思科系列丛书之 Packet Tracer 经典案例篇

Packet Tracer 经典案例之路由交换入门篇

刘彩凤 编著

电子工业出版社
Publishing House of Electronics Industry
北京·BEIJING

内 容 简 介

本书基于 Cisco Packet Tracer 模拟器开发了大量经典、实用的教学案例，以启发读者思考，激发读者灵感，让读者快乐学习。其特色是：案例设计，源于实践不拘一格；内容编排，融网络技术于生活；表现形式，集知识、趣味于一体。

全书共 8 章。第 1 章访问网络设备，介绍了 CTY、VTY、TTY 和 Web 四种访问方法；第 2 章管理网络设备，介绍了路由器和交换机的基本配置、口令恢复、IOS 及配置文件的备份与恢复、DHCP 和 CDP 协议等；第 3 章学习 VLAN 技术，介绍了 VLAN 的配置、Trunk、DTP 和 VTP 协议的应用等；第 4 章学习直连路由，介绍了路由器和三层交换机的功能特点、设备型号及支持的模块类型、路由器物理接口类型及识别、逻辑接口类型、VLAN 间路由的 4 种方法及路由表等；第 5 章学习静态路由，介绍了静态路由的特点及应用场合、标准静态路由的配置、7 种特殊静态路由的配置、企业网综合配置等；第 6 章学习 RIP 协议，介绍了协议的特点、版本、原理及多场景网络的配置和问题解决方案；第 7 章 RIP 网络实战，介绍了网络故障排错、企业网综合配置和园区网的规划等；第 8 章 Multiuser 分布式多用户案例，介绍了 Multiuser 的功能、特点及常见的连接故障、PT 实例的连接及其配置等。

本书既可作为思科网络技术学院及相关大、中专院校的网络技术专业教材，也可作为相关本科院校的实验实训指导书，还可作为高等院校网络方向的教辅用书和计算机网络技能比赛的训练手册，同时也是初学者理想的学习指南。

未经许可，不得以任何方式复制或抄袭本书之部分或全部内容。
版权所有，侵权必究。

图书在版编目（CIP）数据

Packet Tracer 经典案例之路由交换入门篇 / 刘彩凤编著. —北京：电子工业出版社，2017.6
（思科系列丛书. Packet Tracer 经典案例篇）
ISBN 978-7-121-31525-1

Ⅰ. ①P… Ⅱ. ①刘… Ⅲ. ①计算机网络—网络设备—教学软件—高等学校—教材 Ⅳ. ①TP393

中国版本图书馆 CIP 数据核字（2017）第 090076 号

策划编辑：宋　梅
责任编辑：宋　梅
印　　刷：北京捷迅佳彩印刷有限公司
装　　订：北京捷迅佳彩印刷有限公司
出版发行：电子工业出版社
　　　　　北京市海淀区万寿路 173 信箱　邮编　100036
开　　本：787×980　1/16　印张：26　字数：599 千字
版　　次：2017 年 6 月第 1 版
印　　次：2024 年 12 月第 11 次印刷
定　　价：78.00 元

凡所购买电子工业出版社图书有缺损问题，请向购买书店调换。若书店售缺，请与本社发行部联系，联系及邮购电话：（010）88254888，88258888。
质量投诉请发邮件至 zlts@phei.com.cn，盗版侵权举报请发邮件至 dbqq@phei.com.cn。
本书咨询联系方式：mariams@phei.com.cn。

序　言

作为思科公司最大和最重要的一项公益事业，思科网络技术学院项目自1998年进入中国，在近20年的时间里，思科公司累计建立了800多所思科网络技术学院，培养了超过40万名优秀的网络技术人才。现如今思科网络技术学院在校学生人数超过6万人，持续为中国信息产业的飞速发展提供了人才储备。中国教育部连续多年为思科公司颁发了"特殊贡献奖"和"最佳合作伙伴奖"。

思科网络技术学院项目在中国所取得的骄人成绩，离不开兢兢业业工作的老师们数十年如一日的刻苦钻研。刘彩凤老师就是其中非常突出的一位。刘老师第一次引起我的注意，缘于2009年那场竞争激烈的比赛——思科公司在成都举办的全国大学生网络技术总决赛。竞赛现场高手林立，网络精英汇聚一堂，一决高低，刘老师带领的烟台职业学院学生团队获得了全国二等奖。放眼全国，计算机专科院校人才济济，本科院校在硬件条件、学生资源等方面更是遥遥领先。相比之下，获胜的烟台职业学院代表队成为了这次比赛的"黑马"。大家在惊叹之余，认为这或许是一次巧合。然而，仅在一年之后，烟台职业学院的两支代表队双双晋级，再现锋芒。这足以证明，烟台职业学院代表队不是歪打正着，而是实力在握。刘老师和她的团队克服了重重困难，成功的背后是他们付出的超乎想象的艰辛和努力，还有她对学生的那种无私的关爱，让人为之动容。更令人钦佩的是，刘老师在思科公司举办的各项比赛中接连获奖，她的教学案例多次在国内和国际上获奖，尤其是在2013年思科公司举行的首届全球教师教学资源设计竞赛（GIR Contest）中，刘老师的英文教学资源"Skills Challenge"一举获得全球第一名，在全球教师队伍中为中国教师赢得了极高的荣誉，这是我们中国教师的骄傲！更为可贵的是，刘老师无偿地把自己呕心沥血设计的教学案例分享给全世界的老师和学生们，赢得了大家的尊重。

基于此，所有人，包括我在内，都期盼着刘老师能够把自己多年来使用Packet Tracer设计的经典教学案例，包括教学经验和总结，汇编成册，分享给奋斗在教学一线的老师和努力学习网络技术的学生们。通过本书，刘老师实现了她又一个梦想——让更多的人从中获得启发，少走弯路。刘老师以一个教师的情怀，诠释了她对教育事业和对学生的热爱。本书是思科网络技术学院项目在中国开展近20年来最重要的成果之一。我谨代表我个人，向刘老师和她的家人、为本书做出贡献的师生们，以及电子工业出版社的宋梅编审，表示由衷的感谢。同时，也衷心祝愿学习这套书的老师们在教学上取得更加丰硕的成果,祝愿学习这套书的同学们能够在攀登网络技术的高峰上取得更大的进步!

韩　江

思科公司大中华区企业社会责任经理

2017 年 6 月 5 日于北京

前　言

十年磨一剑。2017 年是我作为思科网络技术学院专职讲师的第 11 个年头，呈现给读者的这本书是我十多年来教学实践的总结与积淀。

写书背景

2011 年 10 月，我有幸受邀参加思科首届大中华区教师论坛北京筹备会议。由于我基于 Cisco Packet Tracer 设计的英文教学案例"Troubleshooting"在亚太区获得一等奖，思科公司大中华区企业社会责任经理韩江先生提议让我写一本 Packet Tracer 教学案例分享给思科网络技术学院的师生们。因繁重的教学工作，加上琐碎的家务，此事一直被搁置。2013 年 9 月，很荣幸基于 Packet Tracer 设计的英文教学资源"Skills Challenge"再一次在亚太区获奖，并获得全球第一名。领导、同事和学生们都强烈要求我写一本教学案例，感恩大家的厚爱，2013 年我开始收集整理。2016 年 12 月，书名由韩江经理正式确定，最后在宋梅编审的鞭策与鼓励下，才将想法最终落笔成书。历时四年，得到了思科公司、学校领导、同事以及学生们的大力支持，非常高兴本书终于可以和读者朋友们见面了。

本书目标

本书基于 Packet Tracer 开发了大量经典、实用、富有趣味性和挑战性的教学案例，其目标是培养读者对计算机网络技术的兴趣，引领读者从此爱上网络技术，主动探索和发现全新的网络世界，让网络技术更好、更广泛地应用于生活，进而打造"互联网+"时代的网络技术精英。

内容组织

案例设计遵循认知规律，由简至繁，从单项到综合。采用从访问网络设备到管理网络设备再到配置网络设备的逻辑架构模式。全书共 8 章，各章简要内容如下。

第 1 章　访问网络设备：主要介绍网络设备常用的访问方法，通过本地控制台的 CTY 访问，通过 Telnet、SSH 的 VTY 访问，通过终端服务器的 TTY 访问，以及通过浏览器的 Web 访问。

第 2 章　管理网络设备：主要介绍设备的基本配置，设备的口令恢复，配置文件的备份与恢复，IOS 的备份与恢复，DHCP 和 CDP 协议的应用等。

第 3 章　学习 VLAN 技术：主要介绍 VLAN 的基本概念，单台交换机、两台交换机，以及多台交换机 VLAN 的配置、Trunk、DTP 和 VTP 协议的应用等。

第 4 章　学习直连路由：主要介绍 Cisco 路由器和交换机的诞生背景，设备功能、分类及型号，路由器与三层交换机的区别与联系，路由器物理接口和网络模块的类型及识别，逻辑接口类型，实现 VLAN 间路由的方法，路由表组成、原理及特点等。

第 5 章　学习静态路由：主要介绍静态路由的特点与应用场合、标准静态路由、ARP 代理，特殊静态路由及企业网综合配置等。

第 6 章　学习 RIP 路由协议：主要介绍 RIP 协议特点、版本、原理及 4 个场景网络的应用配置，包括有类网络、不连续网络、等长子网掩码的网络以及采用 VLSM 和 CIDR 的网络。

第 7 章　RIP 网络实战：主要介绍 RIP 协议的综合应用，包括网络故障分析、定位及故障排除，企业网综合配置以及园区网规划等。

第 8 章　Multiuser 分布式多用户案例：主要介绍 Packet Tracer Multiuser 的功能特点，常见连接故障，以及 Multiuser 两用户、三用户和多用户 PT 实例的连接及配置等。

本书特色

教学案例丰富，达 80 多个。案例设计力求创新，设计思路循序渐进，环环相扣。案例形式新颖活泼而不失严谨务实，内容简洁清晰而不失深刻厚重。让读者在仿真环境中快乐地学习，启发思考，激发灵感。

案例设计，源于实践不拘一格

案例独特、类型多样。有来自课堂、实验室的，也有来自企业、园区网的；有配置类的，也有排错类的；有单机单用户类的，也有团队协作类的；有探索发现类的，也有逻辑推理类的；有师生互动类的，也有自主完成类的；有实验验证类的，也有技能强化类的；有课内学做类的，也有课外拓展类的；有基础训练类的，也有挑战闯关类的。

内容编排，融网络技术于生活

把枯燥知识生活化和故事化，借助生活场景，让读者体会学习就是生活。例如，第 1 章有关远程登录的场景，让读者对访问网络设备充满了好奇；第 4 章通过一段浪漫传奇的爱情故事，引出路由器伟大而神圣的诞生背景；通过 IT 部门技术面试闯关场景，整合了前面所学知识；第 5 章通过课堂师生互动场景，掌握各种静态路由的配置，让读者开阔眼界，茅塞顿开；第 7 章通过实验室小组故障排错场景，让读者身临其境，轻松理清排错思路。

表现形式，集知识趣味于一体

把枯燥知识趣味化，借助趣味案例，让读者感到学习就是乐趣。例如，第 2 章探索网络拓扑结构的场景将带领大家一起发现网络的奥秘，深刻体会学习网络技术所带来的乐趣，让学习如同玩一场游戏；第 8 章 Multiuser 多用户场景是对传统实验课的颠覆，它倡导团队协作，共同进步，如同联机游戏；课后挑战闯关场景，有的设置了重重障碍，让读者必须开动脑筋，冥思苦想，只有熟练掌握技术，做到活学活用，才能顺利闯关。

案例开发历经了边设计、边实施、边诊断、边改进、边优化的一系列加工过程，在长期实践中取得了意想不到的教学效果，打开了学生的思维，带领大家走上一条探索学习的道路。

读者对象

本书既可作为思科网络技术学院及相关大、中专院校的网络技术专业教材，也可作为相关本科院校的实验实训指导书，还可作为高等院校网络方向的教辅用书和计算机网络技能比赛的训练手册，同时也是初学者理想的学习指南。

阅读建议

建议读者按顺序阅读本书，因为案例设计循序渐进，环环相扣。对于有基础、有经验的读者，可以直接选读某些章节内容。特别声明：在本书逻辑拓扑图中，设备间的连线若是虚线则表示采用交叉线。

Packet Tracer 是思科网络技术学院的教学工具。思科网络技术学院的教师、学生及校友都可以使用该工具辅助学习 IT 基础、CCNA 路由和交换、CCNA 安全、物联网、无线网络等课程。您可以通过以下链接注册成为"Packet Tracer 101"课程的学生并下载最新版 Packet Tracer 软件：https://www.netacad.com/about-networking-academy/packet-tracer/。

本教材配套有教学资源课件，如有需要，请登录电子工业出版社华信教育资源网（www.hxedu.com.cn），注册后免费下载。

致谢

本书由刘彩凤编写并统稿，韩茂玲参与编写了第 3 章，参加编写工作的还有崔玉礼、于洋和王笑娟。感谢烟台汽车工程职业学院副院长房培玉教授（原烟台职业学院信息工程系主任）引导我走进思科网络技术学院，开启了我的网络教学生涯；感谢原思科网络技术学院全球技术总监 John Lim 及其团队，使我有机会参加全球设计竞赛，吸收国内外先进教学理念，提升设计能力；感谢思科大中华区公共事务部总监练沛强先生，本书编写工作得到了练总大力支持；感谢思科网络技术学院全球产品经理刘亢，给了我参与 Cisco Packet Tracer 测试的机会，鼓励我参加基于 Packet Tracer 的教学案例设计竞赛，让我不断提高；感谢思科公司大中华区企业社会责任经理韩江先生，让我有幸参与思科校企案例项目的开发，积累素材，坚定我的创作信念，也正因为韩江经理的提议才诞生了此书；感谢电子工业出版社宋梅编审，没有宋老师的鞭策和鼓励，本书与读者见面将会遥遥无期，正是宋老师的加班工作，才加快了本书的出版进程；感谢思科网络技术学院大中华区技术经理李涤非老师多年来对我专业的指导和经验传授，让我在写作上少走了很多弯路；感谢思科公司徐如滢经理，让我有机会与思科合作院校进行交流，一路携手，使我不断总结提高；感谢思科公司熊露颖经理，让我有机会参与思科授课计划的制订及 PT 考试系统的开发，为本书编写奠定了基础。

感谢烟台职业学院教务处长原宪瑞教授在教育理念和整体架构上给予我的指导和影响；感谢烟台职业学院信息工程系王作鹏主任在本书创作过程中给予的支持和指导；感谢国家精品资源共享课程负责人薛元昕教授，在课程建设及资源开发方面给予我的帮助；感谢原烟台红十字会副秘书长王吉永老先生，咫尺近邻迟志邦老师，烟台职业学院徐言超老师、张津铭老师和郝志宏老师，广州市黄埔职业技术学校何力老师，吉林铁道职业技术学院王爱华老师，中国石油大学肖军弼老师和曹绍华老师等对本书编写工作的大力支持。

感谢我的学生团：王雪蕾（XL 15NET2）、尹翠红（11NET1）、汲海伦（10NET1）、卜云霞（11NET1）、李璇（LX 14NET1）、胡颖（HY 15NET2）、谢颂根（XSG 13NET）、赵柳玉（07NET1）、黎振（08NET2）、宋术君（09NET1）、王顺顺（11NET2）、孙梦雪（12NET）、门英达（12NET）、高海华（14NET2）、王美超（WMC 15NET1）、杜文超（DWC 15NET2）、张彬（ZHB 15NET2）等同学，对本书的编写提出了很多宝贵意见，并对相关技术细节进行了验证，他们对本书贡献巨大。尤其感谢我的学生雪蕾（XL），是她陪伴我完成了本书创作。最后，感谢思科公司、感谢思科网络技术学院，以及对本书寄予厚望的老师、历届的学生们，是你们给了我无限动力。

感言

本书创作过程非常艰辛，写作周期长，设计的案例也要在实践中不断验证。为潜心创作，我需要阶段性封闭。在封闭期间，通信工具时常会中断，感谢家人、朋友、同事对我的支持、理解和包容。尽管创作艰辛，但我很享受设计灵感一次次迸发的过程，期盼与大家一起分享这份设计成果。虽然尽了最大努力，但鉴于本书是笔者的处女作，水平和视野有限，书中难免存在纰漏和不足之处，愿读者朋友们给予建议指正，我将不胜感激，并不断修改加以完善。

电子邮件地址：yantaicfl@126.com

刘彩凤
2017 年 5 月于烟台

目　　录

第 1 章　访问网络设备 1

1.1　CTY 访问网络设备 2
1.1.1　认识 Console 端口 2
1.1.2　场景一：通过 Console 访问路由器 3
1.1.3　配置 Console 线路 6
1.2　VTY 访问网络设备 6
1.2.1　认识远程登录 6
1.2.2　场景二：配置 Telnet 7
1.2.3　场景三：配置 SSH 10
1.3　TTY 访问网络设备 13
1.3.1　认识终端服务器 13
1.3.2　场景四：连接终端服务器 14
1.3.3　配置终端服务器 17
1.3.4　接入终端服务器 21
1.4　Web 访问网络设备 23
1.4.1　认识 GUI 界面 23
1.4.2　场景五：准备访问 GUI 24
1.4.3　通过 Web 接入 26
1.5　挑战过关练习 28

第 2 章　管理网络设备 30

2.1　管理交换机 31
2.1.1　交换机的基本命令 31
2.1.2　场景一：交换机的基本配置 33
2.1.3　场景二：交换机的口令恢复 37
2.1.4　场景三：交换机机配置文件的备份与恢复 39
2.1.5　文件备份命令总结 44
2.2　管理路由器 45
2.2.1　路由器的基本命令 45

	2.2.2	场景四：路由器的基本配置	46
	2.2.3	场景五：路由器的口令恢复	50
	2.2.4	场景六：路由器配置文件的备份与恢复	52
	2.2.5	动态主机配置协议 DHCP	54
	2.2.6	场景七：配置路由器的 DHCP 服务	55
2.3	管理设备 IOS		58
	2.3.1	认识网络设备 IOS	58
	2.3.2	认识 IOS 命名规则	59
	2.3.3	场景八：升级交换机的 IOS	59
	2.3.4	场景九：恢复路由器的 IOS	65
2.4	探索网络拓扑		69
	2.4.1	认识思科发现协议	69
	2.4.2	场景十：应用 CDP 协议	69
	2.4.3	场景十一：探索网络拓扑结构	76
2.5	挑战过关练习		80

第 3 章 学习 VLAN 技术 ... 84

3.1	认识 VLAN 技术		85
	3.1.1	VLAN 的概念	85
	3.1.2	VLAN 的优点	86
	3.1.3	VLAN 的分类	86
	3.1.4	VLAN 的配置	87
3.2	多场景 VLAN 配置		91
	3.2.1	场景一：单交换机 VLAN 配置	91
	3.2.2	场景二：两台交换机 VLAN 配置	93
	3.2.3	场景三：多交换机 VLAN 配置	95
3.3	认识 Trunk 技术		99
	3.3.1	认识 Trunk 干道	100
	3.3.2	认识 DTP 协议	103
	3.3.3	分析 Trunk 结果	107
	3.3.4	排查 Trunk 故障	108
	3.3.5	场景四：应用 Trunk 技术	110
3.4	认识 VTP 技术		116
	3.4.1	认识 VTP 协议	117

	3.4.2	认识 VTP 修剪	119
	3.4.3	分析 VTP 通告	120
	3.4.4	学习 VTP 配置	122
	3.4.5	场景五：应用 VTP 技术	127
3.5	挑战过关练习		132
	3.5.1	挑战过关练习一	132
	3.5.2	挑战过关练习二	133
	3.5.3	挑战过关练习三	135

第4章 学习直连路由 138

4.1	认识 Cisco 路由设备		139
	4.1.1	Cisco 路由器因爱而生	139
	4.1.2	认识路由器	140
	4.1.3	Cisco 交换机诞生背景	143
	4.1.4	认识交换机	143
	4.1.5	路由设备对比	146
4.2	认识物理接口直连路由		147
	4.2.1	认识路由器物理接口	147
	4.2.2	实验一：配置路由器物理接口	149
	4.2.3	场景一：采用传统路由器实现 VLAN 间路由	152
	4.2.4	认识三层交换机物理接口	156
	4.2.5	实验二：配置三层交换机物理端口	157
	4.2.6	场景二：三层交换机物理接口 VLAN 间路由	159
4.3	认识逻辑接口直连路由		161
	4.3.1	认识路由器子接口	161
	4.3.2	场景三：单臂路由器实现 VLAN 间路由	162
	4.3.3	认识三层交换机 SVI 接口	167
	4.3.4	场景四：三层交换机 SVI 实现 VLAN 间路由	168
	4.3.5	认识设备环回接口	171
4.4	分析路由表		172
	4.4.1	认识路由表	172
	4.4.2	认识直连路由	173
	4.4.3	分析路由表结构	173
	4.4.4	剖析路由表原理	177

            4.4.5 路由表的层次化 ··· 178
    4.5 网络工程师技术面试闯关 ··· 180
    4.6 挑战过关练习 ··· 189
        4.6.1 挑战过关练习一 ··· 189
        4.6.2 挑战过关练习二 ··· 190

第 5 章 学习静态路由 ·· 193
    5.1 认识静态路由 ··· 194
        5.1.1 静态路由的特点 ··· 194
        5.1.2 静态路由的应用 ··· 194
        5.1.3 静态路由的类型 ··· 195
        5.1.4 静态路由的语法 ··· 195
    5.2 配置标准静态路由 ··· 195
        5.2.1 场景一：配置下一跳静态路由 ··· 195
        5.2.2 场景二：配置出接口静态路由 ··· 200
        5.2.3 场景三：下一跳与出接口对 ARP 表的影响 ·· 203
        5.2.4 认识 ARP 代理 ··· 206
    5.3 配置特殊静态路由 ··· 211
        5.3.1 场景四：配置默认路由 ··· 211
        5.3.2 场景五：配置主机路由 ··· 219
        5.3.3 场景六：配置汇总路由 ··· 222
        5.3.4 场景七：配置黑洞路由 ··· 226
        5.3.5 场景八：配置浮动路由 ··· 229
        5.3.6 场景九：配置等价路由 ··· 232
        5.3.7 场景十：配置递归路由 ··· 234
        5.3.8 特殊静态路由小结 ··· 237
    5.4 企业网综合配置案例 ··· 237
        5.4.1 任务背景 ··· 237
        5.4.2 网络拓扑 ··· 238
        5.4.3 配置任务 ··· 238
        5.4.4 任务实施 ··· 239
        5.4.5 连通测试 ··· 243
        5.4.6 案例小结 ··· 244
    5.5 拓展思维案例 ··· 244

5.5.1 案例一：网络配置挑战 ········· 244

5.5.2 案例二：潜精研思挑战 ········· 245

5.5.3 案例三：开拓创新挑战 ········· 246

5.5.4 案例四：面试闯关挑战 ········· 247

第 6 章 学习 RIP 路由协议 ········· 249

6.1 认识 RIP 路由协议 ········· 250

    6.1.1 认识动态路由协议 ········· 250

    6.1.2 RIP 路由协议的特点 ········· 251

    6.1.3 RIP 路由协议的版本 ········· 251

    6.1.4 RIP 路由协议的原理 ········· 252

    6.1.5 RIP 路由协议的配置 ········· 255

6.2 配置多个场景的网络 ········· 256

    6.2.1 场景一：配置有类网络 ········· 256

    6.2.2 场景二：配置不连续网络 ········· 258

    6.2.3 场景三：配置等长掩码网络 ········· 262

    6.2.4 场景四：配置 VLSM 与 CIDR 网络 ········· 264

    6.2.5 总结一：收发更新原则 ········· 266

6.3 RIPv2 解决 RIPv1 存在的问题 ········· 267

    6.3.1 问题一：解决不连续网络问题 ········· 267

    6.3.2 问题二：解决 VLSM&CIDR 问题 ········· 269

    6.3.3 总结二：两个版本的区别与联系 ········· 272

6.4 验证 RIP 协议特性 ········· 272

    6.4.1 实验一：验证用跳数度量最佳路径 ········· 272

    6.4.2 实验二：验证最大负载均衡路径数 ········· 278

    6.4.3 实验三：验证度量的最大跳数限制 ········· 280

    6.4.4 实验四：验证更新的最大路由条数 ········· 286

    6.4.5 实验五：验证两个版本更新的不同 ········· 294

6.5 挑战闯关训练 ········· 298

    6.5.1 挑战任务 ········· 298

    6.5.2 挑战闯关练习拓扑 ········· 299

第 7 章 RIP 网络实战 ········· 301

7.1 RIP 网络故障排错案例 ········· 302

- 7.1.1 故障原因分析 ... 302
- 7.1.2 场景一：故障排错案例 ... 303
- 7.1.3 故障排错总结 ... 306

7.2 企业网综合配置案例 ... 307
- 7.2.1 任务一：配置接口地址 ... 310
- 7.2.2 任务二：配置路由协议 ... 310
- 7.2.3 任务三：关闭自动汇总 ... 312
- 7.2.4 任务四：配置静态路由 ... 314
- 7.2.5 任务五：传播默认路由 ... 316
- 7.2.6 任务六：配置路由注入 ... 317
- 7.2.7 任务七：配置被动接口 ... 318
- 7.2.8 任务八：设置计时参数 ... 319
- 7.2.9 任务九：测试网络连通 ... 320

7.3 园区网络规划案例 ... 323
- 7.3.1 园区网络规划方案 A ... 323
- 7.3.2 园区网络规划方案 B ... 326
- 7.3.3 园区网络规划方案 C ... 330
- 7.3.4 园区网络规划方案 D ... 333
- 7.3.5 园区网规划方案总结 ... 337
- 7.3.6 园区网络优化方案 A ... 337

7.4 课外拓展训练 ... 340
- 7.4.1 训练一：分析能力技能挑战 ... 340
- 7.4.2 训练二：综合应用技能挑战 ... 341
- 7.4.3 训练三：奇思妙想技能挑战 ... 342
- 7.4.4 训练四：洞察分析技能挑战 ... 342

第8章 Multiuser 分布式多用户案例 ... 344

8.1 PT Multiuser 实验案例 ... 345
- 8.1.1 PT Multiuser 功能简介 ... 345
- 8.1.2 PT Multiuser 案例背景 ... 346
- 8.1.3 PT Multiuser 实例创建 ... 348
- 8.1.4 PT Multiuser 实例连接 ... 350
- 8.1.5 PT Multiuser 连通测试 ... 353
- 8.1.6 PT Multiuser 故障排错 ... 355

8.2 PT Multiuser 综合案例 ··· 357
8.2.1 PT Multiuser 任务背景 ··· 357
8.2.2 PT Multiuser 任务分解 ··· 359
8.2.3 PT Multiuser 任务实施 ··· 360
8.2.4 PT Multiuser 实例连接 ··· 367
8.2.5 PT Multiuser 互联结果 ··· 384
8.2.6 PT Multiuser 连通测试 ··· 385
8.2.7 PT Multiuser 冗余测试 ··· 388
8.2.8 PT Multiuser 中断示意图 ··· 393
8.3 PT Multiuser 拓展案例 ··· 394
8.3.1 PT Multiuser 网络拓扑 ··· 394
8.3.2 PT Multiuser 地址规划 ··· 396
8.3.3 PT Multiuser 项目需求 ··· 397

参考文献 ··· 399

第1章 >>>

访问网络设备

本章要点

- CTY 访问网络设备
- VTY 访问网络设备
- TTY 访问网络设备
- Web 访问网络设备
- 挑战过关练习

网络设备是组网的关键设备，如何接入这些设备并为后期配置和管理做准备，这是我们本章探讨的主要问题。通过设置的 5 个应用场景带领你轻松完成学习，设计的 1 个综合过关挑战练习是对本章所掌握知识的考验。网络设备常用接入方法有：通过本地控制台 CTY 接入，通过 Telnet、SSH 的 VTY 接入，通过终端服务器的 TTY 接入以及通过浏览器的 Web 方式接入。根据 OSI RM，我们将网络设备定义为一层物理层设备（如中继器、集线器）、二层数据链路层设备（如网桥、交换机）、三层网络层设备（如路由器、三层交换机）等。一层设备处理比特流，二层设备处理数据帧，三层设备处理数据包。本章主要介绍对二层及以上网络设备的接入。二层及以上设备均有 CPU 和内存，没有键盘和显示器，这就需要借助终端 PC 作为其仿真终端对设备进行控制。

网络设备出厂时通常是空配置（Cisco 最新设备已有一些初始配置以便远程登录），初始设备通常用 Console 端口进行连接，在配置 IP 地址和口令等参数后，便可以采用其他方法接入设备。

1.1 CTY 访问网络设备

1.1.1 认识 Console 端口

Console 控制端口是网络设备与计算机或终端设备进行连接时的常用端口，利用终端仿真程序（如 Windows 下的"超级终端"）对路由器或交换机等网络设备进行本地配置。路由器的 Console 端口多为 RJ-45 端口，控制台会话又称为 CTY 行。控制台采用低速串行连接方式，终端直接连接到路由器或交换机的控制台端口，即 Console 端口。控制台端口是一种管理端口，常用于网络设备的初始连接或设备灾难恢复（如 IOS 恢复或口令恢复）时的访问。

几乎所有 Cisco 网络设备都有一个串行控制台端口（即 Console 端口）。一些型号较新的路由器，如 Cisco 1941 集成多业务路由器（ISR）还带有 USB 控制台端口。路由器 Console 端口如图 1-1 所示。

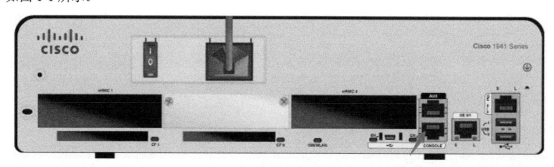

图 1-1　路由器 Console 端口

Console 线缆又名反转线缆，即线缆两端 RJ-45 接头上线序相反。其作用是连接交换机或路由器等网络设备的 Console 端口至计算机的 COM 口。目前大多数 Console 线缆一端的 RJ-45 接头经"RJ-45 to DB9"转换器转换为 DB9 后，再连接到计算机的 COM 口上。带 DB9 接头的 Console 线缆（反转线缆）如图 1-2 所示。

图 1-2　带 DB9 接头的 Console 线缆（反转线缆）

目前，在 Packet Tracer 中，本地控制台的连接仅支持计算机通过 COM 口连接路由器或交换机的 Console 端口。

1.1.2　场景一：通过 Console 访问路由器

接下来，我们将通过场景一熟悉通过 CTY 控制台访问网络设备的方法。

> 🔍 **场景一**：YTVC 思科网络实验室最近采购了一批全新 Cisco 1941 型号的 ISR 路由器，目前正准备验收。实验室负责人请 CFL 老师从 2015 级网络技术专业学生中挑选 6 位同学来帮忙测试。测试要求检查 Cisco 1941 路由器能否正常启动。为了对比模拟器和真机间的差距，老师充分利用了这次学习机会让同学们做好精心准备。她要求同学先在 Packet Tracer 模拟器上熟练操作路由器的初始配置，然后再去实验室测试。

同学们打开 Packet Tracer，拖出了一台 PC 和一台 Cisco 1941 的路由器，如图 1-3 所示。

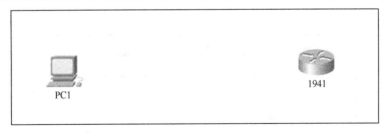

图 1-3　Console 接入前期准备

下面，让我们跟随老师一起通过 Console 线缆，实现对 Cisco 1941 路由器的 CTY 访问。

步骤一：用 Console 线缆连接 PC 和路由器

初次访问网络设备须要使用控制线缆，鼠标单击 Packet Tracer 模拟器左下角的"Connections"图标，点击线缆类型中的"Console"线缆，如图 1-4 所示。

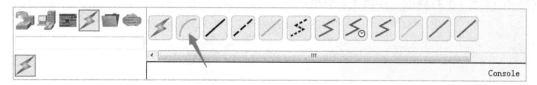

图 1-4　选择 Console 线缆

Console 线缆一端连接 PC 的 COM 口（RS-232 是接口标准），如图 1-5 所示。

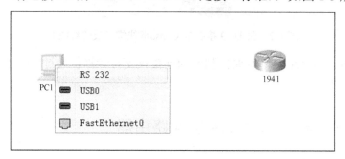

图 1-5　Console 线缆连接 PC 串口

Console 线缆另一端连接 Cisco 1941 路由器的 Console 端口，PC 通过 Console 端口接入路由器，如图 1-6 和图 1-7 所示。

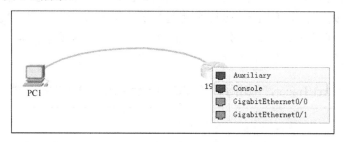

图 1-6　Console 线缆连接路由器 Console 端口

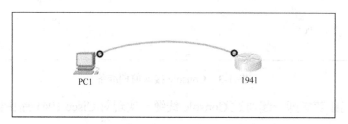

图 1-7　PC 通过 Console 端口接入路由器

通过以上 Console 线缆将 PC1 的 COM 端口与路由器的 Console 端口做好了物理连接。

步骤二：通过终端软件设置串口参数

通过配置 PC 终端软件"Terminal"来完成对串口通信参数的设置，如图 1-8 所示。

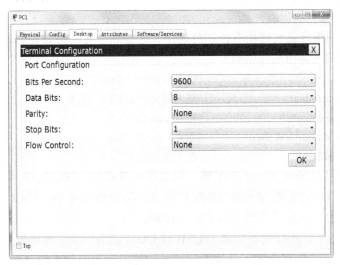

图 1-8　配置终端软件

Console 端口默认参数配置为端口波特率为 9 600 bps、数据位为 8 位、无奇偶校验，停止位是 1 位、无流量控制。设置完毕，单击"OK"按钮，按"Enter"键，终端窗口就会显示路由器提示符，如图 1-9 所示。

图 1-9　接入路由器

现在，我们实现了通过 CTY 访问路由器。为了保护 Console 端口，我们需要对其做相应的安全配置。

1.1.3 配置 Console 线路

默认情况下，网络设备允许所有控制台连接，这显然会对网络设备的安全构成威胁，所以我们应该对控制端口的访问加以限制。

Router>**enable**	//进入路由器特权模式
Router#**configure terminal**	//进入路由器全局配置模式
Router(config)#**line console 0**	//配置 Console 线路
Router(config-line)#**password cisco**	//配置 Console 口令
Router(config-line)#**login**	//配置登录检查

以上是我们使用 Packet Tracer 模拟器，通过终端以 CTY 方式来访问网络设备的方法。但在实验室里，面对真机的我们却不能如此轻松接入。有少数几台 PC 带 COM 端口，可以直接接入路由器；但有的同学自带笔记本电脑想连接路由器，他们发现笔记本电脑竟然没有 COM 端口。在老师的指导下，几位同学领到了 USB 转 COM 的转换器，需要先安装驱动程序，否则 PC 会没有 COM 端口。更有个别同学找不到终端软件，因为有的系统自带"超级终端"，有的系统没有，这就需要额外安装终端软件，如"SecureCRT"、"Xshell"等。

1.2 VTY 访问网络设备

1.2.1 认识远程登录

远程登录提供了访问远程设备的功能，使本地用户可以通过 TCP 连接登录到远程设备，就像控制本地主机一样。远程登录采用客户端/服务器（C/S）模式，客户端和服务器分别执行各自的操作系统。该机制允许客户端程序和服务器程序协商双方能进行身份验证的相关参数，并依此建立 TCP 连接。

服务器可以应付多个用户发起的 TCP 并发连接，并为每个连接请求生成一个对应的进程。每个用户称为一个虚拟终端（VTY），第一个用户为 VTY1，第二个用户为 VTY2，依此类推。在 Cisco 的不同系列产品中，都有一定数量的 VTY 线路可用，具体数目不尽相同。有些网络设备只有 5 条线路可用（VTY 0～4），有些设备提供了十多条，甚至上千条，默认情况下不一定全部启用。

支持远程登录的协议有 SSH（Secure Shell）和 Telnet。SSH 和 Telnet 都是用来管理远程连接的，SSH 使用 TCP 的 22 号端口，Telnet 使用 TCP 的 23 号端口。 SSH 即安全外壳，是一种提供远程安全管理连接协议，对用户名、口令以及在通信设备间传输数据进行强加密，以确保远程连接的安全；而 Telnet 采用明文传输，不能确保安全连接。作为一种最佳实践，只要可能，就应该用 SSH 代替 Telnet。大多数新版 IOS 包含 SSH 服务器，在某些设备中，服务默认启用。

与本地控制台接入不同，远程登录要求网络设备启用网络服务且至少有一个活动接口配置 IP 地址。出于安全考虑，IOS 要求远程会话使用口令，并作为一种最低的身份验证手段。

1.2.2 场景二：配置 Telnet

🔑**场景二**：YTVC 网络中心采购了一台全新的 Cisco 2911 路由器，网管 LX 将其安装在设备间机柜里。因为 LX 的工作地点不在设备间，为便于后期对设备进行配置与管理，LX 决定在设备间先通过控制台完成对路由器的基本配置并对其实现远程管理，如图 1-10 所示。

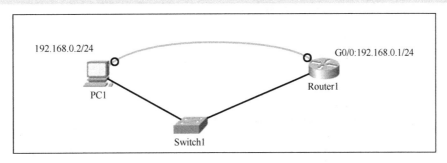

图 1-10　通过 Telnet 远程访问网络设备

请跟网管 LX 学习如何在路由器上配置 Telnet 服务来实现远程访问。

步骤一：配置路由器以太网接口 IP 地址

```
Router>enable
Router#configure terminal
Router(config)#hostname Router1                              //配置路由器的主机名
Router1(config)#interface gigabitEthernet 0/0                //进入路由器的以太网 gigabitEthernet 0/0
Router1(config-if)#ip address 192.168.0.1 255.255.255.0      //配置接口的 IP 地址
Router1(config-if)#no shutdown                               //激活接口，路由器接口默认关闭
```

步骤二：配置路由器 VTY

```
Router1(config-if)#line vty 0 4              //进入 vty 线路模式
Router1(config-line)#password ytvc           //配置 vty 口令
Router1(config-line)#login                   //配置登录检查
```

步骤三：配置路由器特权口令

Router1(config)#**enable password 15net**

步骤四：配置 PC 的 IP 地址

单击 PC1 的"Desktop"，点击"IP configuration"，配置其 IP 地址为 192.168.0.2/24，如图 1-11 所示。

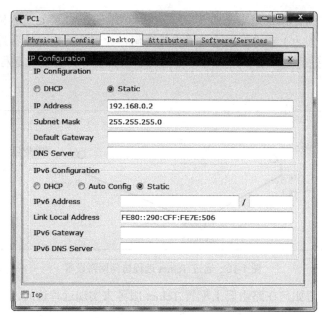

图 1-11　配置 PC1 的 IP 地址

步骤五：通过 Telnet 访问路由器

Telnet 远程登录之前，我们先需要测试 PC 与路由器的连通性。单击 PC1 的"Desktop"，点击"Command Prompt"选项，进入 DOS 命令行窗口，执行如下 ping 命令：

C:\>**ping 192.168.0.1**

Pinging 192.168.0.1 with 32 bytes of data:

Reply from 192.168.0.1: bytes=32 time<1ms TTL=255
Reply from 192.168.0.1: bytes=32 time<1ms TTL=255
Reply from 192.168.0.1: bytes=32 time<1ms TTL=255
Reply from 192.168.0.1: bytes=32 time<1ms TTL=255

Ping statistics for 192.168.0.1:
    Packets: Sent = 4, Received = 4, Lost = 0 (0% loss),
Approximate round trip times in milli-seconds:
    Minimum = 0ms, Maximum = 0ms, Average = 0ms

上述表明，PC 能够 ping 通路由器的以太网口，实现了设备间的连通性。接下来，我们需要通过 PC1 远程登录到路由器，如图 1-12 所示。

```
C:\>telnet 192.168.0.1
Trying 192.168.0.1 ...Open


User Access Verification

Password:                    //输入 vty 的口令
Router1>enable
Password:                    //输入路由器的特权口令
Router1#
```

图 1-12 PC 成功 Telnet 到路由器

从以上输出可见，PC1 通过 Telnet 成功远程登录到了路由器，实现远程接入。

1.2.3 场景三：配置 SSH

> 🔍**场景三**：YTVC 网络安全社团最近在"闲鱼网站"团购了一台二手 Cisco 2901 路由器用于社团开展网络安全实验。部长 XL 同学提议将其安装在 E509 机柜里，供社团成员远程实验。社团最近举办网络安全宣传活动，现要求 XL 先通过控制台完成对路由器的基本配置，如图 1-13 所示。本台设备安装在机柜里，XL 建议今后必须通过 SSH 对其远程管理。请跟 XL 学习如何在路由器上配置 SSH 服务，并通过 PC 对其远程控制。

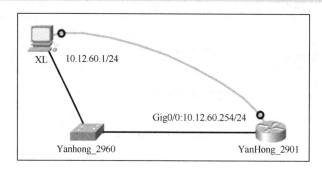

图 1-13 通过 SSH 远程访问网络设备

下面，我们一起看 XL 同学是如何配置 SSH，实现远程访问的。

步骤一：配置路由器主机名

```
Router>enable
Router#configure terminal
Router(config)#hostname YanHong_2901
YanHong_2901 (config)#
```

SSH 要求路由器不使用默认主机名 Router，主机名很可能已经配置。若没有，请在全局配置模式下使用 hostname 命令进行配置，但 Telnet 不做要求。出于习惯，我们最好配置。

步骤二：配置网络的 IP 域名

```
YanHong_2901 (config)#ip domain-name ytvc.com      //创建域名 ytvc.com
```

SSH 必须有域名，服务才能被启用。

步骤三：配置路由器以太网接口 IP 地址

```
YanHong_2901 (config)#interface gigabitEthernet 0/0
YanHong_2901 (config-if)#ip address 10.12.60.254 255.255.255.0
YanHong_2901 (config-if)#no shutdown
```

步骤四：生成 RSA 非对称秘钥

```
YanHong_2901 (config)#crypto key generate rsa           //生成 RSA 加密密钥
The name for the keys will be: YanHong_2901.ytvc.com
Choose the size of the key modulus in the range of 360 to 2048 for your
  General Purpose Keys. Choosing a key modulus greater than 512 may take a few minutes.
How many bits in the modulus [512]: 1024                //密钥设置为 1024 位
% Generating 1024 bit RSA keys, keys will be non-exportable...[OK]
YanHong_2901 (config)#
*3? 1 0:1:12.69:   %SSH-5-ENABLED: SSH 1.99 has been enabled
```

生成 RSA 秘钥对时，SSH 服务将自动启用；删除 RSA 秘钥对时，SSH 服务将自动禁用。生成 RSA 秘钥时，系统会自动提示管理员指定秘钥的长度。秘钥长度可指定为 360～2 048 位。作为一种最佳实践，Cisco 建议秘钥长度不要低于 1 024 位。

步骤五：配置本地身份验证和 vty

```
YanHong_2901 (config)#username XL password yanhong0208  //创建本地用户名和口令
YanHong_2901 (config)#line vty 0 15
YanHong_2901 (config-line)#transport input ssh          //指定 vty 的登录模式为 ssh
YanHong_2901 (config-line)#login local                  //使用本地身份验证
```

VTY 线路默认同时支持 SSH 和 Telnet，即默认命令是 transport input all，我们须要根据实际情况来指定 VTY 线路的登录方式。

步骤六：配置路由器特权口令

```
YanHong_2901 (config)#enable secret ytvc                //基于安全考虑，最好配置密文特权口令
```

步骤七：通过 SSH 访问路由器

SSH 远程登录之前，我们需要先测试 PC 与路由器的连通性，确保两者互通。然后进入 PC 的 DOS 命令行窗口，了解 SSH 的登录模式。

```
C:\>SSH -?                                              //在 PC 上查看 SSH 登录的格式
Packet Tracer PC SSH

Usage: SSH -l username target
```

从以上输出可见，ssh 命令有一个固定参数 –l，其作用是指定以用户名登录。username 参数是本地创建的用户名，target 是远程登录设备的目标 IP 地址。其应用如下：

从图 1-14 可见，PC 通过 SSH 成功登录到了路由器，实现了对设备的访问。

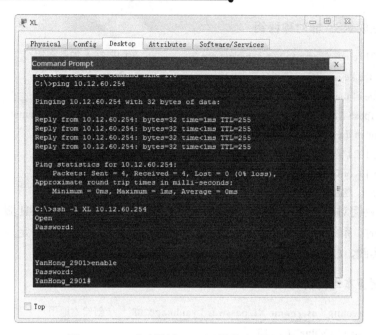

图 1-14　PC 成功通过 SSH 远程登录到了路由器

SSH 的客户端可以是终端 PC 也可以是交换机或路由器等网络设备，即一台网络设备既可以作为服务器又可以作为客户机，如下所示是通过一台 Cisco 2960 交换机 SSH 到 Cisco 2901 路由器：

Yanhong_2960#**ssh -l XL 10.12.60.254**
Open
Password:　　　//输入用户口令 yanhong0208

YanHong_2901>**enable**
Password:　　　//输入特权口令 ytvc
YanHong_2901#

显然，我们可以从一台网络设备 SSH 到另一台网络设备。要在路由器中启用 SSH 服务，必须配置的参数有：主机名、域名、用户名、非对称密钥、enable 密码和用户登录密码。可选的配置参数包括版本、超时时间、重试次数。Cisco 设备默认超时时间最大值为 120 s，重试次数为 3 次。超时机制的作用是可以终止长时间不活动的连接，重试次数作用是防止恶意猜测等非法行为，从而进一步提高连接的安全性。

例如，在路由器上配置 SSH 协议的版本号为 2，超时设置为 20 s，重试次数为 2 次，其对应配置如下：

YanHong_2901 (config)#**ip ssh version 2**
YanHong_2901 (config)#**ip ssh time-out 20**
YanHong_2901 (config)#**ip ssh authentication-retries 2**

我们可以通过命令 show ip ssh 来检验当前 SSH 的配置。

YanHong_2901 #**show ip ssh**
SSH Enabled - **version 2.0**
Authentication timeout: **20 secs**; Authentication retries: **2**

相比 Telnet 而言，SSH 确实更加安全。不同的 VTY 线路，可以配置不同的远程登录协议。例如，可以在 VTY 0 上配置 Telnet，而在 VTY 1 上配置 SSH，这样当 SSH 用户登录时，系统会让 VTY 0 空闲，而使用 VTY 1 进行连接。

1.3　TTY 访问网络设备

TTY 设备包括虚拟控制台、串口和伪终端设备。TTY 是 TeleTYpe 的缩写。TeleType 又名 TeleTypewriters，即电传打字机，可以作为"实时"的输入/输出设备。最终，电传打字机被键盘和显示器终端所取代，但在终端或 TTY 接插的地方，操作系统仍然需要一个程序来监视串行端口。

TTY 终端接入利用路由器作为终端接入发起方，将服务器作为 TCP 连接的接收方，路由器负责在其连接的网点终端和服务器之间，实现数据的透明传输。

TTY 终端接入解决方案不仅实现了固定终端号的基本功能，而且还提供多业务动态切换、屏幕实时存储、终端复位、数据加密等许多增强功能。同时，在服务器上还提供了专业的终端管理软件，在丰富功能的同时，简化了管理。终端接入和路由器的融合提供了一个组建多功能、高效率网络的解决方案，使网点办公和 IP 电话轻松实现。

1.3.1　认识终端服务器

现实生活中由于实验环境需要经常改变以组建不同的网络拓扑，而经常插拔 Console 线缆会对 Console 端口造成极大的损坏，因此需要配置一台终端服务器来解决上述矛盾。我们可以选用任何一台带异步接口的路由器进行终端服务器的配置。通过 CAB-OCTAL-ASYNC 电缆（简称 Octal，俗称八爪线），在一个异步接口上引出 8 条线缆连接到 8 台设备的 Console 端口，可使终端访问服务器避免在配置多台路由器时因频繁插拔 Console 线而损坏 Console 端口。

被调试设备的 Console 端口通过"八爪线"中的一条直接连接到终端服务器的某个异步串行接口。这样，我们就可以在终端服务器上反向 Telnet 到被调试设备。反向 Telnet 技术使用 Telnet 从异步串行接口向外建立连接。为建立一条反向 Telnet 连接，需要远程登录到终端服务器的任何激活的 IP 地址，并附上 20xx 值，xx 是希望访问的线路号。例如，telnet 10.1.1.1 2001，2001 就是指第一条异步线路，依次类推。

1.3.2 场景四：连接终端服务器

> 🔑**场景四**：YTVC 思科网络实验室连续发生了几起在实验中损坏设备 Console 端口的现象，这引起了设备中心的高度重视。经开会讨论决定，为避免频繁插拔 Console 线缆造成 Console 端口损坏，现要求对实验室所有网络设备安装上架，为新买的 6 台 Cisco 1941 路由器安装异步模块，采购 6 根八爪线（Octal），使每一台 Cisco 1941 路由器通过八爪线连接到 8 台网络设备。请将全班同学分成 6 组，分别配置 Cisco 1941 路由器为终端服务器，通过终端服务器访问网络设备，并在不同设备间切换。配置终端服务器的网络拓扑如图 1-15 所示。

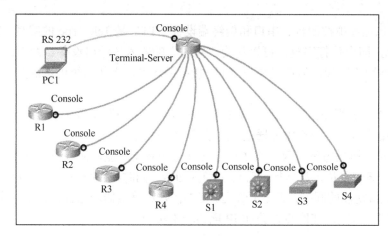

图 1-15 配置终端服务器的网络拓扑

下面让我们一起跟随 CFL 老师，先完成通过终端服务器连接路由器和交换机的工作，再进行相应配置，最后通过终端服务器 TTY 访问这些网络设备。请记录下实验过程以便独立完成实验。

具体步骤如下所述。

步骤一：为 Cisco 1941 终端服务器添加异步串口模块

① 关闭 Cisco 1941 路由器（终端服务器）电源，为其添加一个 HWIC-8A 的异步串口模块到相应的插槽。该模块提供了 8 个异步串口可以连接 8 台设备的 Console 端口，如图 1-16 所示。

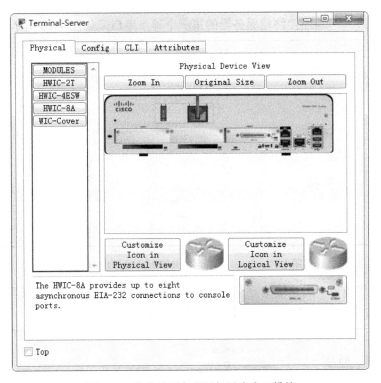

图 1-16　为终端服务器添加异步串口模块

② 选择 Console 线缆，连接 PC 至终端服务器 Cisco 1941 的 Console 端口。

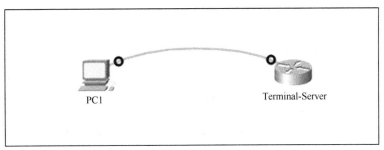

图 1-17　连接 PC 至终端服务器的 Console 端口

③ 选择 Octal（八爪线缆），准备连接终端服务器到其他 8 台网络设备，如图 1-18 所示。

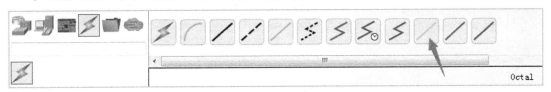

图 1-18　选择"Connections"中的 Octal（八爪线）

④ 将 Octal 线缆一端连接终端服务器的第一个异步串口，如图 1-19 所示。

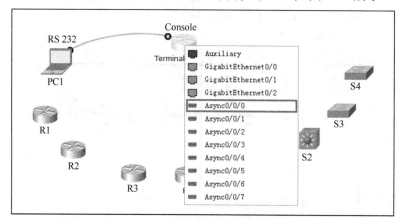

图 1-19　Octal 一端连接终端服务器异步串口

⑤ 将 Octal 线缆另一端连接第一台网络设备，按照从左到右的顺序连接网络设备，先连接 R1，然后连接 R2、R3 和 R4，最后连接 S1、S2、S3 和 S4。

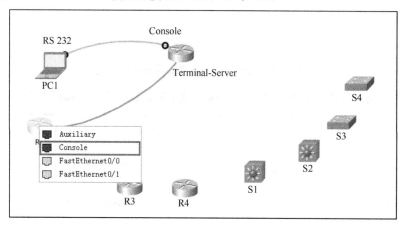

图 1-20　将 Octal 的另一端连接到网络设备的 Console 口

⑥ 使用 Octal 线缆依次连接终端服务器的异步串口至 8 台网络设备的 Console 端口，结果如图 1-21 所示。

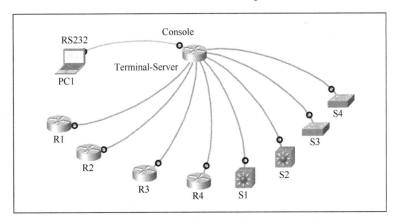

图 1-21 Octal 线路连接终端服务器至 8 台网络设备

到此为止，我们已经顺利完成通过终端服务器连接 8 台网络设备。接下来的任务就是如何配置终端服务器，实现通过 TTY 访问 8 台网络设备。

1.3.3 配置终端服务器

下面我们继续跟随 CFL 老师学习如何配置终端服务器，请详细记录实验步骤。

步骤一：服务器基本配置

```
Router(config)#hostname Terminal-Server              //配置终端服务器主机名
Terminal-Server(config)#enable secret ytvc           //配置特权模式加密口令
Terminal-Server(config)#no ip domain-lookup          //关闭域名解析，防止输错命令长久等待
```

步骤二：配置环回接口

```
Terminal-Server(config)#interface loopback0
Terminal-Server(config-if)#ip address 1.1.1.1 255.255.255.255   //创建环回口 loopback 0 并配置 IP 地址
```

在路由器上最多可以创建 2 147 483 648 个 loopback 环回接口，我们可以通过 ? 来查看。

```
Terminal-Server(config)#interface loopback?
  <0-2147483647>  Loopback interface number
```

该接口属于路由器上的逻辑接口，默认状态为 up，常被用于网络测试。

步骤三：配置 TTY 线路

在配置 TTY 线路之前，我们须要了解当前路由器 TTY 线路的编号，可以通过 show line 命令查看，如图 1-22 所示。

```
Terminal-Server#show line
   Tty Line Typ     Tx/Rx      A Roty AccO AccI   Uses   Noise   Overruns
*    0    0 CTY                -   -    -    -      0      0      0/0
     1    1 AUX    9600/9600   -   -    -    -      0      0      0/0
 0/0/0    3 TTY    9600/9600   -   -    -    -      0      0      0/0
 0/0/1    4 TTY    9600/9600   -   -    -    -      0      0      0/0
 0/0/2    5 TTY    9600/9600   -   -    -    -      0      0      0/0
 0/0/3    6 TTY    9600/9600   -   -    -    -      0      0      0/0
 0/0/4    7 TTY    9600/9600   -   -    -    -      0      0      0/0
 0/0/5    8 TTY    9600/9600   -   -    -    -      0      0      0/0
 0/0/6    9 TTY    9600/9600   -   -    -    -      0      0      0/0
 0/0/7   10 TTY    9600/9600   -   -    -    -      0      0      0/0
   388  388 VTY                -   -    -    -      0      0      0/0
   389  389 VTY                -   -    -    -      0      0      0/0
   390  390 VTY                -   -    -    -      0      0      0/0
   391  391 VTY                -   -    -    -      0      0      0/0
   392  392 VTY                -   -    -    -      0      0      0/0
   393  393 VTY                -   -    -    -      0      0      0/0
   394  394 VTY                -   -    -    -      0      0      0/0
   395  395 VTY                -   -    -    -      0      0      0/0
   396  396 VTY                -   -    -    -      0      0      0/0
   397  397 VTY                -   -    -    -      0      0      0/0
   398  398 VTY                -   -    -    -      0      0      0/0
   399  399 VTY                -   -    -    -      0      0      0/0
   400  400 VTY                -   -    -    -      0      0      0/0
   401  401 VTY                -   -    -    -      0      0      0/0
   402  402 VTY                -   -    -    -      0      0      0/0
   403  403 VTY                -   -    -    -      0      0      0/0
Line(s) not in async mode -or- with no hardware support:
11-387
Terminal-Server#
```

图 1-22　通过 show line 查看 TTY 线路（一）

通过以上输出，我们查到了 TTY 线路的编号是 3～10，于是我们可以做如下配置：

Terminal-Server(config-if)#**line tty 3 10**

Terminal-Server(config-line)#**transport input all**

Terminal-Server(config-line)#**password abc**

Terminal-Server(config-line)#**login**

Terminal-Server(config-line)#**logging synchronous**

//开启日志输出同步，避免用户输入中途被日志信息隔开

Terminal-Server(config-line)#**exec-timeout 0 0**　　　　　　　　　　//抑制登录超时

因为异步串口模块在路由器插槽上位置的不同，会导致路由器 TTY 线路编号的不同，所以配置之前，必须确定终端服务器的 TTY 线路编号，才能正确配置 TTY。图 1-22 所示查看 TTY 线路的编号就与图 1-23 所示的不同。

```
Terminal-Server#show line
   Tty Line Typ    Tx/Rx      A Roty AccO AccI    Uses   Noise   Overruns   Int
*    0    0 CTY               -  -    -    -       0      0       0/0       -
     1    1 AUX    9600/9600  -  -    -    -       0      0       0/0       -
 0/1/0   19 TTY    9600/9600  -  -    -    -       0      0       0/0       -
 0/1/1   20 TTY    9600/9600  -  -    -    -       0      0       0/0       -
 0/1/2   21 TTY    9600/9600  -  -    -    -       0      0       0/0       -
 0/1/3   22 TTY    9600/9600  -  -    -    -       0      0       0/0       -
 0/1/4   23 TTY    9600/9600  -  -    -    -       0      0       0/0       -
 0/1/5   24 TTY    9600/9600  -  -    -    -       0      0       0/0       -
 0/1/6   25 TTY    9600/9600  -  -    -    -       0      0       0/0       -
 0/1/7   26 TTY    9600/9600  -  -    -    -       0      0       0/0       -
   388  388 VTY               -  -    -    -       0      0       0/0       -
   389  389 VTY               -  -    -    -       0      0       0/0       -
   390  390 VTY               -  -    -    -       0      0       0/0       -
   391  391 VTY               -  -    -    -       0      0       0/0       -
   392  392 VTY               -  -    -    -       0      0       0/0       -
   393  393 VTY               -  -    -    -       0      0       0/0       -
   394  394 VTY               -  -    -    -       0      0       0/0       -
   395  395 VTY               -  -    -    -       0      0       0/0       -
   396  396 VTY               -  -    -    -       0      0       0/0       -
   397  397 VTY               -  -    -    -       0      0       0/0       -
   398  398 VTY               -  -    -    -       0      0       0/0       -
   399  399 VTY               -  -    -    -       0      0       0/0       -
   400  400 VTY               -  -    -    -       0      0       0/0       -
   401  401 VTY               -  -    -    -       0      0       0/0       -
   402  402 VTY               -  -    -    -       0      0       0/0       -
   403  403 VTY               -  -    -    -       0      0       0/0       -
Line(s) not in async mode -or- with no hardware support:
3-18, 27-387
Terminal-Server#
```

图 1-23 通过 show line 查看 TTY 线路（二）

此时，我们 TTY 的配置应该改为下面的配置行：

Terminal-Server(config)#**line tty 19 26**

Terminal-Server(config-line)#**password abc**

Terminal-Server(config-line)#**login**

步骤四：配置 VTY 线路

Terminal-Server(config-line)#**line vty 0 15**   //配置 VTY 线路

Terminal-Server(config-line)#**no login**   //取消登录检查，便于登录

Terminal-Server(config-line)#**logging synchronous**

Terminal-Server(config-line)#**exec-timeout 0 0**

为方便后期通过终端服务器远程管理网络设备，建议大家采用主机名登录，因为 IP 地址太长，难以记忆。我们可以在 Octal 线缆上贴上所连接的网络设备名字标签，便于访问。如何通过主机名登录访问设备，我们须要设置主机名与线路的映射，线路对应端口号，即线路号加2000，具体实现命令如下：

Terminal-Server(config-line)#**exit**   //退出线路模式

Terminal-Server(config)#**ip host r1 2003 1.1.1.1**   //配置主机名到线路的映射

Terminal-Server(config)#**ip host r2 2004 1.1.1.1**

Terminal-Server(config)#**ip host r3 2005 1.1.1.1**
Terminal-Server(config)#**ip host r4 2006 1.1.1.1**
Terminal-Server(config)#**ip host s1 2007 1.1.1.1**
Terminal-Server(config)#**ip host s2 2008 1.1.1.1**
Terminal-Server(config)#**ip host s3 2009 1.1.1.1**
Terminal-Server(config)#**ip host s4 2010 1.1.1.1**

接下来，我们可以通过 show hosts 命令进一步查看已经建立的映射关系。映射的主机名对大小写敏感，登录时需要注意。

Terminal-Server#**show hosts**
Default Domain is not set
Name/address lookup uses domain service
Name servers are 255.255.255.255

Codes: UN - unknown, EX - expired, OK - OK, ?? - revalidate
       temp - temporary, perm - permanent
       NA - Not Applicable None - Not defined

Host	Port	Flags	Age	Type	Address(es)
r1	2003	(perm, OK)	0	IP	1.1.1.1
r2	2004	(perm, OK)	0	IP	1.1.1.1
r3	2005	(perm, OK)	0	IP	1.1.1.1
r4	2006	(perm, OK)	0	IP	1.1.1.1
s1	2007	(perm, OK)	0	IP	1.1.1.1
s2	2008	(perm, OK)	0	IP	1.1.1.1
s3	2009	(perm, OK)	0	IP	1.1.1.1
s4	2010	(perm, OK)	0	IP	1.1.1.1

当通过终端服务器以 TTY 方式登录到远程设备时，需要在不同设备间进行切换，建议大家一定要记住表 1-1 所总结的终端服务器操作命令。

表 1-1 终端服务器操作命令

命令功能	具体命令	命令注释
从终端服务器登录到调试设备	telnet *ip address* 端口号 或 telnet *hostname* 或 *hostname*	若用主机名，则须提前用命令 ip host 做映射，hostname 是与 IP 地址及端口映射的主机名
从调试设备返回到终端服务器	Ctrl+Shift+6，X	同时按[Ctrl+Shift+6]，松手，再按 X 键，Telnet 会话依然保留，并未释放
清除 TTY 线路到空闲状态	clear line tty *line-number*	关闭 Telnet 会话，line-number 是异步线路号，如 clear line tty 3
查看任何线路的状态信息	show line	查看终端服务器的连接状态信息

1.3.4 接入终端服务器

配置完终端服务器，同学们已经迫不及待地想通过 TTY 登录到远程网络设备，进而远程控制路由器和交换机。下一步怎么做呢？请继续跟随 CFL 老师一起来吧。

输入相应主机名 r1 就可以达到 telnet 1.1.1.1 2003 的作用。2003 对应第一条线路，第 1 路 TTY 端口号是 2003，以此类推……，第 8 路 TTY 端口号为 2010。具体登录方式见下面对应的格式：

格式一：**telnet** *ip-address port-ID*
Terminal-Server#**telnet 1.1.1.1 2003**
Trying 1.1.1.1 ...Open

User Access Verification

Password:abc
Press RETURN to get started!
R1>

格式二：**telnet** *hostname*

Terminal-Server#**telnet r1**
Trying 1.1.1.1 ...Open

User Access Verification

Password: abc
Press RETURN to get started!
R1>

格式三：*hostname*
Terminal-Server#**r1**
Trying 1.1.1.1 ...Open

User Access Verification

Password: abc
Press RETURN to get started!
R1>

若用格式一登录成功后，请按[Ctrl+Shift+6]，松手再按下 X 键，则返回终端服务器模式，可以继续尝试格式二的登录方式，显示结果如下：

```
Terminal-Server#telnet r1

Trying 1.1.1.1 ...% Connection refused by remote host
```

显然第二种登录方式被远程主机拒绝，因为当前 Telnet 会话依然保留，没有被释放。接下来我们清除 TTY 线路使之处于空闲状态。

```
Terminal-Server#clear line tty 3
[confirm]
 [OK]
Terminal-Server#

[Connection to 1.1.1.1 closed by foreign host]
```

若当前 8 条 TTY 线路全部开启，我们可以在终端服务器上输入"1~8"，就可以在不同网络设备上进行切换，如下所示。

```
Terminal-Server#1                              //切换到第 1 条 TTY 线路，即控制路由器 R1
[Resuming connection 1 to 1.1.1.1 ... ]

R1(config)#                                    //按 Ctrl+Shift+6,X 退出，返回终端服务器
Terminal-Server#2                              //切换到第 2 条 TTY 线路，即控制路由器 R2
[Resuming connection 2 to 1.1.1.1 ... ]

R2(config)#
Terminal-Server#3                              //切换到第 3 条 TTY 线路，即控制路由器 R3
[Resuming connection 3 to 1.1.1.1 ... ]

R3(config)#
Terminal-Server#8                              //切换到第 8 条 TTY 线路，即控制交换机 S4
[Resuming connection 8 to 1.1.1.1 ... ]
S4(config)#

Terminal-Server#5                              //切换到第 5 条 TTY 线路，即控制交换机 S1
[Resuming connection 5 to 1.1.1.1 ... ]

S1(config)#
```

从以上操作可以看出，配置终端服务器确实可以方便网络设备的配置与管理，同时可以避免因反复插拔线缆造成对 Console 端口的损坏，避免不必要的损失。

1.4 Web 访问网络设备

1.4.1 认识 GUI 界面

访问网络设备的界面通常有 CLI（Command Line Interface，命令行界面）界面和 GUI（Graphic User Interface，图形化用户界面）界面两种。前面我们讲过访问网络设备可以通过控制台、Telnet 和 SSH 三种方式访问，且都是对网络设备采用 CLI 界面方式的访问，在该界面下，用户必须在提示符前通过命令行控制网络设备，如图 1-24 所示。

图 1-24　CLI 界面

但某些网络设备不具有控制台端口，不支持 CLI 界面，仅有 GUI 界面。与 CLI 命令行界面相比，GUI 图形界面对于用户来讲在视觉上更易于接受，简单易操作，如图 1-25 所示。

图 1-25 GUI 界面

下面我们将通过场景五来学习客户端如何通过 Web 方式访问设备的 GUI 界面。

1.4.2 场景五：准备访问 GUI

> 场景五：绮丽大厦每个实体商铺的收益都出现了严重滑坡，客流量较以往减半。究其原因主要是因为电商的兴起，使消费者的消费观念发生了巨大的转变。为适应互联网发展，迫使商家采取实体销售和网上销售双向并行的营销模式。绮丽大厦各网点在中秋节前夕，自发组织团购了一批 Cisco WRT300N 无线路由器。下面请你来绮丽大厦指导各位业务网点的业主完成对 Cisco WRT300N 网络设备的访问，以便后期大家可以通过产品手册自行安装。准备访问网络设备如图 1-26 所示。

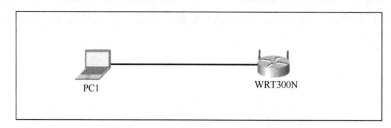

图 1-26 准备访问网络设备

下面就请你详细指导大家完成对 Cisco WRT300N 无线路由器的接入访问。请大家记录操作步骤。

步骤一：通过网线连接个人 PC 至无线路由器

打开产品包装盒，里面提供了一根网线，用该线缆连接计算机至无线路由器局域网口 Ethernet1～Ethernet4 的任何一个接口，如图 1-27 所示。

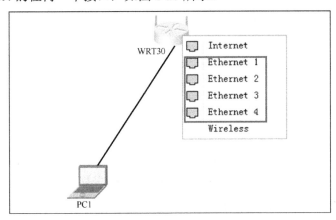

图 1-27　连接网络设备

步骤二：配置 PC DHCP 自动获取地址

点击 PC "Desktop"，并点击 "DHCP"，如图 1-28 所示，PC 将会等待动态获取无线路由器分配的地址。

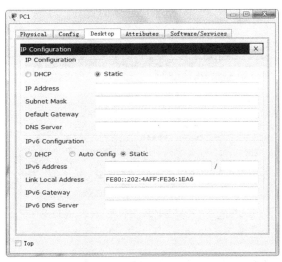

图 1-28　准备动态获取地址

稍等片刻，你会发现 PC 成功获取到了 WRT300N 分配的地址，如图 1-29 所示。

图 1-29 PC1 成功获取 IP 地址参数

PC 获取到的地址是 192.168.0.100，无线路由器的地址是 192.168.0.1。下面我们就准备通过路由器的 IP 地址访问无线路由器。

1.4.3 通过 Web 接入

打开 Web Browser，在 URL 里输入 http://192.168.0.1，然后点击 Go，就会弹出 1-30 所示认证对话框。

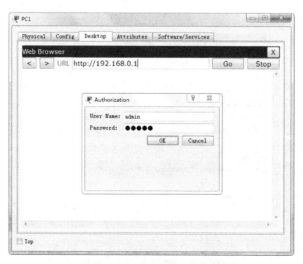

图 1-30 以 Web 方式接入网络设备

在图 1-30 所示对话框里输入产品手册提供的账号 admin，密码 admin，点击 OK 按钮，你就会进入图 1-31 所示的 GUI 界面，说明我们已经成功访问网络设备。

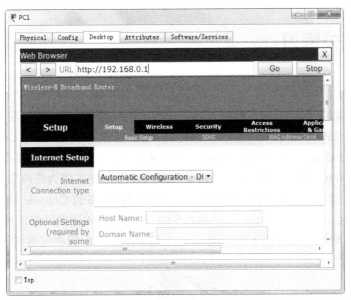

图 1-31　访问网络设备的 GUI 界面

接下来，可以拖动图 1-31 右面的滚动条实现对无线路由器的基本配置，如图 1-32 所示，这就是 GUI 图形化配置界面。

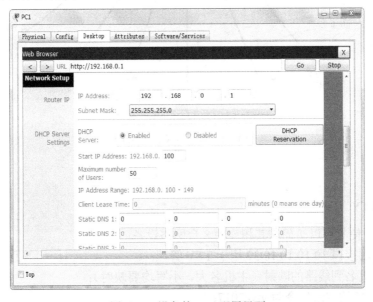

图 1-32　设备的 GUI 配置界面

我们已经顺利完成了通过个人 PC，以 Web 方式访问无线路由器的 GUI 配置界面，想要进行详细配置可以参考产品手册的配置向导。

1.5 挑战过关练习

1. 任务背景

YTVC 思科实验室新购了一批网络设备，搭建的网络拓扑如图 1-33 所示。其中包含一台 Cisco 2911 路由器（R1），5 台 Cisco 1941 路由器（R2~R6），两台 Cisco Catalyst 2960 交换机（S1、S2）以及一台 Cisco 千兆交换机（S3）。为了保护设备，图中所示设备全部安置在机柜中，新购买的 Cisco HWIC-8A 模块安装在 2911 路由器 R1 上，使其成为终端服务器，通过购买的八爪线缆连接到其余设备，以免反复插拔 Console 线损坏设备的 Console 端口。所有网络设备的千兆端口都连接到千兆交换机 S3 上，交换机 S3 所连接设备的逻辑网络地址为 192.168.0.0/24，各设备分配的接口地址如拓扑图 1-33 所示。请按照如下要求配置相应网络设备。

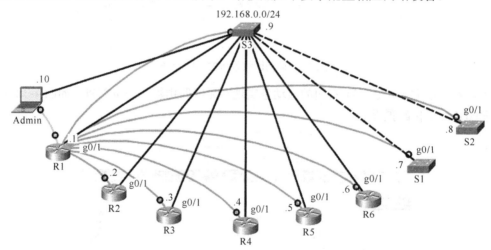

图 1-33 挑战过关练习拓扑

配置要求

（1）配置 TTY
- 配置 R1 使其可以访问与其相连的 8 台网络设备；
- 创建环回接口 10.10.10.10/32 用于 TTY 管理；
- 配置主机名到线路的映射（主机名大、小写均要映射）；
- 设置 TTY 的登录口令为 ytvc2017；

- 开启日志输出同步；
- 关闭域名解析功能。

（2）配置 SSH

- 在 6 台路由器上开启远程 SSH 管理；
- 设置域名为 ytvc.lab；
- 设置用户名为 Admin，口令为 ciscopass；
- 最多允许 9 个用户同时登录。

（3）配置 Telnet

- 在 3 台交换机上开启 Telnet；
- 最多允许 9 个用户同时登录；
- 设置 VTY 口令为 ytvcpass；
- 配置交换机的管理 IP 地址，可以通过如下命令实现：

(switch-config)#**interface vlan 1**
(switch-config-if) #**ip address** *w.x.y.z netmask*
(switch-config-if)#**no shutdown**

注意：所有设备的 enable 口令为 cisco；console 口令为 jsjwl，主机名按照图 1-33 所示进行配置。

到此为止，本章即将结束。我们通过设置的 5 个应用场景，带领大家学习了访问网络设备的几种方法——通过本地控制台的 CTY 访问，通过 Telnet、SSH 的 VTY 访问，通过终端服务器的 TTY 访问以及通过 Web 访问多种访问方法。对于支持 CLI 界面的网络设备，初始化配置和设备的灾难性恢复，如误删 IOS 或丢失口令等，采用控制台访问网络设备是唯一途径。为保护 Console 端口或实现远程管理，可以采用 VTY 或 TTY 方式，我们会在第 2 章详细探讨。对于只支持 GUI 界面而不支持 CLI 界面的网络设备，Web 方式则是另一途径。目前，有些网络设备既支持 CLI 界面又支持 GUI 界面，我们可以根据实际需要灵活选择相应的访问方式。本章最后设置的综合挑战过关练习，目的是加深对知识的理解，并做到对所学知识的活学活用。

第2章 >>>
管理网络设备

本章要点

- 管理交换机
- 管理路由器
- 管理设备 IOS
- 探索网络拓扑
- 挑战过关练习

在第 1 章访问网络设备中,我们介绍了通过终端访问网络设备的几种常用方法。接入设备之后如何管理这些设备,这是我们本章要学习的内容。我们将重点介绍二层设备交换机以及三层设备路由器的管理,先从设备的基本配置开始,再学习设备的口令恢复、配置文件的备份与恢复,以及 IOS 的备份与恢复等内容。本章设置了 11 个生动的场景,带你身临其境,轻松完成学习任务。其中"探索网络拓扑"场景将带领大家一起发现网络的奥秘,深刻体会学习网络技术所带来的乐趣,让学习如同玩一场游戏。最后设计的综合挑战过关练习,概括了本章所学知识,案例设置了重重障碍,我们必须要开动脑筋,冥思苦想,且熟练掌握技术,才能顺利通过挑战闯关。

2.1 管理交换机

2.1.1 交换机的基本命令

表 2-1 为交换机基本配置命令一览表,表 2-2 为常用 show 命令一览表,表 2-3 为命令格式说明一览表。

表 2-1 交换机基本配置命令一览表

配 置 命 令	命 令 功 能	命 令 备 注
hostname	配置主机名	命令环境为全局配置模式
interface vlan <vlan id> ip address ip address subnetmask no shutdown ip default-gateway ip address	配置交换机虚拟接口(SVI)	SVI 接口是交换机管理 VLAN 的接口,SVI 接口须配 IP 地址和子网掩码,交换机才可以被远程访问;若配置默认网关,则可以跨网段被远程访问。管理 VLAN 默认为 1,且须激活才能启用
enable password password enable secret password	配置特权口令	用户在访问特权执行(EXEC)模式前要进行身份验证。命令环境为全局配置模式,前者设置的是明文口令,后者是密文口令,若两者共存,前者失效;若均不配置,IOS 不允许用户远程登录访问其特权 EXEC 模式
line console 0 password password login	配置 Console 口令	命令环境为全局配置模式,Console 0 表示路由器的第一个控制台接口且唯一。login 的作用是当登录时要进行口令验证
line vty [0-15] [1-15] password password login	配置 VTY 口令	命令环境为全局配置模式,可批量配置 VTY 线路,最多支持 16 条,如 line vty 0 15、line vty 0 4,也可以逐条配置 VTY 线路,如 line vty 0、line vty 1 等,Cisco 设备默认支持 5 条
service password-encryption	口令加密	在查看配置文件时防止将口令显示为明文,其作用是对配置文件中的明文口令进行弱加密,一旦加密,即使取消加密服务,也不会消除加密效果
banner motd # message #	配置登录标语	MOTD(Message Of The Day):当日消息,通常用于发布法律通知,声明仅授权人员才可以访问设备,通知会向连接的所有终端显示,"# #"为消息的定界符

续表

配置命令	命令功能	命令备注
copy running-config startup-config 或 write	保存配置	将内存中的运行配置文件 running-config 保存到 NVRAM 中作为启动配置文件的 startup-config 中，copy running-config startup-config 与 write 命令功能一样
delete flash:config.text 或 erase startup-config	删除配置	将交换机的启动配置文件删除

表 2-2　常用 show 命令一览表

常用 show 命令	命令功能解释
show running-config	查看设备当前配置信息
show startup-config	查看设备启动配置文件
show interfaces	查看当前设备所有接口的详细信息
show interfaces *Port ID*	查看当前设备指定接口的详细信息
show ip interface *Port ID*	查看当前设备指定接口的三层详细信息
show ip interface brief	查看当前设备所有接口的三层简要信息
show flash	查看 flash，包括 IOS、启动配置文件、数据库文件等
show version	查看 IOS 的版本信息，包括内存、闪存、接口等

表 2-2 所列出的 show 命令不仅适用于交换机，同样也适用于路由器。

表 2-3　命令格式说明一览表

符号约定	命令符号描述
粗体	需要用户原样输入的命令或关键字
斜体	需要由用户提供，是可变的参数
<X>	尖括号内为必选项（X 为关键字或参数）
[X]	方括号内为可选项（X 为关键字或参数）
[X|Y]	方括号内由竖线分割的关键字或参数表示可选项，即 X 和 Y 二选一
{ X|Y }	花括号内由竖线分割的关键字或参数表示必选项，即 X 和 Y 二选一

要管理交换机，首先要配置交换机。交换机有的工作在二层，有的工作在三层，本章我们只研究二层交换机。交换机的管理是通过 IP 地址实现的。交换机一旦分配了 IP 地址，主机就可以通过 Telnet、SSH 或 Web 方式远程访问该交换机，这在第 1 章中介绍过。交换机的 IP 地址实际上是分配给管理 VLAN 的，详见第 3 章。管理 VLAN 默认只有一个，ID 号为 1，名称为 default，该 VLAN 不可以删除，也不能被修改。

2.1.2 场景一：交换机的基本配置

🔍**场景一**：网络课上，老师要求同学们根据本堂课所学的交换机基本配置命令，依据图 2-1 所示拓扑完成对 Switch1 的基本配置。交换机的主机名为 SW1，特权口令为 ytvc，Console 口令为 cisco，VTY 口令为 ccna，且最多允许 5 个用户 Telnet，为确保安全，要求对明文口令进行加密，登录标语为"Authorized Access Only!"。最后提醒大家：做完实验后，一定要保存配置文件，为避免遗漏，建议通过查看配置文件来检验配置结果。

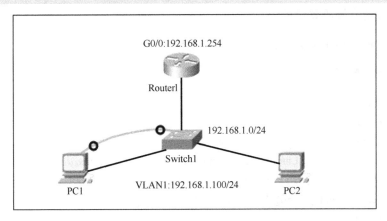

图 2-1 交换机的基本配置

本堂课结束前，老师分享了 15NET2 班 XL 小组提交的实验过程文档，如下所述。

步骤一：配置交换机主机名

```
Switch>enable
Switch#configure terminal
Switch(config)#hostname SW1
```

步骤二：配置虚拟局域网接口（SVI）

```
SW1(config)#interface vlan 1
SW1(config-if)#ip address 192.168.1.100 255.255.255.0
SW1(config-if)#no shutdown
SW1(config-if)#exit
SW1(config)#ip default-gateway 192.168.1.254
```

步骤三：配置交换机口令

```
SW1(config)#enable secret ytvc
SW1(config)#line console 0
```

```
SW1(config-line)#password cisco
SW1(config-line)#login
SW1(config-line)#exit
SW1(config)#line vty 0 4
SW1(config-line)#password ccna
SW1(config-line)#login
SW1(config-line)#exit
SW1(config-)#service password-encryption
```

步骤四：配置交换机登录标语

```
SW1(config)#banner motd #Authorized Access Only!#
```

步骤五：保存配置

```
SW1(config)#end
SW1#copy running-config startup-config
%LINK-5-CHANGED: Interface Vlan1, changed state to up
%LINEPROTO-5-UPDOWN: Line protocol on Interface Vlan1, changed state to up
%SYS-5-CONFIG_I: Configured from console by console
Destination filename [startup-config]?
Building configuration...
[OK]
SW1#
```

步骤六：查看配置文件

```
SW1#show running-config
Building configuration...
Current configuration : 1231 bytes
!
version 12.2
no service timestamps log datetime msec
no service timestamps debug datetime msec
service password-encryption
!
hostname SW1
!
enable secret 5 $1$mERr$PCtyL3OOvlTmQK5Q9sm9H0          //特权模式的密文口令
!
spanning-tree mode pvst
```

```
!
interface FastEthernet0/1
interface FastEthernet0/2
interface FastEthernet0/3
interface FastEthernet0/4
interface FastEthernet0/5
interface FastEthernet0/6
interface FastEthernet0/7
interface FastEthernet0/8
interface FastEthernet0/9
interface FastEthernet0/10
interface FastEthernet0/11
interface FastEthernet0/12
interface FastEthernet0/13
interface FastEthernet0/14
interface FastEthernet0/15
interface FastEthernet0/16
interface FastEthernet0/17
interface FastEthernet0/18
interface FastEthernet0/19
interface FastEthernet0/20
interface FastEthernet0/21
interface FastEthernet0/22
interface FastEthernet0/23
interface FastEthernet0/24
interface GigabitEthernet0/1
interface GigabitEthernet0/2
!
interface Vlan1
  ip address 192.168.1.100 255.255.255.0
!
ip default-gateway 192.168.1.254
!
banner motd ^CAuthorized Access Only!^C        //交换机登录标语
!
line con 0
  password 7 0822455D0A16                      //Console 口令 cisco 已经被加密
  login
!
line vty 0 4
```

```
  password 7 08224F4008                //vty 口令 ccna 已经被加密
  login
 line vty 5 15
  login
 !
 end
 SW1#
```

show running-config 命令后面可以接符号"|"，其作用类似一个筛选符号，后边可以跟关键字 begin（以……开始）、include（包含）、exclude（排除）。例如：

```
SW1#show running-config | begin line     //显示运行配置中以命令 line 开始的配置内容
 line con 0
  password 7 0822455D0A16
  login
 !
 line vty 0 4
  password 7 08224F4008
  login
 line vty 5 15
  login
 !
 End
```

```
SW1#show running-config | include line   //显示运行配置中包含命令 line 的配置内容
 line con 0
 line vty 0 4
 line vty 5 15
SW1#
```

```
SW1#show running-config | exclude line   //显示运行配置中不包括命令 line 的配置内容
Building configuration...

Current configuration : 1231 bytes
!
version 12.2
no service timestamps log datetime msec
no service timestamps debug datetime msec
service password-encryption
!
hostname SW1
```

```
!
enable secret 5 $1$mERr$PCtyL3OOvlTmQK5Q9sm9H0
!
spanning-tree mode pvst
!
-----------------------Output omitted-----------------------
!
interface Vlan1
 ip address 192.168.1.100 255.255.255.0
!
ip default-gateway 192.168.1.254
!
banner motd ^CAuthorized Access Only!^C
!
 password 7 0822455D0A16
 login
!
 password 7 08224F4008
 login
 login
!
End
```

2.1.3 场景二：交换机的口令恢复

Cisco 交换机和路由器的口令恢复方法差别较大，且不同型号交换机的恢复方法也会有所不同。从 Packet Tracer 7.1 开始交换机已经支持口令恢复。在实验室里我们经常会碰到忘记密码的问题，掌握设备的口令恢复非常重要。

> 🔑场景二：15NET2 班同学们在做实验时，发现有的交换机已经设置了口令影响了本节课的实验，这是上节课 15NET1 班同学做交换机基本配置实验时遗留下来的，实验结束后他们没有按照老师的要求对交换机恢复出厂设置。因为有关交换机口令恢复的知识同学们还没有学，所以请求老师的帮助，并记录恢复过程。如图 2-2 所示。

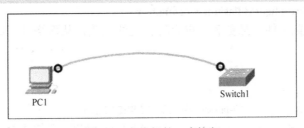

图 2-2 交换机的口令恢复

老师给出的交换机口令恢复过程如下所述。

步骤一：进入交换机口令恢复环境

拔掉交换机电源线，按住面板上的[Mode]按钮不放，而后接通电源，重启交换机，终端显示如下：

switch:

此时，松开[MODE]按钮。

步骤二：初始化交换机 flash 信息

在命令提示符前，输入 **flash_init** 和 **load_helper** 命令来初始化 flash 信息。

switch: **flash_init**
Initializing Flash...
flashfs[0]: 3 files, 0 directories
flashfs[0]: 0 orphaned files, 0 orphaned directories
flashfs[0]: Total bytes: 64016384
flashfs[0]: Bytes used: 4672046
flashfs[0]: Bytes available: 59344338
flashfs[0]: flashfs fsck took 1 seconds.
...done Initializing Flash.
Setting console baud rate to 9600...
switch: **load_helper**

步骤三：更改交换机启动配置文件名

switch: **dir flash:**
Directory of flash:/

4 -rw- 4670455   \<date\>   c2960-lanbase-mz.122-25.SEE1.bin
2 -rw- 975   \<date\>   config.text   //config.text 为交换机的启动配置文件
3 -rw- 616   \<date\>   vlan.dat
59344338 bytes available (4672046 bytes used)

switch: **rename flash:config.text flash:new-config.text**

交换机的启动配置文件一旦更名，重启时就无法加载，从而绕过了口令检查。

switch: **dir flash:**
Directory of flash:/

4 -rw- 4670455   \<date\>   c2960-lanbase-mz.122-25.SEE1.bin
2 -rw- 975   \<date\>   new-config.text   //交换机启动配置文件被更名为 new-config.text
3 -rw- 616   \<date\>   vlan.dat

59344338 bytes available (4672046 bytes used)

步骤四：引导系统重启绕过口令检查

Switch:**boot**

请输入 boot 命令，静待交换机重启，当屏幕出现如下提示：

Would you like to terminate autoinstall?[yes]:**n**　　　　//请回答 n

到此为止，交换机口令恢复已经完成，你可以开始后续实验，若想设置新口令，则须转向步骤五的流程。

步骤五：改回交换机启动配置文件名

Switch#**rename flash:new-config.text flash:config.text**

在特权模式下将启动配置文件恢复为原来的名字 config.text。

步骤六：将启动配置文件加载至内存

Switch:**copy flash:config.text runnig-config**

步骤七：更改各口令，保存配置文件

Cisco 系列交换机口令恢复的方法并不完全一样，但基本思路一致，都需要系统下次重启时，不再加载启动配置文件，绕过口令检查。

2.1.4 场景三：交换机机配置文件的备份与恢复

🔎**场景三**：网络课上，老师向同学们强调要养成一个好的习惯，设备配置完后一定要及时做好备份，必须保存到当前设备，最好再备份到远程主机，以防不测。这与我们经常把计算机上的数据备份到云端是一样的。那么就请你接上次课场景一把配置文件备份到远程服务器，再制造一个交换机配置文件被误删的故障，看你能否通过远程服务器进行恢复。如图 2-3 所示。提示：你可以选择远程 FTP 服务，也可以选择 TFTP 服务。

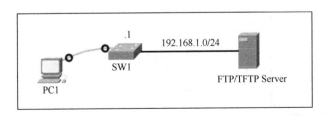

图 2-3　配置文件的备份

XL 同学顺利完成了用 FTP 服务实现交换机配置文件的备份与恢复，下面我们一起来分享

她的实验过程。

步骤一：交换机基本配置

```
Switch(config)#hostname SW1
SW1(config)#interface vlan 1
SW1(config-if)#ip address 192.168.1.1 255.255.255.0
SW1(config-if)#no shutdown
```

步骤二：配置 FTP 客户端

```
SW1(config)#ip ftp username ccna
SW1(config)#ip ftp password ccnp
```

用 FTP 协议在设备之间传输文件，是在 Cisco IOS Release 12.0 之后开始引进的，作为一个面向连接的 TCP 应用，在传输文件之前，FTP 须要提供登录的用户名和口令。我们设置登录 FTP 的用户名为 ccna，口令为 ccnp。

步骤三：配置 FTP 服务器

开启 FTP 服务，如图 2-4 所示，在 FTP 服务器上，创建用户名为 ccna，口令为 ccnp，并为该用户设置相应的权限。

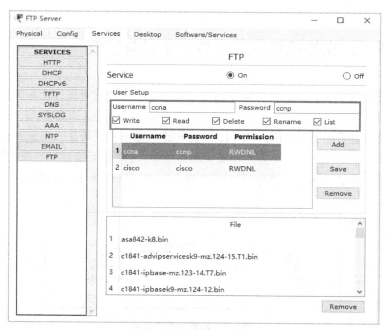

图 2-4　配置 FTP 服务

如图 2-5 所示，配置 FTP 服务器的 IP 地址，确保 FTP 服务器与 FTP 客户端（交换机）在同一网段。

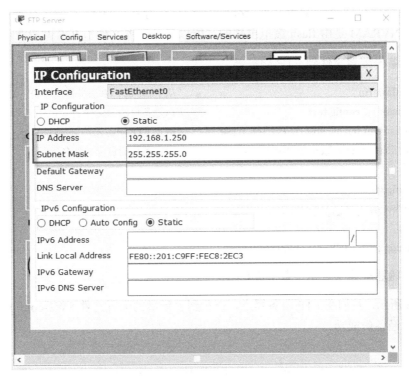

图 2-5　配置 FTP 服务器的 IP 地址

步骤四：测试连通性

```
SW1#ping 192.168.1.250

Type escape sequence to abort.
Sending 5, 100-byte ICMP Echos to 192.168.1.250, timeout is 2 seconds:
!!!!!
Success rate is 100 percent (5/5), round-trip min/avg/max = 0/1/4 ms
```

从以上输出我们可以看出，FTP 客户端 SW1 与 FTP 服务器已经可以 ping 通。

步骤五：查看交换机配置文件

```
SW1#show flash
Directory of flash:/

    1  -rw-     4414921          <no date>    c2960-lanbase-mz.122-25.FX.bin
```

```
    2  -rw-         1099              <no date>  config.text
64016384 bytes total (59599628 bytes free)
```

我们发现交换机的启动配置文件 config.text 已经存在。注意 Cisco 交换机没有真正的 NVRAM，其 NVRAM 是由 flash 虚拟出来的，所以启动配置文件与 IOS 共存于 flash 中。

步骤六：备份配置文件到 FTP 服务器

```
SW1#copy flash ftp                                     //从 flash 到 FTP 服务器的备份
Source filename []? config.text                        //指定备份的源文件名
Address or name of remote host []? 192.168.1.250       //指定 FTP 服务器的 IP 地址
Destination filename [config.text]? SW1-config         //指定备份的目标文件名

Writing config.text...
[OK - 1099 bytes]

1099 bytes copied in 0.078 secs (14000 bytes/sec)
```

步骤七：检验配置文件是否备份成功

如图 2-6 所示，我们在 FTP 服务器上，查看到了新增的配置文件 SW1-config，因此交换机的配置文件已经备份成功。

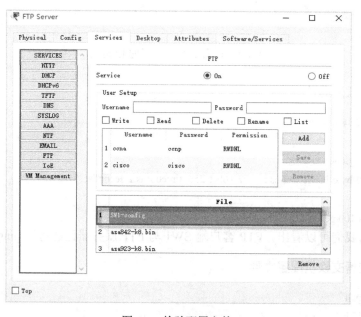

图 2-6 检验配置文件

通过以上实验，我们将交换机的配置文件通过 FTP 协议备份到了远程服务器进行保存。若当前交换机的配置不慎丢失，我们可以通过远程 FTP 服务器进行恢复。

接下来，我们人为制造一个配置文件被误删的故障，将交换机恢复出厂设置。

步骤一：删除配置文件

```
SW1#delete flash:config.text
Delete filename [config.text]?
Delete flash:/config.text? [confirm]
```

通过 delete 命令，我们将 flash 中的配置文件删除，用 show flash 命令进一步确认配置文件 config.text 已被删除。

```
SW1#show flash
Directory of flash:/

    1  -rw-    4414921           <no date>   c2960-lanbase-mz.122-25.FX.bin

64016384 bytes total (59601463 bytes free)
```

重启交换机，发现交换机的提示符为"Switch>"，显然恢复到了出厂设置。

```
SW1#reload
Proceed with reload? [confirm]
Switch>
```

步骤二：配置交换机 SVI

配置交换机管理 VLAN 的 IP 地址，确保与 FTP 服务器同一网段，并完成连通性测试。

步骤三：恢复配置文件

```
Switch#copy ftp flash      //从 FTP 服务器到 flash 的备份
Address or name of remote host []? 192.168.1.250
Source filename []? SW1-config
Destination filename [SW1-config]? config.text

Accessing ftp://192.168.1.250/SW1-config...
[OK - 1099 bytes]

1099 bytes copied in 0.018 secs (61055 bytes/sec)
```

注意：当恢复配置文件时，一定要采用启动配置文件的名字 config.text。通过 show flash 我们发现配置文件已经被成功恢复。

```
Switch#show flash
```

```
Directory of flash:/

    4  -rw-        1099         <no date>   config.text
    1  -rw-     4414921         <no date>   c2960-lanbase-mz.122-25.FX.bin

64016384 bytes total (59600364 bytes free)
```

步骤四：将配置文件加载到内存

```
Switch#copy startup-config running-config
Destination filename [running-config]?

1099 bytes copied in 0.416 secs (2504 bytes/sec)
SW1#
```

这一步非常关键，但往往会被我们忽视。假设我们在恢复配置文件之后，没有将其调入内存，接下来对设备进行配置并保存，那么，启动配置文件中原有的配置就会被覆盖。所以，我们切记要先将启动配置文件调入内存，在此基础上才可以继续进行相应配置。

2.1.5 文件备份命令总结

表 2-4 为文件备份命令一览表。

表 2-4 文件备份命令一览表

文件备份命令	备份命令解释
copy running-config startup-config	把运行配置文件复制到 NVRAM 中保存
copy running-config flash	把运行配置文件复制到 flash 中保存
copy running-config tftp	把运行配置文件复制到 TFTP 服务器中保存
copy running-config ftp	把运行配置文件复制到 FTP 服务器中保存
copy startup-config running-config	把启动配置文件复制到 RAM 中运行
copy startup-config flash	把启动配置文件复制到 flash 中保存
copy startup-config tftp	把启动配置文件复制到 TFTP 服务器中保存
copy startup-config ftp	把启动配置文件复制到 FTP 服务器中保存
copy flash running-config	把 flash 中的配置文件复制到 RAM 中运行
copy flash startup-config	把 flash 中的配置文件复制到 NVRAM 中保存
copy flash tftp	把 flash 中的文件复制到 TFTP 服务器中
copy flash ftp	把 flash 中的文件复制到 FTP 服务器中
copy ftp running-config	把 FTP 服务器中的配置文件复制到 RAM 中
copy ftp startup-config	把 FTP 服务器中的配置文件复制到 NVRAM 中

文件备份命令	备份命令解释
copy ftp flash	把 FTP 服务器中的文件复制到 flash 中
copy tftp running-config	把 TFTP 服务器中的配置文件复制到 RAM 中
copy tftp startup-config	把 TFTP 服务器中的配置文件复制到 NVRAM 中
copy tftp flash	把 TFTP 服务器中文件复制到 flash 中

2.2 管理路由器

2.2.1 路由器的基本命令

路由器基本配置命令一览表如表 2-5 所示。

表 2-5 路由器基本配置命令一览表

配 置 命 令	命 令 功 能	命 令 备 注
hostname	配置主机名	命令环境为全局配置模式
interface *XX* ip address *ip address subnetmask* no shutdown	配置路由器接口 IP 地址	*XX* 可以是局域网接口、广域网接口、子接口等，路由器接口默认被关闭
description *message*	配置接口描述	命令环境为接口配置模式
enable password *password* enable secret *password*	配置特权口令	同交换机基本配置
line console 0 password *password* login	配置 Console 口令	同交换机基本配置
line vty [0-15] [1-15] password *password* login	配置 VTY 口令	同交换机基本配置
service password-encryption	口令加密	同交换机基本配置
banner motd # *message* #	配置登录标语	同交换机基本配置
copy running-config startup-config 或 write	保存配置	同交换机基本配置
erase startup-config	删除配置	同交换机基本配置

2.2.2 场景四：路由器的基本配置

🔑**场景四**：今天的网络课，老师带领大家学习了路由器的基本配置命令。同学们一致反映特别简单，和前期学习交换机的基本配置几乎一样。老师向大家反复强调路由器工作在三层，交换机主要工作在二层。路由器是网间互联设备，它能分割广播域，且每个接口都连接一个网段，每个接口都可以配置一个地址，但是交换机只能配置一个管理 IP 地址。今天老师给大家布置的作业是：依据图 2-7 所示拓扑完成对路由器的基本配置。路由器的主机名为 R1，两个接口分别连接 15NET1 和 15NET2 两个班的 LAN。为管理方便，要求配置所在网段的接口描述，特权密文口令为 ytvc，Console 口令为 cisco，VTY 口令为 ccna，且最多允许 5 个用户 Telnet。为确保安全，要求对明文口令进行加密，登录标语为 "Authorized Access Only!"。请保存配置，进一步检验配置结果，确保两个网段间彼此可以通信。

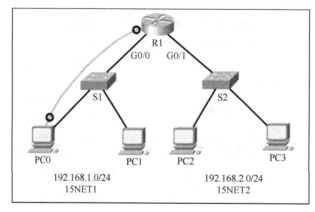

图 2-7　路由器的基本配置

实验结束后，老师分享了 ZHB 小组提交的实验过程文档，我们一起来学习。

步骤一：配置路由器主机名

Router>**enable**
Router#**configure terminal**
Router(config)#**hostname R1**

步骤二：配置路由器接口 IP 地址

R1(config)#**interface GigabitEthernet 0/0**
R1(config-if)#**ip address 192.168.1.254 255.255.255.0**
R1(config-if)#**description Link to 15NET1**
R1(config-if)#**no shutdown**

```
R1(config-if)#exit
R1(config)#interface GigabitEthernet 0/1
R1(config-if)#ip address 192.168.2.254 255.255.255.0
R1(config-if)#description Link to 15NET2
R1(config-if)#no shutdown
```

步骤三：配置路由器相关口令

```
R1(config)#enable secret ytvc
R1(config)#line console 0
R1(config-line)#password cisco
R1(config-line)#login
R1(config-line)#exit
R1(config)#line vty 0 4
R1(config-line)#password ccna
R1(config-line)#login
R1(config-line)#exit
R1(config-)#service password-encryption
```

步骤四：配置路由器登录标语

```
R1(config)#banner motd #Authorized Access Only!#
```

步骤五：保存路由器配置

```
R1(config)#end
R1#copy running-config startup-config
Destination filename [startup-config]?
Building configuration...
[OK]
R1#
```

步骤六：查看配置文件

```
R1#show running-config
Building configuration...

Current configuration : 757 bytes
!
version 15.1
no service timestamps log datetime msec
no service timestamps debug datetime msec
```

```
service password-encryption
!
hostname R1
!
enable secret 5 $1$mERr$PCtyL3OOvlTmQK5Q9sm9H0    //特权模式的密文口令
!
ip cef
no ipv6 cef
!
license udi pid CISCO1941/K9 sn FTX1524IQF4
!
spanning-tree mode pvst
!
interface GigabitEthernet0/0
 description Link to 15NET1
 ip address 192.168.1.254 255.255.255.0
 duplex auto
 speed auto
!
interface GigabitEthernet0/1
 description Link to 15NET2
 ip address 192.168.2.254 255.255.255.0
 duplex auto
 speed auto
 shutdown
!
interface Vlan1
 no ip address
 shutdown
!
ip classless
!
ip flow-export version 9
!
banner motd ^CAuthorized Access Only!^C    //路由器登录标语
!
line con 0
 password 7 0822455D0A16    //console 口的口令 cisco 已经被加密
```

```
  login
 !
 line aux 0
 !
 line vty 0 4
  password 7 08224F4008    //vty 的口令 ccna 已经被加密
  login
 !
 end

R1#
```

步骤七：测试网间互通

```
C:\>ipconfig

FastEthernet0 Connection:(default port)

   Link-local IPv6 Address.........: FE80::207:ECFF:FE34:6816
   IP Address..................  .........: 192.168.1.1
   Subnet Mask..................  ......: 255.255.255.0
   Default Gateway................ ..: 192.168.1.254
```

我们选出 15NET1 班的一台 PC，其 IP 地址为 192.168.1.1

```
C:\>ping 192.168.2.1

Pinging 192.168.2.1 with 32 bytes of data:

Reply from 192.168.2.1: bytes=32 time<1ms TTL=127
Reply from 192.168.2.1: bytes=32 time=11ms TTL=127
Reply from 192.168.2.1: bytes=32 time=13ms TTL=127
Reply from 192.168.2.1: bytes=32 time=16ms TTL=127

Ping statistics for 192.168.2.1:
    Packets: Sent = 4, Received = 4, Lost = 0 (0% loss),
Approximate round trip times in milli-seconds:
    Minimum = 0ms, Maximum = 16ms, Average = 10ms
```

通过测试，我们发现 15NET1 班的网段 192.168.1.0/24 与 15NET2 班的网段 192.168.2.0/24 实现了互通。

2.2.3 场景五：路由器的口令恢复

Cisco 路由器口令恢复和交换机口令恢复的相同之处在于都要绕过口令检查，两者都是控制设备启动时不加载启动配置文件。交换机是通过更改启动配置文件的文件名，让交换机启动时找不到启动配置文件，从而无法加载启动配置文件，绕过口令检查。路由器则是通过修改配置寄存器的值，达到不加载启动配置文件的目的。路由器的配置寄存器指定位的值决定了路由器是加载还是忽略 NVRAM 中配置信息。

> 🔑 **场景五**：今天上午网络课学习的内容是路由器的口令恢复。老师告知同学们用自己的笔记本电脑打开 Packet Tracer 就可以完成，不必去实验室，因为模拟器支持。本节课老师让大家通过自学，依据 2.1.3 场景二中交换机的口令恢复过程整理路由器口令恢复的步骤。同时，老师给大家群发了一个 PKA 文件，如图 2-8 所示，图中路由器被老师设置了 enable 口令，要求同学们来破解。接下来，同学们开始分组讨论，学习热情高涨，HY 同学很快就完成了任务。老师让他依据交换机的口令恢复步骤，整理了一份路由器口令恢复的文档，和大家一起来分享。

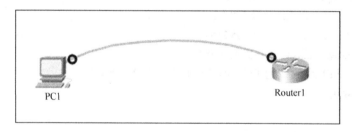

图 2-8 路由器的口令恢复

下面让我们一起依据 HY 同学的文档来学习路由器口令恢复的过程。

步骤一：进入路由器口令恢复环境

关闭路由器电源，重启路由器，开机 60 s 内按【Ctrl+Break】或【Ctrl+C】键中断路由器的启动，进入 rommon 监视模式，具体如下：

System Bootstrap, Version 15.1(4)M4, RELEASE SOFTWARE (fc1)
Technical Support: http://www.cisco.com/techsupport
Copyright (c) 2010 by cisco Systems, Inc.
Total memory size = 512 MB - On-board = 512 MB, DIMM0 = 0 MB
CISCO1941/K9 platform with 524288 Kbytes of main memory
Main memory is configured to 64/-1(On-board/DIMM0) bit mode with ECC disabled

```
Readonly ROMMON initialized

program load complete, entry point: 0x80803000, size: 0x1b340
program load complete, entry point: 0x80803000, size: 0x1b340
IOS Image Load Test
_____
Digitally Signed Release Software
program load complete, entry point: 0x81000000, size: 0x2bb1c58
Self decompressing the image :
###########################
monitor: command "boot" aborted due to user interrupt
rommon 1 >
```

步骤二：修改路由器配置寄存器值

```
rommon 1 >confreg 0x2142
```

修改配置寄存器的值为 0x2142，当路由器下一次重启时，不会加载 NVRAM 中的配置文件 startup-config，从而也就绕过了口令检查。

步骤三：引导系统重启绕过口令检查

```
rommon 2 > reset
Output omitted

        --- System Configuration Dialog ---

Continue with configuration dialog? [yes/no]: no      //退出 setup 模式，返回 CLI 界面


Press RETURN to get started!
Router>
```

步骤四：将启动配置文件加载至内存

```
Router>enable
Router#copy startup-config running-config
Destination filename [running-config]?
657 bytes copied in 0.416 secs (1579 bytes/sec)
```

将启动配置文件 startup-config 从 NVRAM 拷贝到 RAM。这样，可以保留原有配置，并在此基础上修改口令。切记：修改后一定要再次保存配置。

步骤五：改回配置寄存器值

Router1#**configure terminal**
Router1(config)#**config-register 0x2102**

把配置寄存器的值从 0x2142 恢复为正常值 0x2102，确保下次重启会加载启动配置文件，提高系统安全性。

步骤六：修改各口令，保存配置

Router1(config)#**enable secret cisco**
Router1#**copy running-config startup-config**
Destination filename [startup-config]?
Building configuration...
[OK]
Router1#**reload**   //重启路由器，检查路由器是否正常

更改特权口令，若还配置其他口令（如 Console、VTY 等）则一起修改，并将新口令更新到启动配置文件。

2.2.4 场景六：路由器配置文件的备份与恢复

> 🔍**场景六**：今天网络课的前 20 分钟，老师让同学们自主学习路由器配置文件的备份与恢复，并采用小组比赛的方式进行。要求比赛不仅要成功，还要写出中间过程，上交文档，这需要小组成员间积极配合。老师说 2.1.4 场景三中交换机配置文件的备份与恢复，我们采用 FTP 服务和 TFTP 服务都可以，今天为节省时间，指定大家采用 TFTP 服务，如图 2-9 所示。

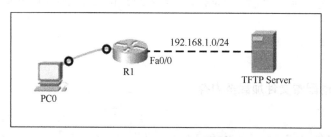

图 2-9 备份路由器配置文件

WMC 同学所在组第一个通过验收，下面我们一起来分享他们组的实验成果吧。

步骤一：路由器基本配置

Router(config)#**hostname R1**
R1(config)#**interface fastEthernet 0/0**
R1(config-if)#**ip address 192.168.1.1 255.255.255.0**
R1(config-if)#**no shutdown**

步骤二：配置 TFTP 服务器

确保 TFTP 服务已经开启，配置 TFTP 服务器的 IP 地址为 192.168.1.250/24。

步骤三：测试连通性

SW1#**ping 192.168.1.250**

Type escape sequence to abort.
Sending 5, 100-byte ICMP Echos to 192.168.1.250, timeout is 2 seconds:
!!!!!
Success rate is 100 percent (5/5), round-trip min/avg/max = 0/1/4 ms

通过以上输出，我们可以看出路由器已经具备与 TFTP 服务器通信的基本条件。

步骤四：备份配置文件到 TFTP 服务器

R1#**copy startup-config tftp:**   //从 NVRAM 到 TFTP 服务器的备份
Address or name of remote host []? **192.168.1.250**
Destination filename [R1-confg]?

Writing startup-config...!!
[OK - 563 bytes]
563 bytes copied in 0.001 secs (563000 bytes/sec)

步骤五：检验 TFTP 服务器是否备份成功

通过检验，我们发现在 TFTP 服务器上，路由器 R1 的配置 R1-confg 已经备份成功。以上实验，我们实现了配置文件的远程备份。接下来，我们再通过远程 TFTP 服务器来恢复当前路由器的配置。

步骤一：删除路由器配置

R1#**erase startup-config**
Erasing the nvram filesystem will remove all configuration files! Continue? [confirm]
[OK]
Erase of nvram: complete

%SYS-7-NV_BLOCK_INIT: Initialized the geometry of nvram

步骤二：恢复路由器出厂设置

使用 reload 命令重启路由器，恢复路由器到出厂设置。

步骤三：配置路由器接口地址

配置路由器接口 IP 地址，使其与 TFTP 服务器在同一网段，确保路由器接口能与 TFTP 服务器 ping 通。

步骤四：恢复路由器配置

Router#**copy tftp startup-config**   //从 TFTP 服务器到 NVRAM 的备份
Address or name of remote host []? **192.168.1.250**
Source filename []? **R1-confg**
Destination filename [startup-config]?

Accessing tftp://192.168.1.250/R1-confg....
Loading R1-confg from 192.168.1.250: !
[OK - 563 bytes]

563 bytes copied in 3.003 secs (187 bytes/sec)

步骤五：加载配置文件至内存

Router#**copy startup-config running-config**
R1#

注意：恢复配置文件，我们也可以先使用命令 copy tftp running-config，而后使用 copy running-config startup-config 或 write 命令来保存配置文件。

2.2.5 动态主机配置协议 DHCP

DHCP（Dynamic Host Configuration Protocol，动态主机配置协议）是为客户端动态分配 IP 地址的协议。服务器能够从预先设置的 IP 地址池里自动给主机分配 IP 地址，不仅能够保证 IP 地址不重复分配，也能及时回收 IP 地址，提高 IP 地址的利用率。DHCP 具有可伸缩性，相对容易管理。DHCP 包括 3 种不同的地址分配机制，如表 2-6 所示，其工作过程和报文格式如表 2-7、表 2-8 所示。

表 2-6　DHCP 三种分配机制

DHCP 分配机制	DHCP 分配机制说明
手工分配（Manual Allocation）	管理员为客户端提供指定预分配的 IP 地址
自动分配（Automatic Allocation）	DHCP 服务器为客户端指定一个永久性 IP 地址
动态分配（Dynamic Allocation）	动态地分配或出租 IP 地址，有租期期限

表 2-7　DHCP 工作过程

DHCP 工作过程	DHCP 工作过程说明
发现阶段	DHCP 客户端发送 DHCP DISCOVER 报文寻找 DHCP 服务器
提供阶段	DHCP 服务器发送 DHCP OFFER 报文提供 IP 地址等相关参数
选择阶段	DHCP 客户端发送 DHCP REQUEST 报文选择某台 DHCP 服务器所提供地址参数
确认阶段	DHCP 服务器发送 DHCP ACK 报文确认所提供的 IP 地址等相关参数
重新续约	DHCP 客户端发送 DHCP REQUEST 报文请求前一次所分配的 IP 地址参数
更新租约	DHCP 服务器发送 DHCP REQUEST 报文更新续租 IP 地址参数

表 2-8　DHCP 报文格式

DHCP 报文格式	DHCP 报文说明
DHCP DISCOVER	客户端通过发送 DHCP DISCOVER 报文寻找网络中的 DHCP 服务器
DHCP OFFER	当 DHCP 服务器收到 DHCP DISCOVER 报文时，DHCP 服务器会以 DHCP OFFER 报文作为响应。DHCP 服务器还会创建一个 ARP 条目，该条目包含请求客户端的 MAC 地址和客户端的租赁地址
DHCP REQUEST	当 DHCP 客户端收到服务器端 DHCP OFFER 报文时，它会发回一条 DHCP REQUEST 报文，目的用于向对方发起租用和租约的更新请求
DHCP ACK	DHCP 服务器收到客户端的 DHCP REQUEST 报文后，它会使用 ICMP ping 工具来检验该地址的租用信息以确保该地址尚未使用，并为 DHCP 客户端创建新的 ARP 条目，以单播 DHCP ACK 报文作为回复

2.2.6　场景七：配置路由器的 DHCP 服务

🔍 **场景七**：XBG 同学向老师提出了一个问题，他在网上看了一句话"DHCP 协议的诞生，是网络管理员的福音"，他问老师这是什么意思。老师笑着回答说："家庭洗衣机的诞生是广大劳动妇女的福音，这大家明白是什么意思吗？"同学们不解地望着老师，老师进一步解释说，有了家庭洗衣机，广大劳动妇女就可以从繁重的手工洗衣中解脱出来，DHCP 协议的诞生也使得网络管理员从烦琐的手工为主机分配地址工作中解脱出来，更重要的是避免手工分配主机地址所带来的不可避免的地址冲突问题。接下来老师给大家演示了让路由器作为 DCHP 服务器给主机动态分配地址的案例，同学们听得很轻松，很快就能理解 DHCP 动态分配地址的作用了，大家一致反映以后不再静态指定地址了，但老师强调网络设备以及服务器的地址必须要手工指定，为方便管理，终端 PC 地址最好采用 DHCP 自动获取。

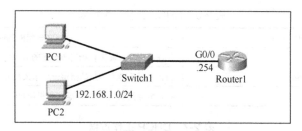

图 2-10 路由器做 DHCP 服务器

下面我们一起来总结一下本节课 DHCP 服务的配置过程，请记住实验步骤。

步骤一：配置路由器为 DHCP 服务器

Router(config)#**hostname Router1**
Router1(config)#**interface G0/0**
Router1(config-if)#**ip address192.168.1.254 255.255.255.0**
Router1(config-if)#**no shutdown**
Router1(config-if)#**exit**
Router1(config)#**ip dhcp excluded-address 192.168.1.1 192.168.1.10**   //排除地址段，根据要求自行决定
Router1(config)#**ip dhcp pool ytvc**   //建立地址池，名称为 ytvc
Router1(dhcp-config)#**network 192.168.1.0 255.255.255.0**   //设置 DHCP 分配的网络子网掩码
Router1(dhcp-config)#**default-router 192.168.1.254**   //设置 DHCP 分配的网关
Router1(dhcp-config)#**dns-server 1.1.1.1**   //设置 DHCP 分配的 DNS

步骤二：配置 PC 为 DHCP 客户机

图 2-11 配置主机自动获取地址

步骤三：测试 DHCP 服务

DHCP 的相应命令如表 2-9 所示。

表 2-9 DHCP 的相应命令

DHCP 相应命令	DHCP 命令说明
ipconfig /renew	更新 IP 地址等相关参数
ipconfig /all	查看 IP 地址等相关参数
ipconfig /release	释放 IP 地址等相关参数
show ip dhcp pool	查看 DHCP 地址池的信息
show ip dhcp binding	查看 DHCP 地址绑定情况

更新主机 IP 地址，并检查 PC 是否获取到地址，如图 2-12 所示。

```
C:\>ipconfig /renew

    IP Address......................: 192.168.1.11
    Subnet Mask.....................: 255.255.255.0
    Default Gateway.................: 192.168.1.254
    DNS Server......................: 1.1.1.1
```

图 2-12 更新主机 IP 地址

经检查验证 PC 已经获取 DHCP 服务器分配来的地址，包括 IP 地址、子网掩码、默认网关等参数，如图 2-13 所示。

```
C:\>ipconfig /all

FastEthernet0 Connection:(default port)

    Connection-specific DNS Suffix..:
    Physical Address................: 0002.4A77.D158
    Link-local IPv6 Address.........: FE80::202:4AFF:FE77:D158
    IP Address......................: 192.168.1.11
    Subnet Mask.....................: 255.255.255.0
    Default Gateway.................: 192.168.1.254
    DNS Servers.....................: 1.1.1.1
    DHCP Servers....................: 192.168.1.254
    DHCPv6 Client DUID..............: 00-01-00-01-2E-C8-C0-C6-00-02-4A-77-D1-58
```

图 2-13 查看主机 IP 地址

通过 ipconfig /all 命令，我们看到了更详细的地址信息，包括 DHCP Server 的地址，进一步确认 DHCP 已经正常工作。通过 ipconfig /release 命令可成功释放当前 PC 获取的地址，如图 2-14 所示。

```
C:\>ipconfig /release

    IP Address......................: 0.0.0.0
    Subnet Mask.....................: 0.0.0.0
    Default Gateway.................: 0.0.0.0
    DNS Server......................: 0.0.0.0
```

图 2-14　释放主机 IP 地址

我们也可以通过图 2-15 来查看 DHCP 地址池信息，包括地址池的名字、排除的地址数、租用的地址数等。

```
Router1#show ip dhcp pool

Pool ytvc :
 Utilization mark (high/low)    : 100 / 0
 Subnet size (first/next)       : 0 / 0
 Total addresses                : 254
 Leased addresses               : 2
 Excluded addresses             : 1
 Pending event                  : none

 1 subnet is currently in the pool
 Current index        IP address range                       Leased/Excluded/Total
 192.168.1.1          192.168.1.1     - 192.168.1.254        2    / 1     / 254
```

图 2-15　查看 DHCP 地址池信息

由图 2-16 我们知道了 DHCP 地址绑定情况，以上 DHCP 的查看命令对我们理解 DHCP 的工作过程是非常有帮助的。

```
Router1#show ip dhcp binding
IP address      Client-ID/              Lease expiration        Type
                Hardware address
192.168.1.11    0002.4A77.D158          --                      Automatic
192.168.1.12    00E0.F70C.7506          --                      Automatic
```

图 2-16　查看 DHCP 地址绑定

2.3　管理设备 IOS

2.3.1　认识网络设备 IOS

思科路由器采用的操作系统软件称为思科 IOS（Internetwork Operating System）。与计算机上的操作系统一样，思科 IOS 会管理路由器的硬件和软件资源，包括存储器分配、进程、安全性和文件系统。思科 IOS 属于多任务操作系统，集成了路由、交换、Internet 网络及电信等功能。思科 IOS 包含不同类型的 IOS 映像，这些映像对应着相应的硬件平台。

虽然思科路由器中的许多 IOS 看似相同，但实际上却是不同类型。IOS 映像是一种包含相应路由器完整 IOS 的文件。思科根据路由器型号和 IOS 内部的功能，创建了许多不同类型的 IOS 映像。通常，IOS 内部的功能越多，IOS 映像就越大，因此就需要越多的闪存和 RAM 来存储和加载 IOS。

2.3.2　认识 IOS 命名规则

Cisco 通用 IOS 命名规则可以表示为 AAAAA-BBBB-CC-DDDD.EE。其中 AAAAA 标识映像运行的硬件平台，BBBB 指定功能集，CC 代表 IOS 的文件格式，DDDD 表示 IOS 软件版本，EE 则代表 IOS 映像的扩展名。

下面我们以 c2800nm-advipservicesk9-mz.124-6.T.bin 和 c1900-universalk9-mz.SPA.152-4.M3.bin 两种 IOS 映像为例解释每部分的含义。

- **c2800nm 和 c1900**：标识映像运行的平台，其中 c2800nm 代表的平台是带网络模块的 Cisco 2800 路由器，c1900 代表的平台是 Cisco1900 路由器。
- **advipservicesk9 和 universalk9**：指定功能集，advipservicesk9 包含高级安全、服务提供商软件包以及 IPv6 高级 IP 服务功能集；universaik 9 包含 IP Base、安全、统一通信和数据 4 个技术包；而 universalk9_npe 不包含强加密，适用于有加密限制的国家 / 地区。
- **mz**：表示映像在何处运行以及文件是否经过压缩。最常见的存储位置和压缩格式标识是 mz，表示是从 RAM 运行，并经过压缩，压缩格式是 zip。存储位置有 m（内存 RAM）、f（flash）、r（ROM）、l（可重定位）；压缩格式也可以是 x，表示 mzip。压缩是为了减小映像大小，因此加载映像到 RAM 中第一个操作就是自我解压。
- **124-6.T 和 152-4.M3**：表示映像的文件格式，前者是映像 12.4（6）T 的文件格式，后者是映像 15.2（4）M3 的文件格式。映像的版本（12.4 和 15.2）信息包括主版本（12 和 15）、次版本（4 和 2）、维护版本号以及维护重建编号（漏洞修复）。T 表示标准维护版，M 表示扩展维护版本。
- **SPA**：表示文件是由思科以数字形式签名的。
- **bin**：文件扩展名，该扩展名表示此文件是二进制可执行文件。

2.3.3　场景八：升级交换机的 IOS

🔎**场景八**：今天，15NET2 班的网络课被老师从实验室调换到教室。上课时，老师因临时调换教室向同学们道歉。接下来，老师告知大家不能去实验室进行真机实验的原因。上节课，15NET1 班在实验室做交换机 IOS 升级实验时，造成多台交换机升级失败，影响了 15NET2 班不能正常进行实验。老师再三强调 IOS 命名规则，第一项就是映像的运行平台，可是真正实验时，有同学竟然将 c2960-lanbasek9-mz.150-2.SE4.bin 映像文件传到了 2950 交换机上，因为 flash 存储空间太小，所以有同学删除了原有的旧版 IOS，重启交换机，造成设备无法正常启动，问题非常严重，老师不得已才临时回教室上课。还好，因为同学们都有自己的笔记本电脑，所以今天真机实验改成了模拟实验。老师布置了本节课任务：打开 Packet Tracer 模拟器，做如图 2-17 所示模拟交换机 IOS 升级实验。同时提醒同学们要养成好习惯，在做 IOS 升级之前，一定要把交换机原有 IOS 进行远程备份，以防不测。老师还提醒大家一定要检查 flash 存储空间大小，把旧版 IOS 和新版 IOS 文件大小做一下比较，若足够容纳两个版本 IOS，那么最好保留两个。如何采用新版 IOS 启动，老师让同学们通过网上查阅资料和小组讨论方式去解决问题。

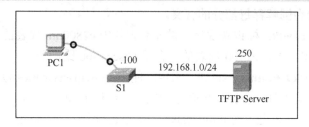

图 2-17　升级交换机 IOS

下面我们来一起分享 15NET2 班 ZHB 小组总结的实验步骤。

步骤一：配置交换机的 SVI 接口

配置交换机的 SVI 接口地址为 192.168.1.100/24，TFTP 服务器地址为 192.168.1.250/24，做 ping 连通性测试，使得交换机和服务器之间能相互 ping 通。

步骤二：备份交换机 IOS 至 TFTP

```
S1#show flash
Directory of flash:/

    1  -rw-     4414921          <no date>    c2960-lanbase-mz.122-25.FX.bin
    2  -rw-        1056          <no date>    config.text

64016384 bytes total (59600407 bytes free)
```

通过 show 命令，我们查看到当前交换机现有 IOS，了解当前交换机 flash 所剩空间大约有 59 MB，而现有 IOS 映像大约是 4 MB。

```
S1#copy flash: tftp
Source filename []? c2960-lanbase-mz.122-25.FX.bin
Address or name of remote host []? 192.168.1.250
```

Destination filename [c2960-lanbase-mz.122-25.FX.bin]?

Writing c2960-lanbase-mz.122-25.FX.bin
...!!
!!!!!!!!
[OK - 4414921 bytes]

4414921 bytes copied in 0.412 secs (291149 bytes/sec)

使用命令 copy flash tftp 将交换机旧版 IOS c2960-lanbase-mz.122-25.FX.bin 成功备份到 TFTP 服务器，当然备份时间长短取决于 IOS 映像的大小。

步骤三：升级交换机的 IOS

升级 IOS 之前，一定要比较 flash 所剩空间与高版本 IOS 映像文件的大小，若 flash 存储空间足够大，则先保留原有 IOS，再升级；若空间不足仅能容纳新版 IOS，则需要删除旧版 IOS，再升级；但若删除旧版 IOS 之后，才发现空间不足以容纳新版 IOS，则无法升级，需要先做 IOS 恢复，切记恢复 IOS 之前，不要重启交换机。

```
S1#copy tftp: flash          //从 TFTP 到 flash 的备份
Address or name of remote host []? 192.168.1.250
Source filename []? c2960-lanbasek9-mz.150-2.SE4.bin      //新版 IOS 的名称
Destination filename [c2960-lanbasek9-mz.150-2.SE4.bin]?

Accessing tftp://192.168.1.250/c2960-lanbasek9-mz.150-2.SE4.bin...
Loading c2960-lanbasek9-mz.150-2.SE4.bin from 192.168.1.250:

!!!!!!!!!!!!!!!!!!!!!!!!!!!!!!!!!!!!!!!!!!!!!!!!!!!!!!!!!!!!!!!!!!!!
[OK - 4670455 bytes]

4670455 bytes copied in 0.256 secs (1466748 bytes/sec)
```

flash 有足够的存储空间，可以容纳多个版本的 IOS，我们可以通过 show flash 命令查看，发现有两个 IOS 存在。

```
S1#show flash
Directory of flash:/

    1  -rw-     4414921          <no date>  c2960-lanbase-mz.122-25.FX.bin
    3  -rw-     4670455          <no date>  c2960-lanbasek9-mz.150-2.SE4.bin
    2  -rw-        1056          <no date>  config.text
```

64016384 bytes total (54929952 bytes free)

我们可以进一步用 show version 命令来查看当前交换机加载的是哪一个版本。

S1#show version
Cisco IOS Software, C2960 Software (C2960-LANBASE-M), Version 12.2(25)FX, RELEASE SOFTWARE (fc1)   //交换机 IOS 版本
Copyright (c) 1986-2005 by Cisco Systems, Inc.   //交换机 IOS 版权相关信息
Compiled Wed 12-Oct-05 22:05 by pt_team
ROM: C2960 Boot Loader (C2960-HBOOT-M) Version 12.2(25r)FX, RELEASE SOFTWARE (fc4)   //引导程序的版本
System returned to ROM by power-on
Cisco WS-C2960-24TT (RC32300) processor (revision C0) with 21039K bytes of memory.
//系统映像文件名称及引导程序加载该文件的位置
24 FastEthernet/IEEE 802.3 interface(s)
2 Gigabit Ethernet/IEEE 802.3 interface(s)   //交换机物理接口的数量及类型
63488K bytes of flash-simulated non-volatile configuration memory   //flash 模拟 NVRAM

Base ethernet MAC Address	: 0009.7C77.7849	//交换机的基 MAC 地址
Motherboard assembly number	: 73-9832-06	//主板组装编号
Power supply part number	: 341-0097-02	//电源部分编号
Motherboard serial number	: FOC103248MJ	//主板序列号
Power supply serial number	: DCA102133JA	//电源序列号
Model revision number	: B0	//模块修订号
Motherboard revision number	: C0	//主板修订号
Model number	: WS-C2960-24TT	//交换机型号
System serial number	: FOC1033Z1EY	//系统序列号
Top Assembly Part Number	: 800-26671-02	//顶级组装零件编号
Top Assembly Revision Number	: B0	//顶级组装修订编号
Version ID	: V02	//版本标识
CLEI Code Number	: COM3K00BRA	//CLEI 代码编号
Hardware Board Revision Number	: 0x01	//硬件板修订号

Switch	Ports	Model	SW Version	SW Image
* 1	26	WS-C2960-24TT	12.2	C2960-LANBASE-M

//交换机默认 VLAN 号、端口数量、型号、映像版本号以及映像名称
Configuration register is 0xF   //配置寄存器值
S1#

当前交换机已经升级了 IOS，用 show version 命令查看发现，当前内存里运行的依然是旧版 12.2 的 IOS。让交换机加载新版 IOS，应采用 boot system 命令来指定加载 IOS 映像名。

```
S1(config)#boot system flash:c2960-lanbasek9-mz.150-2.SE4.bin
S1(config)#end
S1#reload
```

交换机经 reload 命令重启后，我们再查看内存运行的 IOS 版本。

```
S1#show version
Cisco IOS Software, C2960 Software (C2960-LANBASEK9-M), Version 15.0(2)SE4, RELEASE SOFTWARE (fc1)
Technical Support: http://www.cisco.com/techsupport    //技术支持的网址
Copyright (c) 1986-2013 by Cisco Systems, Inc.
Compiled Wed 26-Jun-13 02:49 by mnguyen


ROM: Bootstrap program is C2960 boot loader
BOOTLDR: C2960 Boot Loader (C2960-HBOOT-M) Version 12.2(25r)FX, RELEASE SOFTWARE (fc4)


Switch uptime is 39 minutes
System returned to ROM by power-on
System image file is "flash:c2960-lanbasek9-mz.150-2.SE4.bin"


This product contains cryptographic features and is subject to United
States and local country laws governing import, export, transfer and
use. Delivery of Cisco cryptographic products does not imply
third-party authority to import, export, distribute or use encryption.
Importers, exporters, distributors and users are responsible for
compliance with U.S. and local country laws. By using this product you
agree to comply with applicable laws and regulations. If you are unable
to comply with U.S. and local laws, return this product immediately.


A summary of U.S. laws governing Cisco cryptographic products may be found at:
http://www.cisco.com/wwl/export/crypto/tool/stqrg.html
//通过网址查看思科产品法律摘要
If you require further assistance please contact us by sending email to
export@cisco.com.        //发送电子邮件获取进一步帮助


cisco WS-C2960-24TT-L (PowerPC405) processor (revision B0) with 65536K bytes of memory.
Processor board ID FOC1010X104
```

```
Last reset from power-on
1 Virtual Ethernet interface
24 FastEthernet interfaces
2 Gigabit Ethernet interfaces
The password-recovery mechanism is enabled.

64K bytes of flash-simulated non-volatile configuration memory.
Base ethernet MAC Address       : 00:17:59:A7:51:80
Motherboard assembly number     : 73-10390-03
Power supply part number        : 341-0097-02
Motherboard serial number       : FOC10093R12
Power supply serial number      : AZS1007032H
Model revision number           : B0
Motherboard revision number     : B0
Model number                    : WS-C2960-24TT-L
System serial number            : FOC1010X104
Top Assembly Part Number        : 800-27221-02
Top Assembly Revision Number    : A0
Version ID                      : V02
CLEI Code Number                : COM3L00BRA
Hardware Board Revision Number  : 0x01


Switch  Ports   Model             SW Version       SW Image
------  -----   -----             ----------       --------
*  1    26      WS-C2960-24TT-L   15.0(2)SE4       C2960-LANBASEK9-M


Configuration register is 0xF
```

从以上输出，我们确认新版 IOS 加载映像成功。命令 boot system flash 让引导程序在闪存中指定加载 IOS 映像。默认情况下，引导程序会加载闪存中第一个有效 IOS 映像。boot system 命令迫使引导程序加载指定映像。执行升级时，闪存中若有新、旧两个版本 IOS 映像，则默认加载旧版（因为它在闪存中位于前面），除非你使用命令 boot system flash 改变或删除旧版 IOS 映像。你还可以让引导程序从 TFTP 服务器加载 IOS 映像，但对于大型 IOS，不推荐这样做，因为速度很慢。也可使用命令 boot system rom 让引导程序加载 ROM 中的简版 IOS，但目前版本的 Packet Tracer 尚不支持该命令。

2.3.4 场景九：恢复路由器的 IOS

🔍**场景九**：今天 15NET1 班的网络课老师选择在教室上课，学习任务是路由器 IOS 的备份与恢复。老师让全班同学打开 Packet Tracer 模拟器，把上节课所采用的交换机换成路由器即可开始今天的实验，如图 2-18 所示。实验前，老师还总结了上次在实验室做交换机 IOS 升级时同学们犯下的错误，提醒同学们用模拟器做实验也必须严格按照规范去做。可是模拟实验并没有像老师想象的那么顺利，班上同学讨论异常激烈。ZHN 同学向老师反映他用 FTP 备份好长时间也没有反应，人为中断若干次，最终还是需要等好久系统才有反应，而他们组 HW 同学用 TFTP 备份速度非常快；ZHP 同学则向老师反映他的路由器 IOS 删除后，还没有做恢复，就被重启，可是路由器的主机名竟然变成了 rommon；XL 同学请老师快过去看看她的路由器 flash 上有两个怪怪的文件名，不知道是做什么用的。老师让三位提问题的同学把实验输出结果保存到记事本上传到教师机。

接下来，老师将问题进行了汇总，结果如下。

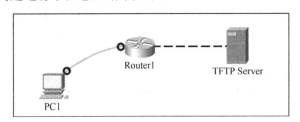

图 2-18　备份路由器 IOS

问题一：ZHN 组的问题

同一 IOS，用 FTP 协议传输用时间为 164.073 s，而 HW 用 TFTP 协议备份，0.717s 就完成，前者为什么那么慢，而后者那么快呢？ZHN 实验输出如下：

R1#**copy flash ftp**
Source filename []? C2900-universalk9-mz.SPA.151-4.M4.bin
Address or name of remote host []? 192.168.1.250
Destination filename [c2900-universalk9-mz.SPA.151-4.M4.bin]?

Writing c2900-universalk9-mz.SPA.151-4.M4.bin…
[OK – 33591768 bytes]
33591768 bytes copied in **164.073 secs (204000 bytes/sec)**

与 ZHN 同组的 HW 同学实验输出如下：

R1#**copy flash: tftp**
Source filename []? C2900-universalk9-mz.SPA.151-4.M4.bin

Address or name of remote host []? 192.168.1.250
Destination filename [c2900-universalk9-mz.SPA.151-4.M4.bin]? c2900-universalk9-mz.SPA.151-4.M4_17_1_14.bin

Writing c2900-universalk9-mz.SPA.151-4.M4.bin
…!!!
!!
!!
!!
!!!

[OK – 33591768 bytes]

33591768 bytes copied in **0.717 secs (4919103 bytes/sec)**

问题二：ZHP 同学遇到的奇怪现象

路由器的主机名怎么变成了 rommon？他提交的文档如下所示（备注：ZHP 同学因生病，路由器口令恢复的课程没有学习）。

Router#**reload**
Proceed with reload? [confirm]
System Bootstrap, Version 15.1(4)M4, RELEASE SOFTWARE (fc1)
Technical Support: http://www.cisco.com/techsupport
Copyright (c) 2010 by cisco Systems, Inc.
Total memory size = 512 MB – On-board = 512 MB, DIMM0 = 0 MB
CISCO1941/K9 platform with 524288 Kbytes of main memory
Main memory is configured to 64/-1(On-board/DIMM0) bit mode with ECC disabled

Readonly ROMMON initialized

Boot process failed…

The system is unable to boot automatically. The BOOT
environment variable needs to be set to a bootable
image.
Rommon 1 >

问题三：XL 同学反馈的问题

路由器 flash 中的两个文件 sigdef-category.xml 和 sigdef-default.xml 作用是什么？XL 提交

的文档内容如下。

```
R1#show flash

System flash directory:
File   Length      Name/status
  3    33591768    c1900-universalk9-mz.SPA.151-4.M4.bin
  2    28282       sigdef-category.xml
  1    227537      sigdef-default.xml
[33847587 bytes used, 221896413 available, 255744000 total]
249856K bytes of processor board System flash (Read/Write)
```

老师当着全班同学的面，表扬了这三位同学，并给他们日常成绩各加了两分。针对这三个问题，老师并没有直接给出答案，而是让全班同学以小组为单位讨论，寻求以互联网的方式去解决问题。通过团队协作，问题很快就有了答案：

HY 小组对 ZHN 提出的问题，回复如下所述。

- FTP（File Transfer Protocol，文件传输协议）：使用 TCP 协议的 20 和 21 端口。FTP 需要建立两个信道（每个信道都需要建立三次握手），信道 1 是控制信道，使用端口 21 来进行登录认证；信道 2 是数据信道，利用 20 号端口传输数据。而 TCP 协议采用了滑动窗口、重传以及确认等机制，确保 FTP 传输的可靠性，因此适合传输大文件，速度慢。
- TFTP（Trivial File Transfer Protocol，简单文件传输协议）：使用 UDP 协议的 69 号端口传输数据，因为 UDP 协议不确保文件传输的可靠性，因此不提供相应机制，传输速度快，适合传输小文件。

DWC 小组解答了 ZHP 提出的问题，如下所述。

"rommon"并不是路由器的主机名，路由器的主机名也根本没有发生变化。"rommon>"说明路由器进入了一种特殊的模式——ROM 监视模式。前面我们所学过的"路由器口令恢复"就是在该模式下进行的。监视模式是路由器进行口令恢复、系统恢复等灾难性恢复的特殊模式。因为路由器重启，而 flash 中又没有 IOS，所以引导程序就加载了 ROM 中固化的微型 IOS 映像，于是进入了监视模式，以待系统恢复。

ZHT 小组回答了 XL 提出的问题，如下所述。

文件 sigdef-category.xml：其作用是表示该设备固件特征类别；而文件 sigdef-default.xml 是签名文件，其作用是鉴别此 IOS 映像的真实性。

下面是本次课的新内容，请同学们在 ZHP 同学的实验环境下恢复 IOS，CFL 老师给出了详细的实验步骤。

步骤一：检验设备的连接

路由器的接口必须采用第一个以太网接口，如 Ethernet0、Fastethernet0/0、Gigabitethernet0/0 等。

步骤二：设置环境变量的值

```
Rommon 1 > IP_ADDRESS=192.168.1.1
rommon 2 > IP_SUBNET_MASK=255.255.255.0
rommon 3 > DEFAULT_GATEWAY=192.168.1.254
rommon 4 > TFTP_SERVER=192.168.1.250
rommon 5 > TFTP_FILE=c2900-universalk9-mz.SPA.151-4.M4.bin
```

温馨提醒：
- 变量名必须大写（大小写敏感）；
- 变量名必须被赋值（变量值仅是临时赋值，不能保存，IOS 系统恢复后自动消失）；
- DEFAULT_GATEWAY 必须被赋值（同网段 IOS 的恢复，变量值可以是任意合法 IP 地址）；
- "=" 左右不能有空格。

步骤三：用 tftpdnld 命令恢复 IOS

```
rommon 6 > tftpdnld    //开始从 TFTP 恢复 IOS

            IP_ADDRESS : 192.168.1.1
        IP_SUBNET_MASK : 255.255.255.0
       DEFAULT_GATEWAY : 192.168.1.254
           TFTP_SERVER : 192.168.1.250
             TFTP_FILE : c2900-universalk9-mz.SPA.151-4.M4_17_1_14.bin
Invoke this command for disaster recovery only.
WARNING: all existing data in all partitions on flash will be lost!

Do you wish to continue? y/n:   [n]:   y       //回答"y"
  Receiving c2900-universalk9-mz.SPA.151-4.M4.bin from 192.168.1.250!!!!!!!!!!!!!!!!!!!!!!!!!!!!!!!!!
!!!!!!!!!!!!!!!!!!!!!!!!!!!!!!!!!!!!!!!!
（此处省略部分输出）
File reception completed.
Validating file c2900-universalk9-mz.SPA.151-4.M4.bin to flash.
Eeeeeeeeeeeeeeeeeeeeeeeeeeeeeeeeeeeeeeeee
//从 TFTP 服务器接收 IOS 并进行差错校验
```

步骤四：重启路由器加载新 IOS

```
rommon 7 > reset
```

步骤五：查看 flash 的新 IOS

以上是路由器在删除 IOS 后，不慎重启进入监视模式，然后采取的通过设置环境变量来恢复 IOS 的办法。核验需要反复练习直到熟练掌握。

2.4 探索网络拓扑

2.4.1 认识思科发现协议

CDP（Cisco Discovery Protocol，思科发现协议）是思科专用的数据链路层协议，其作用是使 Cisco 网络设备能够发现直连邻居的二层及二层以上的 Cisco 设备。因此，使用不同网络层协议的 Cisco 设备也可以获得对方的信息。CDP 协议最初在 Cisco IOS 10.3 版中引入，CDP 使用组播来传播信息，如果收到 CDP 消息，说明数据链路层工作正常。所以，CDP 协议是网络排错的一个有效工具。在支持 CDP 的设备上，CDP 默认被启用，Cisco 设备默认每 60 s 发送一次通告给直连的邻居，通告中包含了自身的基本信息，例如，主机名、硬件型号、软件版本，以及 CDP 通告的有效时间（默认为 180 s）。默认情况下，Cisco 设备不会将收到的 CDP 消息转发给其他设备，仅会将邻居信息保存在自己的邻居列表中。

CDP 协议常用命令一览表如表 2-10 所示。

表 2-10　CDP 协议常用命令一览表

CDP 命令	CDP 命令功能
show cdp	查看 CDP 状态
show cdp interface	查看各个接口的 CDP 配置
show cdp neighbors	查看所有直连设备的简明信息
show cdp neighbors detail	查看所有直连设备的详细信息
show cdp entry *	查看所有直连设备的详细信息
show cdp entry *Device ID*	查看指定直连设备的详细信息
cdp run/no cdp run	全局模式下，启用/禁用 CDP 协议
cdp enable/no cdp enable	接口模式下，启用/禁用 CDP 协议

2.4.2 场景十：应用 CDP 协议

> 场景十：课堂上老师让大家打开 Packet Tracer，搭建如图 2-19 所示拓扑，并要求同学们只需要配置设备的主机名和相应接口地址。接下来的任务是使用上面表中的 show 命令来学习各自的作用，并分析相应 show 命令的结果。通过在不同模式下开启或禁用 CDP 协议来理解相应命令的作用。

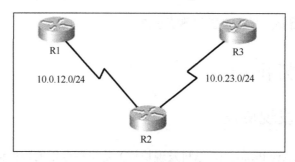

图 2-19 应用思科发现协议

15NET1 班 MFL 小组提交的实验过程文档让老师比较满意,因此拿来和大家一起分享。

步骤一:路由器基本配置

(1)配置路由器 R1

Router(config)#**hostname R1**
R1(config)#**interface serial 0/0/0**
R1(config-if)#**ip address 10.0.12.1 255.255.255.0**
R1(config-if)#**no shutdown**

(2)配置路由器 R2

Router(config)#**hostname R2**
R2(config)#**interface serial 0/0/0**
R2(config-if)#**ip address 10.0.12.2 255.255.255.0**
R2(config-if)#**no shutdown**
R2(config-if)#**exit**
R2(config)#**interface serial 0/0/1**
R2(config-if)#**ip address 10.0.23.2 255.255.255.0**
R2(config-if)#**no shutdown**

(3)配置路由器 R3

Router(config)#**hostname R3**
R3(config)#**interface serial 0/0/1**
R3(config-if)#**ip address 10.0.23.3 255.255.255.0**
R3(config-if)#**no shutdown**

步骤二:使用 show 命令探索

(1)show cdp

R1#**show cdp**

```
Global CDP information:
        Sending CDP packets every 60 seconds
        Sending a holdtime value of 180 seconds
        Sending CDPv2 advertisements is enabled
```

这里显示 CDP 计时器的更新时间是 60 s，保持时间是 180 s，即路由器每 60 s 向邻居发送一次 CDP 信息，如果 60 s 没有收到相邻路由器的 CDP 信息，也不会马上删除这条邻居信息，而是要等待 180 s 后仍然没有收到，才会删除。

（2）show cdp interface

```
R1#show cdp interface
Vlan1 is administratively down, line protocol is down
    Sending CDP packets every 60 seconds
    Holdtime is 180 seconds
GigabitEthernet0/0 is administratively down, line protocol is down
    Sending CDP packets every 60 seconds
    Holdtime is 180 seconds
GigabitEthernet0/1 is administratively down, line protocol is down
    Sending CDP packets every 60 seconds
    Holdtime is 180 seconds
Serial0/0/0 is up, line protocol is up
    Sending CDP packets every 60 seconds
    Holdtime is 180 seconds
Serial0/0/1 is administratively down, line protocol is down
    Sending CDP packets every 60 seconds
    Holdtime is 180 seconds
```

使用 show cdp interface 命令可以显示每个接口的 CDP 信息，包括每个接口的物理层和数据链路层状态、CDP 发送消息的时间间隔和保持时间。

（3）show cdp neighbors

```
R1#show cdp neighbors
Capability Codes: R - Router, T - Trans Bridge, B - Source Route Bridge
                 S - Switch, H - Host, I - IGMP, r - Repeater, P - Phone
Device ID      Local Intrfce    Holdtme    Capability    Platform    Port ID
R2             Ser 0/0/0        173        R             C1900       Ser 0/0/0
```

命令显示的参数解释如下。

- **Device ID**：邻居设备标识，也就是邻居设备的主机名，如 R1 的邻居 ID 是 R2。
- **Local Intrfce**：本地接口，即当前路由器 R1 的接口，如 R1 通过接口 Ser0/0/0 连接 R2。

- **Holdtme**：保持时间，如当前路由器 R1 与邻居 R2 的保持时间还有 173 s。从 180 s 倒计时。
- **Capability**：设备性能，R 代表路由器，S 代表交换机等，如邻居 R2 是路由器。
- **Platform**：硬件平台，如邻居 R2 的设备型号是 C1900。
- **Port ID**：端口标识，指邻居路由器的端口标识，如当前路由器 R1 通过本地接口 Ser0/0/0 连接到了邻居 R2 的 Ser0/0/0 口上。

由此可见，CDP 协议可以发现邻居的很多信息，从而探索发现网络。

（4）show cdp neighbors detail

R2#**show cdp neighbors detail**

Device ID: R1
Entry address(es):
　　IP address : 10.0.12.1
Platform: cisco C1900, **Capabilities**: Router
Interface: Serial0/0/0, **Port ID** (outgoing port): Serial0/0/0
Holdtime: 131

Version :
　Cisco IOS Software, C1900 Software (C1900-UNIVERSALK9-M), Version 15.1(4)M4, RELEASE SOFTWARE (fc2)
　Technical Support: http://www.cisco.com/techsupport
　Copyright (c) 1986-2012 by Cisco Systems, Inc.
　Compiled Thurs 5-Jan-12 15:41 by pt_team

advertisement version: 2
Duplex: full

Device ID: R3
Entry address(es):
　　IP address : 10.0.23.3
Platform: cisco C1900, **Capabilities**: Router
Interface: Serial0/0/1, **Port ID** (outgoing port): Serial0/0/1
Holdtime: 133

Version :

Cisco IOS Software, C1900 Software (C1900-UNIVERSALK9-M), Version 15.1(4)M4, RELEASE SOFTWARE (fc2)

Technical Support: http://www.cisco.com/techsupport

Copyright (c) 1986-2012 by Cisco Systems, Inc.

Compiled Thurs 5-Jan-12 15:41 by pt_team

advertisement version: 2

Duplex: full

R2#

（5）show cdp entry *

R2#**show cdp entry ***

Device ID: R1
Entry address(es):
   IP address : 10.0.12.1
Platform: cisco C1900, **Capabilities**: Router
Interface: Serial0/0/0, **Port ID** (outgoing port): Serial0/0/0
Holdtime: 155

Version :
Cisco IOS Software, C1900 Software (C1900-UNIVERSALK9-M), Version 15.1(4)M4, RELEASE SOFTWARE (fc2)

Technical Support: http://www.cisco.com/techsupport

Copyright (c) 1986-2012 by Cisco Systems, Inc.

Compiled Thurs 5-Jan-12 15:41 by pt_team

advertisement version: 2

Duplex: full

Device ID: R3
Entry address(es):
   IP address : 10.0.23.3
Platform: cisco C1900, Capabilities: Router
Interface: Serial0/0/1, Port ID (outgoing port): Serial0/0/1
Holdtime: 157

Version :

Cisco IOS Software, C1900 Software (C1900-UNIVERSALK9-M), Version 15.1(4)M4, RELEASE SOFTWARE (fc2)

Technical Support: http://www.cisco.com/techsupport

Copyright (c) 1986-2012 by Cisco Systems, Inc.

Compiled Thurs 5-Jan-12 15:41 by pt_team

advertisement version: 2
Duplex: full

R2#

由以上输出可见，命令 show cdp neighbors detail 与 show cdp entry *显示结果一样，都是显示邻居的详细信息，其命令功能相同。显示内容包含直连邻居的设备标识、IP 地址、硬件平台、设备性能、本地发送 CDP 消息的接口，以及对方接收 CDP 消息接口、保持时间和 IOS 的版本号等。

（6）show cdp entry *Device ID*

R2#show cdp entry **R1**

Device ID: R1
Entry address(es):
    IP address : 10.0.12.1
Platform: cisco C1900, **Capabilities**: Router
Interface: Serial0/0/0, Port ID (outgoing port): Serial0/0/0
Holdtime: 167

Version :

Cisco IOS Software, C1900 Software (C1900-UNIVERSALK9-M), Version 15.1(4)M4, RELEASE SOFTWARE (fc2)

Technical Support: http://www.cisco.com/techsupport

Copyright (c) 1986-2012 by Cisco Systems, Inc.

Compiled Thurs 5-Jan-12 15:41 by pt_team

advertisement version: 2
Duplex: full
R2#

本命令只显示指定直连设备的详细信息。

步骤三：禁用/启用 CDP 协议

（1）全局模式下禁用／启用 CDP 协议

R2#**configure terminal**
R2(config)#**no cdp run**
在全局模式下，禁用 CDP 协议，并用 show 命令进行验证。

R2(config)#**end**
R2#**show cdp neighbors**
% CDP is not enabled
我们发现，CDP 协议已经被禁用，因此也就无法发现邻居。

R2#**configure terminal**
R2(config)#**cdp run**
在全局模式下，启用 CDP 协议，并用 show 命令进行验证。

R2#**show cdp neighbors**
Capability Codes: R - Router, T - Trans Bridge, B - Source Route Bridge
                  S - Switch, H - Host, I - IGMP, r - Repeater, P - Phone

Device ID	Local Intrfce	Holdtme	Capability	Platform	Port ID
R1	Ser 0/0/0	153	R	C1900	Ser 0/0/0
R3	Ser 0/0/1	155	R	C1900	Ser 0/0/1

从以上输出我们发现，如果在全局模式下禁用或启用 CDP 协议，那么该设备所有接口的 CDP 协议均被禁用或启用。

（2）接口模式下禁用／启用 CDP 协议

R2(config)#**interface serial 0/0/0**
R2(config-if)#**no cdp enable**
在接口模式下，禁用 CDP 协议并用 show 命令进行验证。

R2#**show cdp neighbors**
Capability Codes: R - Router, T - Trans Bridge, B - Source Route Bridge
                  S - Switch, H - Host, I - IGMP, r - Repeater, P - Phone

Device ID	Local Intrfce	Holdtme	Capability	Platform	Port ID
R1	Ser 0/0/0	160	R	C1900	Ser 0/0/0
R3	Ser 0/0/1	162	R	C1900	Ser 0/0/1

在路由器 R2 的 serial 0/0/0 接口下已经禁用 CDP 协议，但是，我们发现其直连的邻居 R1 依然存在，似乎协议的禁用没有生效。

R2#**show cdp neighbors**

Capability Codes: R - Router, T - Trans Bridge, B - Source Route Bridge
             S - Switch, H - Host, I - IGMP, r - Repeater, P - Phone

Device ID	Local Intrfce	Holdtme	Capability	Platform	Port ID
R1	Ser 0/0/0	0	R	C1900	Ser 0/0/0
R3	Ser 0/0/1	123	R	C1900	Ser 0/0/1

我们继续不断使用 show cdp neighbors 命令查看，结果发现 R1 的"Holdtme"值始终在递减，最后到达 0，但 R3 的却再一次出现了反弹，然后再递减。

R2#**show cdp neighbors**
Capability Codes: R - Router, T - Trans Bridge, B - Source Route Bridge
             S - Switch, H - Host, I - IGMP, r - Repeater, P - Phone

Device ID	Local Intrfce	Holdtme	Capability	Platform	Port ID
R3	Ser 0/0/1	122	R	C1900	Ser 0/0/1

我们进一步查看结果发现，接口 serial 0/0/0 上的邻居 R1 彻底从邻居列表中消失，这说明我们在接口下禁用 CDP 协议是有效果的，但是这种现象如何解释？答案是 CDP 协议邻居的保持时间为 180 s，而且计时器采用倒计时，若从 180 s 至 0 s，接口没有收到来自邻居的 CDP 消息，则就会从邻居列表中删除该邻居。邻居 R3 出现的保持时间递减然后反弹至 180 s，接下来又递减至 120 s，再反弹回 180 s，是因为 CDP 协议会每 60 s 向直连的邻居发送 CDP 消息，收到 CDP 消息的设备会重新刷新"Holdtme"值为 180 s，从而动态维持邻居关系。

在接口模式下，启用 CDP 协议并用 show 命令进行验证。

R2(config)#**int s0/0/0**
R2(config-if)#**cdp enable**

我们发现，当前路由器 R2 通过接口 Serial 0/0/0，已经接收到来自邻居 R1 的 CDP 消息。

R2#**show cdp neighbors**
Capability Codes: R - Router, T - Trans Bridge, B - Source Route Bridge
             S - Switch, H - Host, I - IGMP, r - Repeater, P - Phone

Device ID	Local Intrfce	Holdtme	Capability	Platform	Port ID
R3	Ser 0/0/1	172	R	C1900	Ser 0/0/1
R1	Ser 0/0/0	170	R	C1900	Ser 0/0/0

深入理解发送消息的时间间隔和保持时间"Holdtime"对我们后续理解其他协议的邻居关系是非常有意义的。

2.4.3 场景十一：探索网络拓扑结构

场景十一：今天的网络课，CFL 老师给同学们布置了一道非常有挑战性的任务：依据图 2-20 所示拓扑，利用上次课所学 CDP 知识来发现 ABC Company 云中的网络拓扑，你可以使用 Packet Tracer 来搭建云中的网络拓扑，也可以在白纸上画出拓扑，标注设备名字和相应接口。ABC Company 全部采用 Cisco 设备，设备均已正确配置好，并且允许远程用户 Telnet，所有设备的特权口令为 class，VTY 口令为 cisco。这项任务让同学们束手无策。HY 同学提示全班同学突破口应从主机 Admin 入手，他已经快完成任务了，老师给 HY 下达一个任务，让他把发现整个网络拓扑的过程记录下来分享给全班同学。

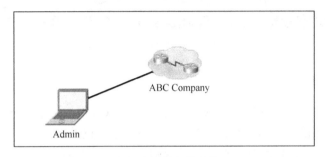

图 2-20　探索网络拓扑结构

下面是 HY 同学提交的发现网络拓扑过程的文档，请大家一起来分享。

步骤一：查看主机 IP 地址

在主机 Admin 上通过 ipconfig 命令，查看到当前主机的网关是 192.168.0.1，如图 2-21 所示。

```
C:\>ipconfig

FastEthernet0 Connection:(default port)

  Link-local IPv6 Address.........: ::
  IP Address......................: 192.168.0.100
  Subnet Mask.....................: 255.255.255.0
  Default Gateway.................: 192.168.0.1
```

图 2-21　查看主机的 IP 地址

步骤二：Telnet 到网关设备

在主机 Admin 上通过 Telnet 登录到网关设备，登录的 VTY 口令为"cisco"，特权口令为"class"，图 2-22 显示已经成功登录到设备 R1 上。

```
C:\>telnet 192.168.0.1
Trying 192.168.0.1 ...Open

User Access Verification

Password:
R1>enable
Password:
R1#
```

图 2-22　Telnet 到网关设备

步骤三：探索网络拓扑

在网关设备 R1 上，通过使用 show version 和 show ip interface brief 命令，我们了解到当前设备是一台 28 系列的路由器，并且仅有两个百兆的以太网接口。

接下来，我们开始使用 CDP 协议来发现网络，探索网络拓扑。

R1#show cdp neighbors
Capability Codes: R - Router, T - Trans Bridge, B - Source Route Bridge
                  S - Switch, H - Host, I - IGMP, r - Repeater, P - Phone

Device ID	Local Intrfce	Holdtme	Capability	Platform	Port ID
S4	Fas 0/1	157	S	2960	Fas 0/1
S1	Fas 0/0	157	S	2960	Fas 0/1

通过 show cdp neighbors 命令，我们发现路由器 R1 只有两个直连的二层邻居，即两台 Cisco 2960 的交换机 S1 和 S4，其连接关系，我们可以很轻松画出来，如图 2-23 所示。

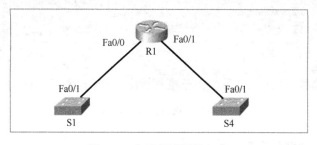

图 2-23　初次探索网络拓扑

接下来，我们的任务是探索路由器左右分支的网络拓扑，也就是设备 S1 和 S4 所连设备的拓扑。因为全网都支持 Telnet，所以我们需要了解 S1 和 S4 的 IP 地址信息。在路由器 R1 上发布 **show cdp neighbors detail** 或 **show cdp entry *** 命令，我们查到的 IP 地址信息如下：

S1: 192.168.0.10、**S4:** 192.168.1.40

从路由器 R1 登录到交换机 S1 上，准备探索左分支网络拓扑，如图 2-24 所示。

图 2-24　Telnet 到 S1

在交换机 S1 上，查看其直连的邻居，发现有 R1、S2 和 S3。

S1#**show cdp neighbors**
Capability Codes: R - Router, T - Trans Bridge, B - Source Route Bridge
　　　　　　　　　S - Switch, H - Host, I - IGMP, r - Repeater, P - Phone

Device ID	Local Intrfce	Holdtme	Capability	Platform	Port ID
S3	Fas 0/3	122	S	2960	Fas 0/1
S2	Fas 0/2	122	S	2960	Fas 0/1
R1	Fas 0/1	122	R	C2800	Fas 0/0

根据 show 的结果，我们可以画出图 2-25 所示的网络拓扑。

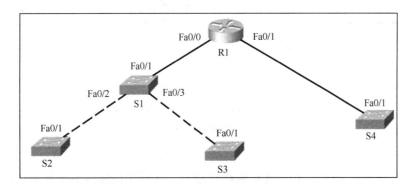

图 2-25　在交换机 S1 上探索网络拓扑

我们继续 Telnet 到设备 S2 和 S3 上探索，左分支网络拓扑最终如图 2-26 所示。

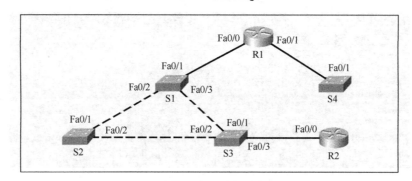

图 2-26 探索左分支网络拓扑

查看右分支设备 S4 的 IP 地址，重复上述探索过程，直至将整个右分支探索完毕，最终发现的网络拓扑如图 2-27 所示。

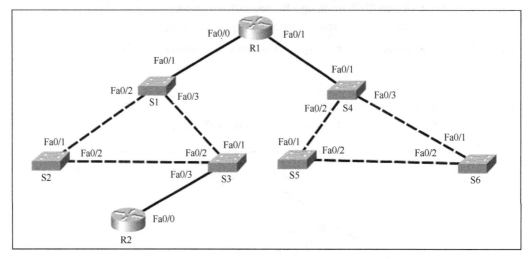

图 2-27 探索完整网络拓扑

这项挑战任务，同学们参与度很高，通过一个简单的 CDP 协议就可以探索出网络拓扑，老师发现同学们对网络技术的学习越来越有兴趣了，有同学反馈这节课就像在玩一场游戏一样，终于通关。

2.5 挑战过关练习

1. 任务背景

请打开 Cisco Packet Tracer 模拟器，参照表 2-11，从设备选区中选择相应设备拖至工作区，更改 Display Name，并按照图 8-28 所示拓扑进行布放。

表 2-11 挑战练习设备一览表

设备型号	设备名称	主机名	设备数量
CISCO 1941	路由器	R1	2
		R2	
CISCO 2811	路由器	R3	2
		R4	
CISCO 2960	交换机	SW1	1
Laptop	终端 PC	Admin	1
Server	服务器	TFTP Server	1

R2

R1

Admin

SW1

TFTP Server

R3

R4

图 2-28 挑战过关练习拓扑

2. 准备工作

- 请打开 TFTP Server，点击 TFTP 服务，如图 2-29 所示，再逐个选中文件单击 Remove File 按钮，如图 2-30 所示，将文件全部清空；
- 删除路由器 R1 的 IOS，不允许关机重启；
- 删除路由器 R3 的 IOS，重启路由器，进入监视模式；
- 在路由器 R4 上输入如下命令：

```
enable
configure terminal
hostname R4
line console 0
password ytvc
```

```
exec-timeout 0 1
login
end
write
```

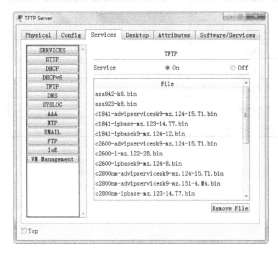

图 2-29 TFTP Server 初始配置

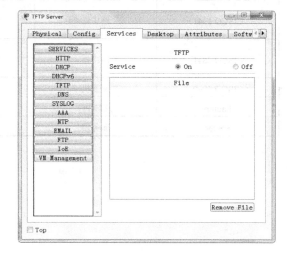

图 2-30 TFTP Server 清空配置

3. 挑战任务

- 恢复路由器 R1 的 IOS。
- 删除路由器 R4 的配置。
- 恢复路由器 R3 的 IOS。
- 对所有设备进行配置：
  - 配置主机名；
  - 配置 Console 口令为 ytvc，特权口令为 cisco；
  - 配置如图 2-31 所示的登录横幅。

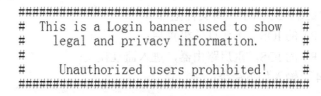

图 2-31 网络设备登录横幅

  - 保存配置，并上传到 TFTP 服务器。
- 将图 2-28 所示的所有设备进行连线，搭建一个局域网。
- 规划并配置地址，使任意两台设备均可以相互 ping 通。

本章内容就此结束，通过设置的 11 个应用场景，让我们轻松完成了对交换机和路由器的基本配置、口令恢复、配置文件以及 IOS 的备份与恢复，最后通过"探索网络拓扑"，让我们深刻体会到熟练掌握网络知识并能灵活应用所学技术给我们所带来的无限乐趣。本章的应用场景实际是课堂的缩影，呈现了课堂上师生互动的场景，汇总了同学们反馈出来的种种问题，带领我们总结经验、深入思考，不断挖掘网络技术内涵。最后的挑战闯关练习凝聚了本章精华，能引领大家从点到面、系统地深入学习，面对出现的种种问题，能积极思考，不断提升随机应变能力，进而提高自行分析和解决问题的能力。

第 3 章

学习 VLAN 技术

本章要点

- 认识 VLAN 技术
- 多场景 VLAN 配置
- 认识 Trunk 技术
- 认识 VTP 技术
- 挑战过关练习

我们已经可以接入网络设备并能对设备进行基本的管理,那又如何配置设备使其实现不同的功能呢?本章"学习 VLAN 技术"将带我们真正迈入设备配置的大门。

VLAN 技术是基于二层交换机的技术,它通过软件方式实现广播域的划分与管理,从而提高组网的灵活性与安全性,进一步提升网络性能。

本章设计的 5 个应用场景——YH 食品有限公司企业网发展的五个阶段,承载了本章所有教学内容。公司不同的发展阶段,需求不同,采用的技术也不尽相同,这需要不断升级和优化网络。通过本章学习,你将会轻松掌握 VLAN、Trunk、DTP、VTP 等知识,做到融会贯通,灵活应用。最后通过精心设计的 3 个挑战过关练习,检验对本章相关知识的理解,相信会让你大开眼界,传递的信息,也一定会给你留下深刻的印象。

3.1 认识 VLAN 技术

3.1.1 VLAN 的概念

VLAN(Virtual Local Area Network),虚拟局域网,是一种二层技术。它是在交换式局域网中将一个较大广播域按照部门、功能等因素分割成较小广播域的技术。VLAN 划分实质上就是将一台交换机逻辑分成了几台交换机,通过软件方式实现逻辑工作组的划分与管理。

默认情况下交换机不能分割广播域,多台交换机互联后依然属于一个广播域,如图 3-1 所示。

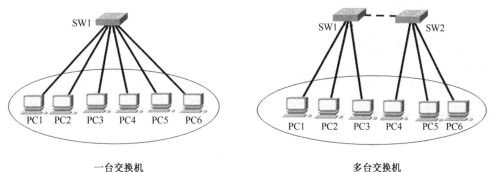

图 3-1 默认一个广播域

但是,在交换机上通过 VLAN 技术,就可以分割广播域,即一个 VLAN 就是一个广播域。VLAN 由交换机的一组物理端口组成,这些物理端口可以属于一台交换机,也可以属于多台交换机。各个广播域之间在逻辑上是分开的,如图 3-2 所示。

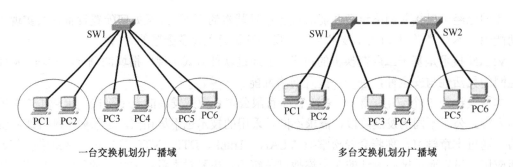

图 3-2　VLAN 技术分割广播域

3.1.2　VLAN 的优点

采用 VLAN 技术，使网络管理更加方便、灵活，从而提高设备使用效率，增强网络适应性。表 3-1 为 VLAN 优点一览表。

表 3-1　VLAN 优点一览表

VLAN 优点	VLAN 优点描述
提高网络安全	可将含敏感数据的用户组与其他组分开，降低泄密的可能性
降低网络成本	现有带宽和上行链路的利用率更高，减少升级需求，节约成本
提高网络性能	基于二层划分多个逻辑分组，可以减少网络中不必要的流量
抑制广播风暴	分割广播域，增加了广播域数量，缩小了广播域的范围
提高网管效率	若增加新设备，添加新 VLAN，简单易实现，管理开销小
简化服务管理	使项目管理或特殊应用处理方便，容易确定网络服务的影响范围

3.1.3　VLAN 的分类

VLAN 按照不同的标准，有多种分类方式，具体如下。

1. 按照 VLAN 创建方式

VLAN 可分为系统默认 VLAN 和用户 VLAN

- **默认 VLAN**：VLAN ID 号和名称是固定的，不能被用户删除或修改。如 default、fddi-default、token-ring-default 等。
- **用户 VLAN**：VLAN ID 号和名称是用户创建的，且能被用户删除或修改。如 VLAN10、VLAN99、Teacher、Student 等。

2. 按照 VLAN 成员添加方式

VLAN 可分为静态 VLAN 和动态 VLAN。

- **静态 VLAN**：通过手工方式将交换机端口添加到某个 VLAN 中，端口属于某个 VLAN 后，除非管理员重新分配，否则该端口永远属于该 VLAN。例如，将端口 Fa0/1 添加到 VLAN10，则该端口只属于 VLAN10，不属于其他 VLAN，除非重新再分配。
- **动态 VLAN**：交换机会根据用户设备信息，如 MAC 地址、IP 地址乃至高层用户信息等，自动将端口分配给某个 VLAN。动态 VLAN 要求网络中必须有一台策略服务器（VMPS），该服务器包含设备信息到 VLAN 的映射关系。例如，当公司有新员工入职、离职或员工更换工作岗位时，需要网管员更新 VMPS 服务器中设备信息和 VLAN 的映射关系。

与静态 VLAN 相比，动态 VLAN 的优点是依据 VMPS 服务器中设备信息和 VLAN 的映射关系，动态将相关端口添加到某个 VLAN 中。动态 VLAN 可以根据交换端口所连接的计算机，随时改变端口所属 VLAN，当计算机变更所连交换端口或交换机时，VLAN 不用重新配置。

VMPS 策略服务器通常由高端交换机充当（如 Cisco Catalyst 6500 交换机），低端交换机作为客户机。当某台主机发送数据帧到网络中时，由客户机向服务器发起查询请求，VMPS 依据设备信息，找到与 VLAN 的对应关系，再依据主机所连接交换机端口动态将其添加到相应 VLAN。

3. 按照 VLAN 承载信息类型不同

VLAN 可分为数据 VLAN、管理 VLAN。

- **数据 VLAN**：承载用户数据，如 VLAN10、VLAN99、Teacher、Student 等。
- **管理 VLAN**：承载用于网络设备管理的流量，如 Telnet、SSH、SNMP 等信息。

数据 VLAN 和管理 VLAN 既可以使用默认 VLAN，也可使用用户 VLAN。从安全角度考虑，建议大家两种 VLAN 都自己创建。

另外，还有其他几种类型的 VLAN，如语音 VLAN、本征 VLAN、黑洞 VLAN 等。

3.1.4 VLAN 的配置

在 Cisco Catalyst 交换机上配置 VLAN 的模式有如下两种。

① 全局配置模式：Switch(config)#。新版 IOS 模式，建议使用。

② 数据库配置模式：Switch(vlan)#。Catalyst 旧版操作系统 CatOS 模式，不推荐使用。配置 VLAN 前，需要先确定可用 VLAN ID，VLAN ID 分为普通范围和扩展范围两部分。

- 普通范围 VLAN ID：1~1005，其中 VLAN1、VLAN1002~VLAN1005 为保留，不可用；

- 扩展范围 VLAN ID：1006~4094。

我们可选择普通范围可用 ID 来配置静态 VLAN，具体配置如下所述。

1. 创建 VLAN

创建 VLAN 可以在全局配置模式，也可以在 VLAN 子模式下进行。

```
Switch>enable
Switch#configure terminal
Switch(config)#vlan 10
Switch(config-vlan)#name Finance
Switch(config-vlan)#vlan 20
Switch(config-vlan)#name Marketing
```

2. 添加 VLAN 成员

（1）添加单个端口

```
Switch(config)#interface f0/10
Switch(config-if)#switchport mode access
Switch(config-if)#switchport access vlan 10
```

（2）添加批量连续端口

```
Switch(config)#interface range f0/1-6
Switch(config-if-range)#switchport mode access
Switch(config-if-range)#switchport access vlan 10
```

（3）添加批量不连续端口

```
Switch(config)#interface range f0/12-14,f0/16
Switch(config-if-range)#switchport mode access
Switch(config-if-range)#switchport access vlan 20
```

一般情况下，我们先创建 VLAN，再添加 VLAN 成员。但是，若没有创建 VLAN，直接添加 VLAN 成员，则系统可以使用默认名称自动创建相关 VLAN，详见如下面命令提示。

```
Switch(config)#interface f0/20
Switch(config-if)#switchport mode access
Switch(config-if)#switchport access vlan 30
% Access VLAN does not exist. Creating vlan 30
```

3. 修改 VLAN 成员

我们可以灵活改变 VLAN 成员，即从一个 VLAN 改变到另外一个 VLAN，只需重新分配相应端口到新 VLAN 即可。例如，连接到 Fa0/10 的用户，原本属于财务部（VLAN10），后因工作需要，调动到市场部（VLAN20），其修改如下：

```
Switch(config)#interface f0/10
Switch(config-if)#switchport access vlan 20
```

4. 删除某个 VLAN

删除某个 VLAN 需要先把 VLAN 成员从该 VLAN 中删除。先查看当前 VLAN 信息，了解 VLAN 成员，即 VLAN 端口信息。

```
Switch#show vlan brief

VLAN Name                          Status     Ports
---- ------------------------------ --------- -------------------------------
1    default                        active    Fa0/7, Fa0/8, Fa0/9, Fa0/11
                                              Fa0/15, Fa0/17, Fa0/18, Fa0/19
                                              Fa0/20, Fa0/21, Fa0/22, Fa0/23
                                              Fa0/24, Gig0/1, Gig0/2
10   Finance                        active    Fa0/1, Fa0/2, Fa0/3, Fa0/4
                                              Fa0/5, Fa0/6, Fa0/10
20   Marketing                      active    Fa0/12, Fa0/13, Fa0/14, Fa0/16
1002 fddi-default                   active
1003 token-ring-default             active
1004 fddinet-default                active
1005 trnet-default                  active
```

若要删除 VLAN 10，则需先删除 VLAN 成员，再删除 VLAN ID，具体操作如下：

```
Switch(config)#interface range f0/1-6,f0/10
Switch(config-if-range)#no switchport access vlan 10        //默认将端口放回 VLAN 1
Switch(config)#no vlan 10
```

再次查看当前 VLAN 信息，我们发现 VLAN 10 的成员目前属于 VLAN 1。

```
Switch#show vlan brief

VLAN Name                          Status     Ports
---- ------------------------------ --------- -------------------------------
1    default                        active    Fa0/1, Fa0/2, Fa0/3, Fa0/4
                                              Fa0/5, Fa0/6, Fa0/7, Fa0/8
                                              Fa0/9, Fa0/10, Fa0/11, Fa0/15
                                              Fa0/17, Fa0/18, Fa0/19, Fa0/20
                                              Fa0/21, Fa0/22, Fa0/23, Fa0/24
```

			Gig0/1, Gig0/2
20	Marketing	active	Fa0/12, Fa0/13, Fa0/14, Fa0/16
1002	fddi-default	active	
1003	token-ring-default	active	
1004	fddinet-default	active	
1005	trnet-default	active	

若删除 VLAN，未先删除其成员，而直接删除的是 VLAN ID，如下所示：

Switch(config)#**no vlan 10**

则删除 VLAN 后，其成员 Fa0/1～Fa0/6 及 Fa0/10 也一并被删除，如下 show 显示，此时我们看不到这些端口。

Switch#show vlan brief

VLAN	Name	Status	Ports
1	default	active	Fa0/7, Fa0/8, Fa0/9, Fa0/11
			Fa0/15, Fa0/17, Fa0/18, Fa0/19
			Fa0/20, Fa0/21, Fa0/22, Fa0/23
			Fa0/24, Gig0/1, Gig0/2
20	Marketing	active	Fa0/12, Fa0/13, Fa0/14, Fa0/16
1002	fddi-default	active	
1003	token-ring-default	active	
1004	fddinet-default	active	
1005	trnet-default	active	

目前，这些端口处于"游离"状态，是不可用的。若想使用这些端口，则必须将其重新添加到相应 VLAN。这种情况会给网络故障排除带来一定难度。

5. 删除所有 VLAN

如果用户需要把某台交换机上所有 VLAN 删除，则无须逐个删除。因为在创建 VLAN 时，系统会在 flash 中自动生成 vlan.dat 文件，它记录了所有 VLAN 信息。所以，删除 vlan.dat 也就删除了所有 VLAN。若恢复交换机出厂设置，flash 中的启动配置文件 config.text 和数据库文件 vlan.dat 需要同时删除，具体操作如下。

SW1#**delete flash:vlan.dat**
Delete filename [vlan.dat]?
Delete flash:/vlan.dat? [confirm]
SW1#**delete flash:config.text**
Delete filename [config.text]?
Delete flash:/config.text? [confirm]

接下来，我们在特权模式下输入 reload 命令重启设备，交换机就恢复到出厂设置。

3.2 多场景 VLAN 配置

3.2.1 场景一：单交换机 VLAN 配置

> 🔑**场景一：** YH 食品有限公司是一家小型干果食品加工公司。目前，公司仅有十几名员工，公司的业务数据承载在一台 Cisco 2960 交换机上。为确保数据传输的安全性，现请你按照工作部门划分三个 VLAN。部门名称分别是财务部（Finance）、市场部（Marketing）和生产部（Production）。

如图 3-3 所示，我们看到没有划分 VLAN 之前，所有用户都在一个广播域即同一网络中，他们使用同一网段 IP 地址；划分 VLAN 之后，不同部门用户被逻辑分割到不同广播域即不同网络中，网络规划如表 3-2 所示。

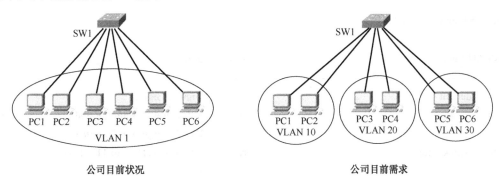

图 3-3 单交换机 VLAN

表 3-2 单交换机 VLAN 规划一览表

VLAN ID	VLAN 名称	端口分配	网络地址	备注
10	Finance	Fa0/1~8	192.168.0.0/24	财务部
20	Marketing	Fa0/9~16	192.168.1.0/24	市场部
30	Production	Fa0/17~24	192.168.2.0/24	生产部

具体实现，操作步骤如下。

步骤一：创建 VLAN

```
SW1(config)#vlan 10
```

```
SW1(config-vlan)#name Finance
SW1(config-vlan)#vlan 20
SW1(config-vlan)#name Marketing
SW1(config-vlan)#vlan 30
SW1(config-vlan)#name Production
```

步骤二：添加 VLAN 成员

```
SW1(config)#interface range f0/1-8
SW1(config-if-range)#switchport mode access
SW1(config-if-range)#switchport access vlan 10
SW1(config-if-range)#interface range f0/9-16
SW1(config-if-range)#switchport access vlan 20
SW1(config-if-range)#interface range f0/17-24
SW1(config-if-range)#switchport access vlan 30
```

步骤三：查看 VLAN 信息

```
SW1#show vlan brief
```

VLAN	Name	Status	Ports
1	default	active	Gig0/1, Gig0/2
10	Finance	active	Fa0/1, Fa0/2, Fa0/3, Fa0/4
			Fa0/5, Fa0/6, Fa0/7, Fa0/8
20	Marketing	active	Fa0/9, Fa0/10, Fa0/11, Fa0/12
			Fa0/13, Fa0/14, Fa0/15, Fa0/16
30	Production	active	Fa0/17, Fa0/18, Fa0/19, Fa0/20
			Fa0/21, Fa0/22, Fa0/23, Fa0/24
1002	fddi-default	active	
1003	token-ring-default	active	
1004	fddinet-default	active	
1005	trnet-default	active	

从以上信息我们可以看出，VLAN ID、名称及成员的配置都是正确的。

步骤四：配置 PC 的 IP 地址，测试网络连通性

此时，同一 VLAN PC 间可以正常通信，不同 VLAN PC 间不能通信，即 VLAN 技术实现了广播域分割。

3.2.2 场景二：两台交换机 VLAN 配置

🔑**场景二**：YH 食品有限公司规模在逐步扩大，目前员工发展到近四十人。办公室分布在两个楼层。每个楼层有一台 Cisco 2960 交换机，两台交换机级联在一起。为确保数据传输安全，公司要求按职能重新规划 VLAN，如图 3-4 所示，VLAN 名称依然按照所属部门命名。三个部门分别是财务部、市场部和生产部。请你根据要求划分 VLAN。

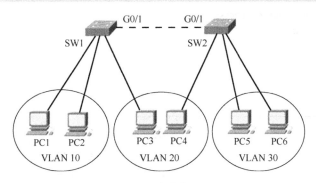

图 3-4 两台交换机 VLAN

根据需求，按照如图 3-4 所示网络拓扑，我们做了如表 3-3 所示的规划。

表 3-3 两台交换机网络规划一览表

VLAN ID	VLAN 名称	端口分配	网络地址	备注
10	Finance	SW1：Fa0/1～16	192.168.0.0/24	财务部
20	Marketing	SW1：Fa0/17～24,G0/1 SW2：Fa0/17～24,G0/1	192.168.1.0/24	市场部
30	Production	SW2：Fa0/1～16	192.168.2.0/24	生产部

经分析，VLAN10、VLAN30 的端口成员均属于一台交换机，而 VLAN20 的端口成员跨越了两台交换机。例如，VLAN20 中的主机 PC3 要发送数据到 PC4，PC3 的数据需经 SW1 的 G0/1 端口转发，由 SW2 的 G0/1 端口接收并转发到 PC4。因此，两台交换机的 G0/1 端口也应属于 VLAN20。

具体配置步骤如下。

步骤一：在 SW1 上创建 VLAN，添加 VLAN 成员

```
SW1(config)#vlan 10
SW1(config-vlan)#name Finance
```

SW1(config-vlan)#**vlan 20**
SW1(config-vlan)#**name Marketing**
SW1(config-vlan)#**exit**
SW1(config)#**interface range f0/1-16**
SW1(config-if-range)#**switchport mode access**
SW1(config-if-range)#**switchport access vlan 10**
SW1(config-if-range)#**interface range f0/17-24,G0/1**
SW1(config-if-range)#**switchport mode access**
SW1(config-if-range)#**switchport access vlan 20**

查看 SW1 的 VLAN 信息。

SW1#**show vlan brief**

VLAN	Name	Status	Ports
1	default	active	Gig0/2
10	Finance	active	Fa0/1, Fa0/2, Fa0/3, Fa0/4
			Fa0/5, Fa0/6, Fa0/7, Fa0/8
			Fa0/9, Fa0/10, Fa0/11, Fa0/12
			Fa0/13, Fa0/14, Fa0/15, Fa0/16
20	Marketing	active	Fa0/17, Fa0/18, Fa0/19, Fa0/20
			Fa0/21, Fa0/22, Fa0/23, Fa0/24
			Gig0/1
1002	fddi-default	active	
1003	token-ring-default	active	
1004	fddinet-default	active	
1005	trnet-default	active	

步骤二：在 SW2 上创建 VLAN，添加 VLAN 成员

SW2(config)#**vlan 20**
SW2(config-vlan)#**name Marketing**
SW2(config-vlan)#**vlan 30**
SW2(config-vlan)#**name Production**
SW2(config-vlan)#**exit**
SW2(config)#**interface range f0/17-24 ,G0/1**
SW2(config-if-range)#**switchport mode access**
SW2(config-if-range)#**switchport access vlan 20**
SW2(config-if-range)#**interface range f0/1-16**

SW2(config-if-range)#**switchport mode access**
SW2(config-if-range)#**switchport access vlan 30**
查看 SW2 的 VLAN 信息。
SW2#**show vlan brief**

VLAN	Name	Status	Ports
1	default	active	Gig0/2
20	Marketing	active	Fa0/17, Fa0/18, Fa0/19, Fa0/20
			Fa0/21, Fa0/22, Fa0/23, Fa0/24
			Gig0/1
30	Production	active	Fa0/1, Fa0/2, Fa0/3, Fa0/4
			Fa0/5, Fa0/6, Fa0/7, Fa0/8
			Fa0/9, Fa0/10, Fa0/11, Fa0/12
			Fa0/13, Fa0/14, Fa0/15, Fa0/16
1002	fddi-default	active	
1003	token-ring-default	active	
1004	fddinet-default	active	
1005	trnet-default	active	

步骤三：配置 PC 的 IP 地址，做连通性测试

至此，同一 VLAN 的 PC 间可以直接通信，不同 VLAN 的 PC 间不能通信。我们通过 VLAN 技术成功实现了逻辑工作组的划分与管理。

3.2.3 场景三：多交换机 VLAN 配置

🔍**场景三**：YH 食品有限公司在 2017 年年初新增一个客服部，公司招聘新员工近 20 人，目前员工总人数接近 70 人。因新增员工，公司又在三楼租赁多间办公室，所以为三楼再新购一台 Cisco 2960 交换机，并连接至二楼交换机。由于公司最近对人员进行了一次大整合，部门有所调整，办公地点也发生了改变。现要求你，针对公司三个楼层的交换机，重新规划部署网络。请先收集部门整合的最新信息，再进行详细规划和设计并进一步实施。公司网络拓扑如图 3-5 所示。

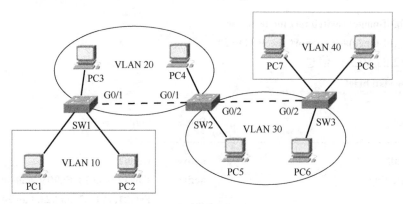

图 3-5　多台交换机 VLAN

经调研得知，财务部在一楼办公，客服部在三楼，市场部分布在一楼和二楼，生产部分布在二楼和三楼。依据部门位置，以及 PC 所连交换机端口位置，我们对网络做了重新规划，如表 3-4 所示。

表 3-4　多台交换机网络规划一览表

VLAN ID	VLAN 名称	端口分配	网络地址	备注
10	Finance	SW1：Fa0/1～Fa15	192.168.0.0/24	财务部
20	Marketing	SW1：Fa0/16～Fa24,G0/1 SW2：Fa0/1～Fa9,G0/1	192.168.1.0/24	市场部
30	Production	SW2：Fa0/10～Fa24,G0/2 SW3：Fa0/1～Fa5,G0/2	192.168.2.0/24	生产部
40	Customer	SW3：Fa0/6～Fa24	192.168.3.0/24	客服部

具体实施过程如下。

步骤一：在 SW1 上创建 VLAN，添加 VLAN 成员

```
SW1(config)#vlan 10
SW1(config-vlan)#name Finance
SW1(config-vlan)#vlan 20
SW1(config-vlan)#name Marketing
SW1(config-vlan)#exit
SW1(config)#interface range f0/1-15
SW1(config-if-range)#switchport mode access
SW1(config-if-range)#switchport access vlan 10
SW1(config-if-range)#interface range f0/16-24,G0/1
SW1(config-if-range)#switchport mode access
SW1(config-if-range)#switchport access vlan 20
```

查看 SW1 的 VLAN 信息。

```
SW1#show vlan brief

VLAN Name                          Status     Ports
---- ------------------------------ --------- -------------------------------
1    default                        active    Gig0/2
10   Finance                        active    Fa0/1, Fa0/2, Fa0/3, Fa0/4
                                              Fa0/5, Fa0/6, Fa0/7, Fa0/8
                                              Fa0/9, Fa0/10, Fa0/11, Fa0/12
                                              Fa0/13, Fa0/14, Fa0/15
20   Marketing                      active    Fa0/16, Fa0/17, Fa0/18, Fa0/19
                                              Fa0/20, Fa0/21, Fa0/22, Fa0/23
                                              Fa0/24, Gig0/1

1002 fddi-default                   active
1003 token-ring-default             active
1004 fddinet-default                active
1005 trnet-default                  active
```

步骤二：在 SW2 上创建 VLAN，添加 VLAN 成员

```
SW2(config)#vlan 20
SW2(config-vlan)#name Marketing
SW2(config-vlan)#vlan 30
SW2(config-vlan)#name Production
SW2(config-vlan)#exit
SW2(config)#interface range f0/1-9,G0/1
SW2(config-if-range)#switchport mode access
SW2(config-if-range)#switchport access vlan 20
SW2(config-if-range)#interface range f0/10-24,G0/2
SW2(config-if-range)#switchport mode access
SW2(config-if-range)#switchport access vlan 30
```

查看 SW2 的 VLAN 信息。

```
SW2#show vlan brief

VLAN Name                          Status     Ports
---- ------------------------------ --------- -------------------------------
1    default                        active
20   Marketing                      active    Fa0/1, Fa0/2, Fa0/3, Fa0/4
                                              Fa0/5, Fa0/6, Fa0/7, Fa0/8
```

30	Production	active	Fa0/9, Gig0/1 Fa0/10, Fa0/11, Fa0/12, Fa0/13 Fa0/14, Fa0/15, Fa0/16, Fa0/17 Fa0/18, Fa0/19, Fa0/20, Fa0/21 Fa0/22, Fa0/23, Fa0/24, Gig0/2
1002	fddi-default	active	
1003	token-ring-default	active	
1004	fddinet-default	active	
1005	trnet-default	active	

步骤三：在 SW3 上创建 VLAN，添加 VLAN 成员

SW3(config)#**vlan 30**
SW3(config-vlan)#**name Production**
SW3(config-vlan)#**vlan 40**
SW3(config-vlan)#**name Customer**
SW3(config-vlan)#**exit**
SW3(config)#**interface range f0/1-5,G0/2**
SW3(config-if-range)#**switchport mode access**
SW3(config-if-range)#**switchport access vlan 30**
SW3(config-if-range)#**interface range f0/6-24**
SW3(config-if-range)#**switchport mode access**
SW3(config-if-range)#**switchport access vlan 40**

查看 SW3 的 VLAN 信息。

SW3#**show vlan brief**

VLAN	Name	Status	Ports
1	default	active	Gig0/1
30	Production	active	Fa0/1, Fa0/2, Fa0/3, Fa0/4 Fa0/5, Gig0/2
40	Customer	active	Fa0/6, Fa0/7, Fa0/8, Fa0/9 Fa0/10, Fa0/11, Fa0/12, Fa0/13 Fa0/14, Fa0/15, Fa0/16, Fa0/17 Fa0/18, Fa0/19, Fa0/20, Fa0/21 Fa0/22, Fa0/23, Fa0/24
1002	fddi-default	active	
1003	token-ring-default	active	

1004 fddinet-default		active
1005 trnet-default		active

步骤四：配置 PC 的 IP 地址，进行连通性测试

完成以上配置后，同一部门的 PC 间可以直接通信，不同部门的 PC 间不能直接通信。

3.3 认识 Trunk 技术

3.2 节中，我们学习了跨越两台交换机同一 VLAN 的配置。在实际工程中，有两个 VLAN 跨越两台交换机的情况，也有更多 VLAN 跨越两台交换机甚至更多交换机的情况。如何保证同一 VLAN 的 PC 间跨交换机通信呢？

一个 VLAN 跨交换机实现同一 VLAN PC 间通信，需一条模式为 Access 的级联链路（共占 2 个交换端口），如图 3-6 所示。

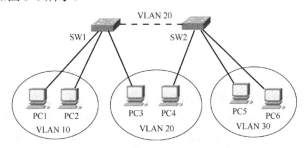

图 3-6 跨交换机一个 VLAN 通信

两个 VLAN 跨交换机实现同一 VLAN PC 间通信，需两条模式为 Access 的链路（共占 4 个交换端口），如图 3-7 所示。

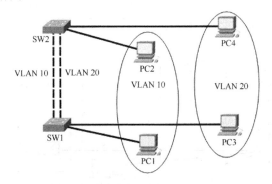

图 3-7 跨交换机两个 VLAN 通信

三个 VLAN 跨交换机实现同一 VLAN PC 间通信，则需要三条模式为 Access 的链路（共占 6 个交换端口），如图 3-8 所示。

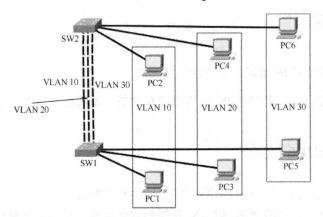

图 3-8 跨交换机三个 VLAN 通信

采用 Access 链路方法虽能实现跨交换机同一 VLAN PC 间的通信，但需要多条 Access 链路的支持，浪费了多个级联端口。为解决这一问题，可以引入 Trunk 技术。

3.3.1 认识 Trunk 干道

1. 干道的基本认识

当有多个 VLAN 跨越两台交换机通信时，交换机间只有一条级联链路就可以实现，我们把这条链路称为干道。干道可以同时传输多个 VLAN 的数据。如图 3-9 所示，▢ 表示 VLAN10 的数据报文，▨ 表示 VLAN20 的数据报文，▩ 表示 VLAN30 的数据报文。当三个 VLAN 的数据跨越交换机访问另一台交换机同一 VLAN 时，都必经干道。此时，这条干道就像高速公路被划分成多车道一样，允许多辆车同时通过。这条干道就是我们本节课要学习的 Trunk 链路，即干道链路。

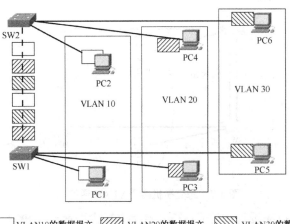

图 3-9 干道链路

2. 干道的应用场合

干道链路通常应用在以下几种场合。

- **交换机与交换机之间**：如图 3-10 所示，SW1 与 SW2 之间的链路。它是 PC1 和 PC2，PC3 和 PC4，以及 PC5 和 PC6 通信的必经之路，因为它需要同时转发三个 VLAN 的数据。
- **交换机与路由器之间**：如图 3-10 所示，SW2 与 Router1 之间的链路。为了实现不同 VLAN 之间的通信，该链路必须允许 3 个 VLAN 的数据通过，才可实现 VLAN 间的互访。
- **交换机与服务器之间**：服务器是多个 VLAN 的共用服务器，但服务器网卡必须支持 Trunk。

若一条链路只允许通过一个 VLAN 的数据，这条链路就不是干道（Trunk），而是访问链路，即接入链路（Access）。接入链路一般是指交换机和主机之间的链路，如图 3-10 所示。

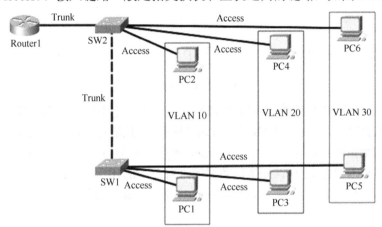

图 3-10　干道链路应用

3. 干道的实现原理

默认情况下，干道允许交换机上所有 VLAN 的数据通过。当交换机转发某一 VLAN 的广播数据帧时，它会向除该接收端口之外且属于同一 VLAN 的其他所有端口转发。学习干道技术后，我们还需要进一步补充，交换机还会向所有干道转发该广播帧。干道如何让所有 VLAN 的数据同时通过呢？又如何区分不同 VLAN 的数据呢？这就是我们本节课要学习的干道原理。

干道怎样实现？交换机又如何识别来自不同 VLAN 的数据？这就是干道 Trunk 要解决的"数据处理"问题。

如图 3-11 所示，若 PC1 要跟 PC2 通信，PC1 就发出一个数据帧，该数据帧沿一条 Access

链路到达 SW1 的 Fa0/1 端口，接下来 SW1 交换机将查看 MAC 地址表，得知该数据帧来自 VLAN10。

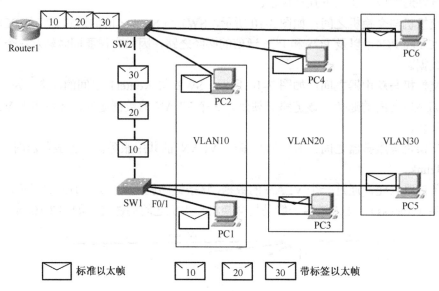

图 3-11　干道链路数据封装

由于干道 Trunk 默认允许所有 VLAN 数据通过，为区别不同 VLAN 的数据，交换机会对这数据帧进行重新封装，打上 VLAN10 的 Tag（标签），然后再发送到 Trunk 链路。

带着 Tag 的数据帧到达对端交换机的 Trunk 端口，该端口就能识别该数据帧。Trunk 端口收到其携带的 VLAN Tag 是 10，知道该数据帧来自 VLAN10。于是接收数据帧并将 VLAN 的 Tag 去掉，再交付给某个属于 VLAN10 的 Access 端口。最后该端口负责把数据帧交付给目标主机，一个单向数据帧的转发工作就此结束。

由此可见，在 Access 链路上传输的数据帧是标准以太帧，而在 Trunk 链路上传输的数据帧往往是经 Trunk 封装协议打上 Tag 的以太帧。

在 Trunk 上怎样对数据重新封装呢？这需要通过 Trunk 协议来完成，即 ISL 或 IEEE 802.1q。这两种协议支持在交换机之间封装和解封装数据帧。为标识数据帧属于哪个 VLAN，则需要对数据帧进行打 Tag 封装，即在以太网帧基础上增加 VLAN 标识。

ISL（Inter Switch Link）是 Cisco 系列交换机专用封装协议，工作在 OSI RM 第二层。ISL 是一个独立协议，能在交换机间封装任何类型数据帧或上层协议的数据。例如，ISL 可以封装令牌环帧、以太帧等。ISL 封装是通过专用集成电路（Application-Specific Integrated Circuits，ASIC）来实现的，因此，ISL 能为 Trunk 链路提供高效性能，其封装如图 3-12 所示。

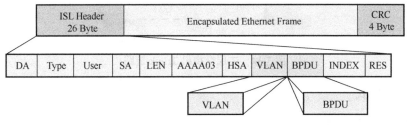

图 3-12 ISL 协议数据封装

IEEE 802.1q 是国际标准协议，能对数据帧附加 VLAN 标识，兼容各厂商交换机。它也是二层协议，但采用的帧格式与 ISL 协议不同。IEEE 802.1q 帧格式如图 3-13 所示。

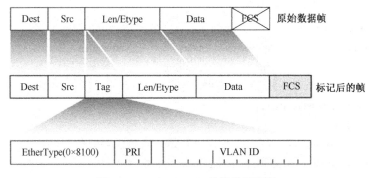

图 3-13 IEEE 802.1q 协议数据封装

IEEE 802.1q 不像 ISL 那样封装原有以太帧，它是修改以太帧。即将一个 4 Byte 的 Tag 字段插入原始以太帧中，原始帧的 FCS（帧检验和）也会重新计算，如图 3-13 所示。Tag 的作用是便于 Trunk 对端交换机识别数据帧所属 VLAN，从而正确转发数据，VLAN 也就可以跨越多台交换机。

另外，ISL 对经过 Trunk 的所有数据帧都打 Tag。IEEE 802.1q 对数据帧可以打 Tag 也可以不打 Tag。本征 VLAN 不打 Tag，一台设备只能有一个本征 VLAN。不打 Tag 的帧中不携带 VLAN 信息，是一个普通以太帧，且要求 Trunk 两端本征 VLAN 必须一致。

3.3.2 认识 DTP 协议

前面我们认识了干道 Trunk，了解了 Trunk 的应用场合，但是，Trunk 是如何形成的呢？这是本节课我们要学习的内容。

Trunk 的形成有两种方式：一种是静态方式，即通过手工指定形成；另一种是动态方式，即通过协议协商形成。

要使 SW1 和 SW2 之间链路成为 Trunk，如图 3-14 所示，我们通过如下操作实现。

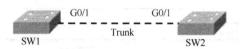

图 3-14 二层交换机 Trunk 链路

1. 静态指定方式

SW1(config)#**interface G0/1**
SW1(config-if)#**switchport mode trunk**
%LINEPROTO-5-UPDOWN: Line protocol on Interface GigabitEthernet0/1, changed state to down
%LINEPROTO-5-UPDOWN: Line protocol on Interface GigabitEthernet0/1, changed state to up

在 SW1 交换机上,查看 Trunk 信息。

SW1#**show interfaces trunk**

Port	Mode	Encapsulation	Status	Native vlan
Gig0/1	on	802.1q	trunking	1

// Gig0/1 端口已经形成 Trunk,封装为 IEEE 802.1q

Port	Vlans allowed on trunk
Gig0/1	1-1005

// 以上显示 Gig0/1 端口的 Trunk 允许 VLAN1~VLAN1005 数据帧通过

Port	Vlans allowed and active in management domain
Gig0/1	1

// 以上显示 Gig0/1 端口的 Trunk 实际允许 VLAN1 的数据帧通过,之所以和上一段不同,是因为当前交换机只有 VLAN 1 存在

Port	Vlans in spanning tree forwarding state and not pruned
Gig0/1	1

// 以上显示 Gig0/1 端口的 Trunk 没有被修剪掉的 VLAN

在 SW2 交换机上,查看 Trunk 信息,结果同 SW1。

SW2#**show interfaces trunk**

Port	Mode	Encapsulation	Status	Native vlan
Gig0/1	on	802.1q	trunking	1

Port	Vlans allowed on trunk
Gig0/1	1-1005

Port	Vlans allowed and active in management domain
Gig0/1	1

Port	Vlans in spanning tree forwarding state and not pruned

Gig0/1   1

通过以上显示，我们发现 Trunk 已经形成，若使用命令 show interfaces trunk，没有信息输出，如下所示：

SW2#**show interfaces trunk**

SW2#

以上结果表明 Trunk 没有形成。确定 Trunk 是否形成，需要分别在两端交换机上查看，若只有一端能看到 Trunk 形成信息，并不能断定 Trunk 已经形成，必须两端都看到 Trunk 形成信息，才能确认。

以上 Trunk 配置命令是在二层交换机上实现的。若要在三层交换机之间实现 Trunk，则必须指定 Trunk 封装协议 IEEE 802.1q 或 ISL，如图 3-15 所示。

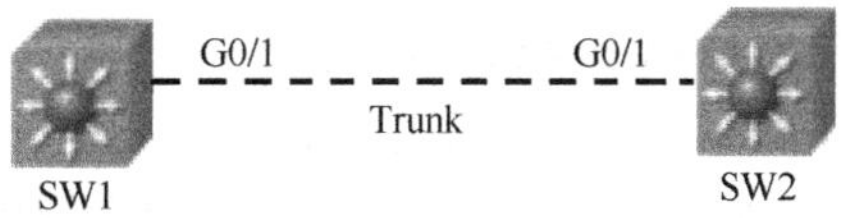

图 3-15 三层交换机 Trunk 链路

在三层交换机上，静态指定 Trunk，正确配置如下：

SW1(config)#**interface G0/1**
SW1(config-if)#**switchport trunk encapsulation dot1q**   //需先指定 Trunk 封装协议
SW1(config-if)#**switchport mode trunk**
%LINEPROTO-5-UPDOWN: Line protocol on Interface GigabitEthernet0/1, changed state to down
%LINEPROTO-5-UPDOWN: Line protocol on Interface GigabitEthernet0/1, changed state to up

若同二层交换机一样，做如下配置：

SW1(config)#**interface G0/1**
SW1(config-if)#**switchport mode trunk**
Command rejected: An interface whose trunk encapsulation is "Auto" can not be configured to "trunk" mode.

则我们会看到错误提示："命令被拒绝：因为 Trunk 封装类型是'Auto'，所以不能被配置成 trunk 模式"，即 Auto 封装类型是不可以形成 Trunk 的。

为何必须在三层交换机上手工指定 Trunk 封装协议？因为 Catalyst 二层交换机默认只支持 IEEE 802.1q 协议，而三层交换机对 ISL 和 IEEE 802.1q 协议均支持，因此，必须手工指定，并确保两端协议一致。

2. 动态协商方式

DTP（Dynamic Trunk Protocol）是二层动态 Trunk 协议，用于两台 Cisco 设备之间协商形成 Trunk，是思科专用协议。DTP 默认启用，也可禁用，DTP 协商会导致网络流量的增加。

只有当相邻交换机端口被配置为某个支持 DTP 的 Trunk 模式时，DTP 才可管理 Trunk 协商。启用 DTP 的交换机间可自动协商 Trunk 链路的形成，也可协商 Trunk 链路的封装类型。DTP 可以配置的端口模式如下所述。

- **Access（Off）**：永久非 Trunk 模式，即关闭 DTP。不管相邻端口是否为 Trunk 模式，最终都无法形成 Trunk。
- **Trunk（On）**：永久 Trunk 模式，即开启 DTP。即使相邻端口不是 Trunk 模式，也会协商成 Trunk。
- **Dynamic Auto**：是目前交换机端口的默认配置。该模式下端口可以响应 DTP 协商报文，但不会主动发送协商报文。所以若相邻两端口都是 auto 模式，因为都不主动发送 DTP 协商报文，最终无法形成 Trunk。只有有一方主动发送 DTP 协商报文才有可能形成 Trunk。
- **Dynamic Desirable**：端口期望同相邻端口形成 Trunk。该模式下端口会主动发送 DTP 协商报文，也会对 DTP 协商报文做出响应。
- **Nonegotiate**：端口为非协商状态，即关闭 DTP 协议。此时端口不会收、发 DTP 协商报文。除非对端和本端被强制置成 Trunk 模式，否则无法形成 Trunk。

表 3-5　为 DTP 端口协商结果一览表。

	dynamic auto	dynamic desirable	trunk	access	nonegotiate
dynamic auto	×	√	√	×	×
dynamic desirable	√	√	√	×	×
trunk	√	√	√	不建议	√
access	×	×	不建议	×	×
nonegotiate	×	×	√	×	×

从表 3-5 相邻端口模式产生的 DTP 协商结果来看，在开启 DTP 协议的情况下，只要有一端端口模式是"desirable"，就能协商成 Trunk；只要一端静态指定为"trunk"，结果肯定是 Trunk；只要有一端静态指定为"access"，就无法形成 Trunk；两端都是"auto"，结果也不能形成 Trunk（均不主动发送 DTP 协商报文，也不会收到 DTP 消息）。若一端关闭 DTP 协议（nonegotiate），除非对端被强制指定为"Trunk"，否则无法形成 Trunk。

开启或关闭 DTP 协议，我们可以采用如下命令实现：

```
SW1(config-if)#switchport nonegotiate          //关闭 DTP 协议
SW1(config-if)#no switchport nonegotiate       //开启 DTP 协议
```

需要注意的是，在关闭 DTP 前，Trunk 形成方式必须设置成手工指定，否则会出现如下提示信息：

```
SW1(config-if)#switchport nonegotiate
Command rejected: Conflict between 'nonegotiate' and 'dynamic' status.
```

以上消息提示，关闭 DTP 失败（非协商与动态状态冲突）。

3.3.3　分析 Trunk 结果

实际工作过程中，能对 Trunk 的结果进行分析非常重要。它可以帮助我们解决很多有关 Trunk 的故障问题。下面我们一起来分析 Trunk 的结果。

```
SW1#show interfaces trunk
Port        Mode            Encapsulation   Status        Native vlan
Gig0/1      on              802.1q          trunking      1

Port        Vlans allowed on trunk
Gig0/1      1-1005

Port        Vlans allowed and active in management domain
Gig0/1      1,10,20,30

Port        Vlans in spanning tree forwarding state and not pruned
Gig0/1      1,10,20,30
```

在交换机 SW1 上查看 Trunk 结果信息，发现其 Gig0/1 端口已形成 Trunk。Mode 为"on"表明 Trunk 是通过手工方式形成的，其封装协议为 IEEE 802.1q，本征 VLAN ID 是 1。默认情况下，Trunk 允许所有 VLAN 数据通过，实际根据本地数据库决定允许通过的 VLAN 是 VLAN 1、VLAN 10、VLAN 20 和 VLAN 30。

在对端交换机 SW2 上，我们看到其端口 Gig0/1，是以 DTP 的"auto"模式与对端协商形成 Trunk 的，其他 Trunk 属性与对端相同。

```
SW2#show interfaces trunk
Port        Mode            Encapsulation   Status        Native vlan
Gig0/1      auto            n-802.1q        trunking      1

Port        Vlans allowed on trunk
Gig0/1      1-1005

Port        Vlans allowed and active in management domain
Gig0/1      1,10,20,30

Port        Vlans in spanning tree forwarding state and not pruned
Gig0/1      1,10,20,30
```

此时 SW1 与 SW2 间的 Trunk 已经正常工作。需要记住的是 Trunk 两端端口都正常，Trunk 才能正常工作。

3.3.4 排查 Trunk 故障

下面的 3 种故障案例都会导致 Trunk 不能正常工作，从而影响网络通信。

1. Trunk 两端本征 VLAN 不一致

默认情况下，Trunk 两端的本征 VLAN 均为 VLAN 1，即本征 VLAN 一致。所谓的本征 VLAN（Native VLAN）是指经 Trunk 转发但不打 VLAN Tag 的特殊 VLAN。

若我们在 SW1 上修改 Trunk 属性中的本征 VLAN，另一端不变，则因 Trunk 链路两端本征 VLAN 不匹配，就会导致网络故障，如下所示：

SW1(config)#**interface G0/1**
SW1(config-if)#**switchport mode trunk**
SW1(config-if)#**switchport trunk native vlan 10**

查看 SW1 的 Trunk 信息，我们发现本征 VLAN ID 为 10。

SW1#**show interfaces trunk**

Port	Mode	Encapsulation	Status	Native vlan
Gig0/1	on	802.1q	trunking	**10**

同时，我们会发现在两台交换机上都有 CDP 信息弹出：

SW1#
%CDP-4-NATIVE_VLAN_MISMATCH: Native VLAN mismatch discovered on GigabitEthernet0/1 (10), with SW2 GigabitEthernet0/1 (1).

交换机 SW1 上的信息显示，在 GigabitEthernet0/1 端口上发现本征 VLAN（10）与 SW2 的 GigabitEthernet0/1 端口的本征 VLAN（1）不匹配，这是一种常见 VLAN 故障。

SW2#
%CDP-4-NATIVE_VLAN_MISMATCH: Native VLAN mismatch discovered on GigabitEthernet0/1 (1), with SW1 GigabitEthernet0/1 (10).

交换机 SW2 上的信息显示，在 GigabitEthernet0/1 端口上发现本征 VLAN（1）与 SW1 的 GigabitEthernet0/1 端口的本征 VLAN（10）不匹配。

由此可知，Trunk 两端报错的信息是一致的。

2. Trunk 两端允许通过的 VLAN 不一致

默认情况下，Trunk 允许所有 VLAN 数据通过。但是，我们若修改 Trunk 链路端口上允许通过的 VLAN，则相关 VLAN 的通信就会受到影响，从而导致网络故障。

下面，我们在交换机 SW1 的 Trunk 链路端口上，禁止 VLAN 10 的数据通过，SW2 的 Trunk

端口不变。操作如下:

```
SW1(config)#interface G0/1
SW1(config-if)#switchport mode trunk
SW1(config-if)#switchport trunk allowed vlan remove 10
```

在交换机 SW1 上查看 Trunk 的结果信息:

```
SW1# show interface trunk
Port        Mode        Encapsulation   Status      Native vlan
Gig0/1      on          802.1q          trunking    1

Port        Vlans allowed on trunk
Gig0/1      1-9,11-1005

Port        Vlans allowed and active in management domain
Gig0/1      1,20,30

Port        Vlans in spanning tree forwarding state and not pruned
Gig0/1      1,20,30
```

从以上输出我们发现,Trunk 已经禁止了 VLAN10 的数据通过。Trunk 的 SW2 端允许 VLAN10 的数据通过,而 SW1 端禁止 VLAN10 的数据通过,则跨越 Trunk 的 VLAN10 主机间的通信就会失败。

3. Trunk 模式不匹配

从以上两个故障案例我们都能看到 Trunk 信息。下面我们一起来分析没有 Trunk 信息的故障。

```
SW1#show interfaces trunk

SW1#
```

在 SW1 交换机上,没有显示出 Trunk 的信息,什么原因导致 Trunk 无法形成呢?接下来,我们要在 SW1 和 SW2 上使用 **show interfaces** *interface-ID* **switchport** 进一步排查。

```
SW1#show interfaces G0/1 switchport
Name: Gig0/1
Switchport: Enabled
Administrative Mode: trunk
Operational Mode: trunk
Administrative Trunking Encapsulation: dot1q
Operational Trunking Encapsulation: dot1q
Negotiation of Trunking: Off
```

Access Mode VLAN: 1 (default)
Trunking Native Mode VLAN: 1 (default)
....

SW1#

SW1 的排查分析：SW1 的 Gig0/1 端口管理模式和运行模式均为 trunk，说明 SW1 是手工指定为 trunk 模式的。通过信息 "Negotiation of Trunking: Off"，可以推断交换机已经关闭 DTP，也就不会收、发 DTP 协商报文。如果 SW1 开启 DTP，则它会向对端 SW2 发送 DTP 协商报文，告知对方自己已是 trunk 模式。

SW2#**show interfaces G0/1 switchport**
Name: Gig0/1
Switchport: Enabled
Administrative Mode: **dynamic auto**
Operational Mode: **static access**
Administrative Trunking Encapsulation: dot1q
Operational Trunking Encapsulation: native
Negotiation of Trunking: On
Access Mode VLAN: 1 (default)
Trunking Native Mode VLAN: 1 (default)
....

SW2#

SW2 的排查分析：SW2 的 Gig0/1 端口管理模式为 **dynamic auto**，运行模式为 **static access**，说明 SW2 已开启 DTP。但是，端口模式为 dynamic auto，即便开启 DTP，也不会主动发送 DTP 协商报文。而对端 SW1 又关闭了 DTP，本端口就收不到 DTP 消息。所以，SW2 的 Gig0/1 端口，只能工作在 Access 模式，也就无法形成 Trunk。

以上是常见的三种 Trunk 故障，其中 Trunk 的本征 VLAN 不一致、Trunk 上允许的 VLAN 不一致，这两种故障，我们都能通过命令 show interfaces trunk 查到故障原因；但 Trunk 模式不匹配导致的故障，需要进一步使用命令 show interfaces *interface-ID* switchport 来详细排查。

3.3.5 场景四：应用 Trunk 技术

🔑**场景四**：YH 食品有限公司随着业务的不断扩大，对数据的存储提出了更高要求。目前公司的 4 个部门依然是财务部、市场部、生产部和客服部。应市场部和生产部的强烈要求，YH 公司为两个部门各配备了一台专用服务器，用于存储公司大数据。现要求你将两台服务器尽快接入公司网络，尽可能不影响网络的正常通信，并确保服务器的安全性。网络拓扑如图 3-16 所示。

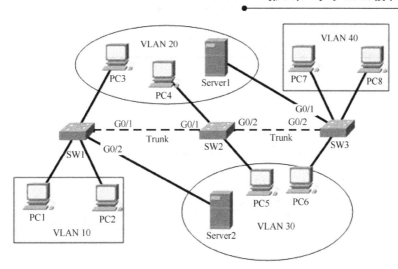

图 3-16 Trunk 链路应用

依据图 3-16 拓扑分析：若市场部 VLAN 20 的主机 PC3 要发送数据到 Server1，需要经过交换机 SW1→SW2→SW3 转发；而生产部 VLAN30 的主机 PC6 要发送数据到 Server2，也需要经过交换机 SW3→SW2→SW1 转发，最终到达目的主机。由此可见，VLAN20 和 VLAN30 的数据都要经链路 SW1—SW2—SW3 承载。而在 3.2.3 节场景三中，SW1—SW2 链路是 VLAN20 的专用链路，SW2—SW3 链路是 VLAN30 的专用链路，如图 3-17 所示。

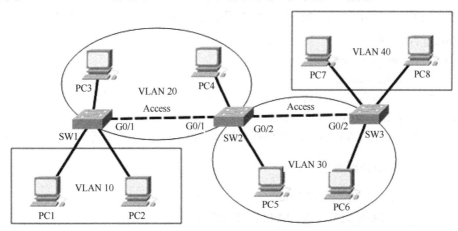

图 3-17 Access 链路应用

经以上对场景四（图 3-16）与场景三（图 3-17）的对比分析，我们发现场景三采用两条 Access 链路是有缺陷的，因为它们是相应 VLAN（VLAN20、VLAN30）的专用通道，显然不适应当前的网络现状。场景四需要进行技术升级，即将 Access 链路更新为 Trunk 链路，使其变成多 VLAN 的公共通道。

现将网络重新规划如表 3-6 所示。

表 3-6 网络重新规划一览表

VLAN ID	VLAN 名称	端口分配	网络地址	备注
10	Finance	SW1：Fa0/1～15	192.168.0.0/24	财务部
20	Marketing	SW1：Fa0/16～24 SW2：Fa0/1～9 SW3：G0/1	192.168.1.0/24	市场部
30	Production	SW1：G0/2 SW2：Fa0/10～24 SW3：Fa0/1～5	192.168.2.0/24	生产部
40	Customer	SW3：Fa0/6～24	192.168.3.0/24	客服部

具体实现过程如下所述。

步骤一：将专用服务器添加到对应 VLAN

```
SW1(config)#vlan 30
SW1(config-vlan)#name Production
SW1(config-vlan)#interface G0/2
SW1(config-if)#switchport mode access
SW1(config-if)#switchport access vlan 30
```

在交换机 SW1 上创建 VLAN 30，将连接专用服务器 Server2 的交换端口添加到 VLAN 30 中。

```
SW3(config)#vlan 20
SW3(config-vlan)#name Marketing
SW3(config-vlan)#interface G0/1
SW3(config-if)#switchport mode access
SW3(config-if)#switchport access vlan 20
```

在交换机 SW3 上创建 VLAN 20，将连接专用服务器 Server1 的交换端口添加到 VLAN 20 中。

步骤二：将级联链路配置成 Trunk

【方案一】 采用静态方式配置 Trunk

在 SW1 上，设置级联链路 SW1-SW2 的 Trunk 端口。

```
SW1(config)#interface G0/1
SW1(config-if)#switchport mode trunk          //静态 Trunk 方式
SW1(config-if)#switchport nonegotiate         //关闭 DTP 协商
```

在 SW2 上，设置级联链路 SW1-SW2 和 SW2-SW3 的 Trunk 端口。

```
SW2(config)#interface range G0/1-2
SW2(config-if-range)#switchport mode trunk
SW2(config-if-range)#switchport nonegotiate
```

在 SW3 上，设置级联链路 SW2-SW3 的 Trunk 端口。

SW3(config)#**interface G0/2**
SW3 (config-if)#**switchport mode trunk**
SW3 (config-if)#**switchport nonegotiate**

查看 SW1 的 Trunk 形成结果。

SW1#**show interfaces trunk**

Port	Mode	Encapsulation	Status	Native vlan
Gig0/1	on	802.1q	trunking	1

Port	Vlans allowed on trunk
Gig0/1	1-1005

Port	Vlans allowed and active in management domain
Gig0/1	1,10,20,30

Port	Vlans in spanning tree forwarding state and not pruned
Gig0/1	1,10,20,30

查看 SW2 的两个 Trunk 形成结果。

SW2#**show interfaces trunk**

Port	Mode	Encapsulation	Status	Native vlan
Gig0/1	on	802.1q	trunking	1
Gig0/2	on	802.1q	trunking	1

Port	Vlans allowed on trunk
Gig0/1	1-1005
Gig0/2	1-1005

Port	Vlans allowed and active in management domain
Gig0/1	1,20,30
Gig0/2	1,20,30

Port	Vlans in spanning tree forwarding state and not pruned
Gig0/1	1,20,30
Gig0/2	1,20,30

查看 SW3 的 Trunk 形成结果。

SW3#**show interfaces trunk**

Port	Mode	Encapsulation	Status	Native vlan
Gig0/2	on	802.1q	trunking	1

Port	Vlans allowed on trunk
Gig0/2	1-1005

Port	Vlans allowed and active in management domain
Gig0/2	1,20,30,40

Port	Vlans in spanning tree forwarding state and not pruned
Gig0/2	1,20,30,40

以上操作，我们是通过手工方式指定的 Trunk，通过检查发现，两条 Trunk 链路已经形成，并能进行正常转发工作。通过 **show interfaces** *interface-ID* **switchport** 进一步查看：

SW1#**show interfaces g0/1 switchport**
Name: Gig0/1
Switchport: Enabled
Administrative Mode: **trunk**
Operational Mode: **trunk**
Administrative Trunking Encapsulation: dot1q
Operational Trunking Encapsulation: dot1q
Negotiation of Trunking: **Off**
Access Mode VLAN: 20 (Marketing)
Trunking Native Mode VLAN: 1 (default)
Voice VLAN: none
………………………………………………

如上显示信息："Negotiation of Trunking: Off"说明 DTP 已经关闭。当网络中有非 Cisco 设备或者考虑到 VLAN 安全时，我们应采用该手工方式指定 Trunk，关闭 DTP 协商。

【方案二】 采用动态方式配置 Trunk

因为交换机端口默认模式是 dynamic auto，所以动态 Trunk 无须在链路两端都配置，只需在一端配置 dynamic desirable 就可以形成 Trunk。我们可以在 SW2 上对两条链路端口做如下配置：

SW2(config)#**interface range G0/1-2**
SW2 (config-if)#**switchport mode dynamic desirable**

接下来，我们先在 SW2 上查看两条链路是否已形成 Trunk。

SW2#**show interfaces trunk**

Port	Mode	Encapsulation	Status	Native vlan
Gig0/1	desirable	n-802.1q	trunking	1
Gig0/2	desirable	n-802.1q	trunking	1

Port	Vlans allowed on trunk

```
Gig0/1          1-1005
Gig0/2          1-1005

Port            Vlans allowed and active in management domain
Gig0/1          1,20,30
Gig0/2          1,20,30

Port            Vlans in spanning tree forwarding state and not pruned
Gig0/1          1,20,30
Gig0/2          1,20,30
```

以上输出结果显示，两条 Trunk 链路已经形成，且已经通过 DTP 协商了封装类型，如结果 "n-802.1q" 所示。

进一步在 SW1 和 SW3 上查看 Trunk 情况。

```
SW1#show interfaces trunk
Port        Mode        Encapsulation   Status      Native vlan
Gig0/1      auto        n-802.1q        trunking    1

Port        Vlans allowed on trunk
Gig0/1      1-1005

Port        Vlans allowed and active in management domain
Gig0/1      1,10,20,30

Port        Vlans in spanning tree forwarding state and not pruned
Gig0/1      1,10,20,30
```

在 SW3 上查看 Trunk 情况。

```
SW3#show interfaces trunk
Port        Mode        Encapsulation   Status      Native vlan
Gig0/2      auto        n-802.1q        trunking    1

Port        Vlans allowed on trunk
Gig0/2      1-1005

Port        Vlans allowed and active in management domain
Gig0/2      1,20,30,40

Port        Vlans in spanning tree forwarding state and not pruned
Gig0/2      1,20,30,40
```

以上输出表明，动态 Trunk 已经配置成功，SW2 两个端口分别以 DTP 模式 dynamic desirable 与 SW1 和 SW3 的端口 dynamic auto 协商形成 Trunk。进一步查看 SW1—SW2 间的 Trunk 端口，结果如下所示：

```
SW1#show interfaces g0/1 switchport
Name: Gig0/1
Switchport: Enabled
Administrative Mode: dynamic auto         //DTP 端口模式 auto
Operational Mode: trunk                   //DTP 协商结果 trunk
Administrative Trunking Encapsulation: dot1q
Operational Trunking Encapsulation: dot1q
Negotiation of Trunking: On               //DTP 协议开启
Access Mode VLAN: 1 (default)
Trunking Native Mode VLAN: 1 (default)
Voice VLAN: none
-------------------------------------------------
```

```
SW2#show interfaces g0/1 switchport
Name: Gig0/1
Switchport: Enabled
Administrative Mode: dynamic desirable    //DTP 端口模式 desirable
Operational Mode: trunk                   //DTP 协商结果 trunk
Administrative Trunking Encapsulation: dot1q
Operational Trunking Encapsulation: dot1q
Negotiation of Trunking: On               //DTP 协议开启
Access Mode VLAN: 1 (default)
Trunking Native Mode VLAN: 1 (default)
Voice VLAN: none
-------------------------------------------------
```

比较 SW1 和 SW2 的端口模式，我们可以进一步确认，在开启 DTP 协议的前提下，两交换机端口自动协商成 Trunk 模式。

3.4 认识 VTP 技术

VTP（VLAN Trunking Protocol）是二层 VLAN 干道协议，即 VLAN 中继协议，是思科专用协议。其作用是管理同一个域内 VLAN 的建立、删除和修改，保证域内 VLAN 信息的同步。

3.4.1 认识 VTP 协议

1. VTP 的必要性

目前 VLAN 技术已非常成熟，且被广泛应用于局域网。如何确保局域网中 VLAN 数据库同步，保证每台交换机都能依据 VLAN 信息正确转发数据，这是本节要学习的内容。

如果企业网中存在大量交换机，且每台都有很多 VLAN，那么添加、修改或删除 VLAN 的维护工作，仅凭手工配置，不但工作量巨大，且容易出错，产生网络故障。在这种情况下，引入了 VTP 技术。

2. VTP 域的规划

VTP 域即 VLAN 管理域，是域名相同且通过 Trunk 链路互连的一组交换机的集合。VTP 实质是将多台交换机划分到一个管理域，域中交换机通过交换 VTP 报文共享 VLAN 信息。一台交换机只能属于一个管理域，不同域中的交换机不能共享 VLAN 信息。

VTP 域划分条件：
- 交换机之间必须通过 Trunk 互联；
- 相同域内交换机域名必须相同；
- 交换机必须相邻，即相邻交换机需要有相同域名。

不满足以上任意一个条件，VLAN 数据库就无法保证正确同步。

在实际工程中，我们可以根据相应情况，灵活划分 VTP 域。接下来，我们以图 3-18 的拓扑为例一起来学习 VTP 域的规划方法。

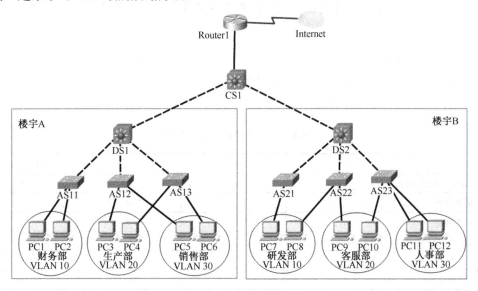

图 3-18 VTP 域的规划

本拓扑中，假设所有交换机都采用 Cisco 设备。我们看到楼宇 A 的部门和楼宇 B 的不同。我们可以把楼宇 A 中所有交换机规划到一个 VTP 域，共享相同 VLAN 信息。同理，将楼宇 B 中的交换机划分在另一 VTP 域。两个 VTP 域，使用不同域名。一个 VTP 域，就是一个交换区块。

同一 VTP 域，主要使用二层技术；不同 VTP 域间，则使用三层技术，即通过三层技术实现 VTP 域间的互通。若把园区网规划到同一 VTP 域，那么整个园区就是一个使用二层技术的交换区块。楼宇交换机 DS1 和 DS2 就不能充分发挥其三层设备的功能，楼宇内 VLAN 间通信要通过核心交换机 CS1 才可以实现，这无形中加重了核心交换机 CS1 的负担，使园区网络性能受到很大影响。因此，不建议把整个园区网规划到一个 VTP 域。

3．VTP 域中交换机的角色

VTP 基于 C/S 工作模式，即客户机／服务器模式。VTP 是 Cisco 交换机间采用的同步 VLAN 信息的协议，域中至少有一台交换机用作服务器。VTP 域中交换机的角色有以下 3 种。

- **Server 模式**：即服务器模式，是 VTP 的默认模式。该模式下交换机能够创建、删除和修改 VLAN，并把 VLAN 信息写入本地数据库文件 vlan.dat 中。Server 模式下，交换机能够产生 VTP 通告，能学习和转发 VTP 通告。
- **Client 模式**：即客户机模式。该模式下交换机不能创建、删除和修改 VLAN，不会更新本地数据库文件 vlan.dat。但能够学习 VTP 通告，并能请求和转发 VTP 通告。
- **Transparent 模式**：即透明模式。该模式下交换机能够创建、删除和更改 VLAN，VLAN 信息也会写入本地数据库文件 vlan.dat 中。不产生、不学习 VTP 通告，但会转发 VTP 通告。该模式下的交换机起到了 VTP 中继的作用，前提条件是交换机也必须加入到 VTP 域。

注意：VTP 域 VLAN 同步后，查看 flash，我们发现即使是 Client 模式的交换机，flash 中也会存在数据库文件 vlan.dat。针对这个问题，我们反复实验，得出的结论与 Cisco 给出的表 3-7 说法并不一致（表中用底纹做了特殊标记）。

表 3-7　VTP 模式比较一览表

	Server	Client	Transparent
能创建、删除、修改 VLAN	√	×	√
能生成 VTP 通告	√	×	×
能转发 VTP 通告	√	√	√
能根据收到的 VTP 通告，更新 VLAN 信息	√	√	×
能保存 VLAN 信息到 NVRAM（flash）	√	×	√
能影响到其他交换机上 VLAN	√	√	×

3.4.2 认识 VTP 修剪

VTP 修剪（VTP Pruning），是思科基于 VTP 协议开发的一项功能。该功能让交换机能动态添加或删除与 Trunk 相关联的 VLAN，从而提高交换网络的可用带宽。

VTP 修剪可以删除 Trunk 链路上不必要的广播、组播以及目标地址未知的单播流量。例如，来自 VLAN 10 主机的通信流，要经交换机 A 转发，交换机 A 依据自己的 VLAN 数据库，不转发给没有 VLAN 10 的邻居交换机 B，而会转发给有 VLAN10 的邻居交换机 C。若没有 VTP 修剪，这些通信流量会在 VTP 域的所有 Trunk 上传输，浪费带宽。

VTP 修剪，仅 VTP V2 支持，默认没有启用。VTP 修剪功能仅会阻止包含在修剪列表中的 VLAN。默认情况下，VLAN 2～VLAN 1001 支持修剪功能。下面我们通过实例来理解 VTP 的修剪功能。

如图 3-19 是一个没有开启 VTP 修剪功能的交换网络。图中 SW1 和 SW4 的端口 Fa0/1 均属于 VLAN 10。若此时，PC1 主机发出一个广播包，经 SW1 泛洪出去，则全网交换机都会收到，即使交换机 SW3、SW5 和 SW6 中没有 VLAN 10 的信息。

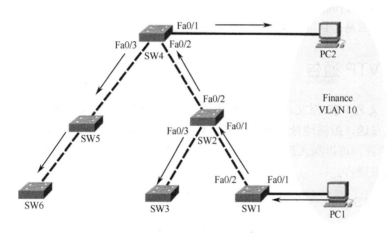

图 3-19 没有开启 VTP 修剪

下面我们来看一下开启 VTP 修剪后，网络的变化，如图 3-20 所示。

如图 3-20 所示开启 VTP 修剪功能后，此时 PC1 的广播流量经交换机 SW1 泛洪，最终不会被转发到 SW3、SW5 和 SW6。因为 SW2—SW3 以及 SW4—SW5 的 Trunk 端口 Fa0/3 已经修剪了 VLAN10 的通信流量。

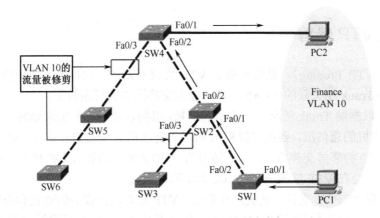

图 3-20 开启 VTP 修剪

注意：只能在具有 VTP Server 模式的交换机上开启 VTP 修剪功能，其他交换机必须处于 VTP Server 或 VTP Client 模式才可以修剪不必要的 VLAN 流量。处于 Transparent 模式的交换机，必须通过手动将 VLAN 从 Trunk 中删除，开启 VTP 修剪功能的命令如下：

Switch(config)#**VTP Prunning**

需要提醒大家的是：目前的 Packet Tracer 尚不支持该命令。

3.4.3 分析 VTP 通告

VTP 通告，又名 VTP 报文，是在交换机间 Trunk 链路上传递 VLAN 信息的数据包。VTP 通告以组播形式发送，组播地址为 01-00-0C-CC-CC-CC，且只能通过 Trunk 传递。

研究 VTP 通告，可以深入理解 VTP 工作原理。VTP 通告主要包括三种类型：汇总通告、子集通告以及请求通告。

1．汇总通告（Summary Advertisements）

默认情况下，Catalyst 交换机会每 5 分钟发送一条汇总通告消息。通知邻近交换机当前 VTP 域名和配置修订号。收到汇总通告消息的交换机，会将将其 VTP 域名与自己 VTP 域名进行比较。若名称不同，则忽略报文；若名称相同，比较两者配置修订号，如果自己的较大，报文会被忽略；如果自己的较小，则会向对方发送通告请求消息。汇总通告如图 3-21 所示。

图 3-21　VTP 汇总通告

2．子集通告（Subset Advertisements）

管理员在某 Server 交换机上添加、删除或修改 VLAN 时，该交换机配置修订号就会增加并发送一条汇总通告消息。随后，它又会发送一条或多条子集通告消息。每条子集通告消息中都包含一个 VLAN 信息列表。若有多个 VLAN，交换机就会请求 Server 交换机发送多条子集通告消息来通告所有这些 VLAN 信息。子集通告如图 3-22 所示。

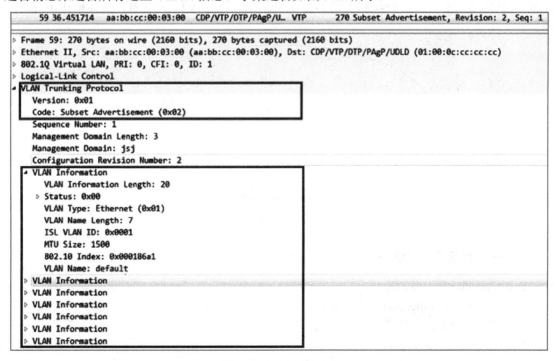

图 3-22　VTP 子集通告

3. 请求通告（Request Advertisements）

在下列情况下，交换机需要发送 VTP 请求通告。
- 交换机重启；
- VTP 域名被修改；
- 交换机收到了一条 VTP 汇总通告消息，且该消息的配置修订号高于其自身修订号。

收到请求通告消息之后，VTP 设备就会发送一条汇总通告消息。在此之后，再发送一条或多条子集通告消息，这样整个 VTP 域中的 VLAN 信息就可以同步。

3.4.4 学习 VTP 配置

前面我们学习了 VTP 协议的作用、域的规划以及域中交换机的角色，了解了 VTP 修剪功能可以优化网络，通过了解 VTP 通告类型，进一步理解了 VTP 工作原理。接下来，我们将学习如何配置 VTP。

VTP 通告只能在 Trunk 上传送，因此配置 VTP，建议顺序是先配置 Trunk，再配置 VTP 域名，接下来指定交换机的 VTP 模式，再在服务器交换机上对 VLAN 进行管理，最后在相应客户交换机上添加 VLAN 成员。除此之外，我们还可以指定 VTP 版本号、配置 VTP 口令、配置 VTP 修剪功能等。下面我们以图 3-23 拓扑为例，学习 VTP 的配置，具体操作步骤如下所述。

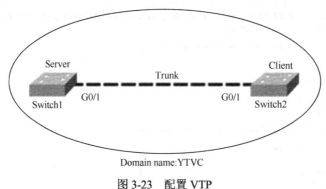

图 3-23　配置 VTP

步骤一：配置 Trunk 链路

Switch1(config)#**interface G0/1**
Switch1(config-if)#**switchport mode trunk**
经查看，Trunk 已经形成。
Switch1#**show interfaces trunk**
Port   Mode   Encapsulation   Status   Native vlan

Gig0/1	on	802.1q	trunking	1

Port	Vlans allowed on trunk
Gig0/1	1-1005

Port	Vlans allowed and active in management domain
Gig0/1	1

Port	Vlans in spanning tree forwarding state and not pruned
Gig0/1	1

步骤二：配置 VTP 域

（1）配置 VTP 服务器

Switch1(config)#**vtp domain YTVC**	//配置 VTP 域名，默认为空（NULL）
Switch1(config)#**vtp mode server**	//配置 VTP 模式
Switch1(config)#**vtp password 15net**	//配置 VTP 口令
Switch1(config)#**vtp version 2**	//配置 VTP 版本

（2）配置 VTP 客户机

Switch2(config)#**vtp domain YTVC**
Switch2(config)#**vtp mode client**
Switch2(config)#**vtp password 15net**

注意：VTP 的版本只能在服务器上指定，客户机不支持指定功能，但其可以通过接收 VTP 通告实现和服务器 VTP 版本的同步。我们可以通过 show vtp status 命令查看 VTP 的配置信息。

（3）查看服务器交换机 Switch1 的 VTP 信息

查看服务器交换机 Switch1 的 VTP 信息，我们看到配置结果完全正确。

Switch1#**show vtp status**	
VTP Version	: 2
Configuration Revision	: 1
Maximum VLANs supported locally	: 255
Number of existing VLANs	: 5
VTP Operating Mode	: Server
VTP Domain Name	: YTVC

VTP Pruning Mode	: Disabled
VTP V2 Mode	: Enabled
VTP Traps Generation	: Disabled
MD5 digest	: 0x07 0x86 0x67 0xDD 0x92 0x81 0xA1 0xB3
Configuration last modified by 0.0.0.0 at 3-1-93 00:19:13	
Local updater ID is 0.0.0.0 (no valid interface found)	

（4）查看客户机交换机 Switch2 的 VTP 信息

查看客户机交换机 Switch2 的 VTP 信息，我们看到结果也正确。

```
Switch2#show vtp status
VTP Version                       : 2
Configuration Revision            : 1
Maximum VLANs supported locally   : 255
Number of existing VLANs          : 5
VTP Operating Mode                : Client
VTP Domain Name                   : YTVC
VTP Pruning Mode                  : Disabled
VTP V2 Mode                       : Enabled
VTP Traps Generation              : Disabled
MD5 digest                        : 0xEB 0xF4 0xE8 0x0B 0x0A 0x7C 0x0C 0x96
Configuration last modified by 0.0.0.0 at 3-1-93 00:04:16
```

（5）对比配置和结果

对比配置和结果，我们发现显示中缺少 VTP 的口令。注意，这需要单独的命令 **show vtp password** 来进行专项查看。

```
Switch1#show vtp password
VTP Password: 15net


Switch2#show vtp password
VTP Password: 15net
```

步骤三：在服务器上创建 VLAN

```
Switch1(config)#vlan 10
Switch1(config-vlan)#name Teacher
Switch1(config-vlan)#vlan 20
Switch1(config-vlan)#name Student
```

在客户端交换机 Switch2 上查看，是否学到了 VLAN 信息。

```
Switch2#show vlan brief

VLAN Name                          Status    Ports
---- -------------------------------- --------- -------------------------------
1    default                       active    Fa0/1, Fa0/2, Fa0/3, Fa0/4
                                             Fa0/5, Fa0/6, Fa0/7, Fa0/8
                                             Fa0/9, Fa0/10, Fa0/11, Fa0/12
                                             Fa0/13, Fa0/14, Fa0/15, Fa0/16
                                             Fa0/17, Fa0/18, Fa0/19, Fa0/20
                                             Fa0/21, Fa0/22, Fa0/23, Fa0/24
                                             Gig0/2
10   Teacher                       active
20   Student                       active
1002 fddi-default                  active
1003 token-ring-default            active
1004 fddinet-default               active
```

在客户机上，通过 show vlan brief 查看 VLAN 信息，我们可以发现，在服务器交换机上创建的 VLAN，已被客户端交换机学习到。

步骤四：VTP 域交换机添加端口成员

（1）给服务器交换机添加 VLAN 成员

```
Switch1(config)#interface range f0/1-12
Switch1(config-if-range)#switchport mode access
Switch1(config-if-range)#switchport access vlan 10
Switch1(config-if-range)#exit
Switch1(config)#interface range f0/13-24
Switch1(config-if-range)#switchport mode access
Switch1(config-if-range)#switchport access vlan 20
```

客户端交换机添加 VLAN 成员。

```
Switch2(config)#interface range f0/1-12
Switch2(config-if-range)#switchport mode access
Switch2(config-if-range)#switchport access vlan 10
Switch2(config-if-range)#exit
Switch2(config)#interface range f0/13-24
Switch2(config-if-range)#switchport mode access
Switch2(config-if-range)#switchport access vlan 20
```

（2）查看客户端交换机已经成功添加的 VLAN 成员

```
Switch2#show vlan brief
VLAN Name                             Status        Ports
---- -------------------------------- ------------- -------------------------------
1    default                          active        Gig0/2
10   Teacher                          active        Fa0/1, Fa0/2, Fa0/3, Fa0/4
                                                    Fa0/5, Fa0/6, Fa0/7, Fa0/8
                                                    Fa0/9, Fa0/10, Fa0/11, Fa0/12
20   Student                          active        Fa0/13, Fa0/14, Fa0/15, Fa0/16
                                                    Fa0/17, Fa0/18, Fa0/19, Fa0/20
                                                    Fa0/21, Fa0/22, Fa0/23, Fa0/24
1002 fddi-default                     active
1003 token-ring-default               active
1004 fddinet-default                  active
1005 trnet-default                    active
```

步骤五：查看 VLAN 是否同步

在服务器和客户端上查看 VTP 信息：

```
Switch1#show vtp status
VTP Version                         : 2                                          //该 VTP 支持版本 2
Configuration Revision              : 5                                          //VTP 配置修订版本号，该值默认为 0
Maximum VLANs supported locally     : 255                                        //VTP 支持的最大 VLAN 数量
Number of existing VLANs            : 7                                          //当前交换机存在的 VLAN 数量
VTP Operating Mode                  : Server                                     //当前交换机的 VTP 模式
VTP Domain Name                     : YTVC                                       //当前 VTP 域名
VTP Pruning Mode                    : Disabled                                   //当前 VTP 禁用修剪功能
VTP V2 Mode                         : Enabled                                    //当前 VTP 已开启版本 2
VTP Traps Generation                : Disabled                                   //不发送 VTP 提示消息至 SNMP 服务器
MD5 digest                          : 0xE3 0xBE 0x8A 0xC0 0xFE 0x3F 0xBE 0xB9   //MD5 值
Configuration last modified by 0.0.0.0 at 3-1-93 00:32:03                        //VTP 配置最近一次修改的时间以及被谁修改
Local updater ID is 0.0.0.0 (no valid interface found)                           //本地修改者的 ID

Switch2#show vtp status
VTP Version                         : 2
Configuration Revision              : 5
Maximum VLANs supported locally     : 255
```

Number of existing VLANs	: 7
VTP Operating Mode	: Client
VTP Domain Name	: YTVC
VTP Pruning Mode	: Disabled
VTP V2 Mode	: Enabled
VTP Traps Generation	: Disabled
MD5 digest	: 0xE3 0xBE 0x8A 0xC0 0xFE 0x3F 0xBE 0xB9
Configuration last modified by 0.0.0.0 at 3-1-93 00:32:03	

比较以上输出表明两台交换机已经实现了 VLAN 同步。比较步骤二 VTP 的信息输出，我们发现当前 **Number of existing VLANs** 为 7，由原来的 5 增加了 2，原因是当前 VTP 域新增了 2 个 VLAN：Teacher 和 Student。**Configuration Revision** 配置修订值增加了 4，原因是 VTP 域新增 4 次 VTP 通告（增加 VLAN 10、命名 Teacher、增加 VLAN 20、命名 Student），也就是服务端对 VLAN 执行了 4 次操作，每操作（包括添加、删除、修改）一次，配置修订值就会增 1；每增 1，就会引发一次 VTP 通告。

配置修订值是 VTP 进行同步消息的一个重要依据。同一 VTP 域中，配置修订值高的交换机的 VLAN 信息能够把配置修订值低的交换机的 VLAN 信息覆盖。因此当使用 VTP 协议实现 VLAN 数据同步时一定要谨慎。

3.4.5 场景五：应用 VTP 技术

🔍**场景五**：YH 食品有限公司因采取了网上店铺和实体销售双向并行营销模式，业务量剧增。目前，财务部、市场部、生产部以及客服部 4 个部门的员工正面临大调整，办公地点也进行了很大改动。财务部的部分人员分别调入到市场部和客服部。网管员面临的很头疼的问题是，每个部门都被分散在三个楼层办公，但不变的是同部门之间还要像往常一样通过公司内网协同工作，部门之间依然实现安全隔离。现公司为每个部门都配备了一台专用服务器，要求集中在三楼设备间统一托管。YH 公司的特点是员工的办公地点经常会发生改变。目前公司委以你重任利用周末双休时间加班重新规划公司企业网，确保下周企业网能正常运行。

网络拓扑如图 3-24 所示。

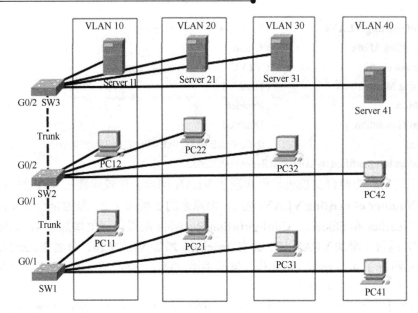

图 3-24 VTP 技术应用

网络重新规划如表 3-8 所示。

表 3-8 网络新规划一览表

VLAN ID	VLAN 名称	端口分配	网络地址	备注
10	Finance	SW1：Fa0/1～F2 SW2：Fa0/1～F2 SW3：Fa0/1～F2	192.168.0.0/24	财务部
20	Marketing	SW1：Fa0/3～F11 SW2：Fa0/3～F11 SW3：Fa0/3～F11	192.168.1.0/24	市场部
30	Production	SW1：Fa0/12～F20 SW2：Fa0/12～F20 SW3：Fa0/12～F20	192.168.2.0/24	生产部
40	Customer	SW1：Fa0/21～F24 SW2：Fa0/21～F24 SW3：Fa0/21～F24	192.168.3.0/24	客服部

步骤一：配置 Trunk

```
SW1(config)#interface G0/1
SW1(config-if)#switchport mode trunk
SW1(config-if)#switchport nonegotiate
```

```
SW2(config)#interface range G0/1-2
SW2(config-if-range)#switchport mode trunk
SW2(config-if-range)#switchport nonegotiate
```
静态指定 trunk 模式，关闭 DTP 协商。

```
SW3(config)#interface G0/2
SW3(config-if)#switchport mode trunk
SW3(config-if)#switchport nonegotiate
```
静态指定两条 Trunk 链路，在 SW2 上查看两条 Trunk 的状态。

```
SW2#show interfaces trunk
Port        Mode         Encapsulation  Status        Native vlan
Gig0/1      on           802.1q         trunking      1
Gig0/2      on           802.1q         trunking      1

Port        Vlans allowed on trunk
Gig0/1      1-1005
Gig0/2      1-1005

Port        Vlans allowed and active in management domain
Gig0/1      1
Gig0/2      1

Port        Vlans in spanning tree forwarding state and not pruned
Gig0/1      1
Gig0/2      1
```
经显示发现，两条 Trunk 链路已形成。

步骤二：配置 VTP

```
SW1(config)#vtp domain YTHW
SW1(config)#vtp mode client
SW1(config)#vtp password ythw
```
配置交换机 SW1 为 VTP 客户机。

```
SW2(config)#vtp domain YTHW
SW2(config)#vtp mode server
SW2(config)#vtp password ythw
```
配置交换机 SW2 为 VTP 服务器。

```
SW3(config)#vtp domain YTHW
```

```
SW3(config)#vtp mode client
SW3(config)#vtp password ythw
```
配置 SW3 为 VTP 客户机。

步骤三：在服务器上创建 VLAN

```
SW2(config)#vlan 10
SW2(config-vlan)#name Finance
SW2(config-vlan)#vlan 20
SW2(config-vlan)#name Marketing
SW2(config-vlan)#vlan 30
SW2(config-vlan)#name Production
SW2(config-vlan)#vlan 40
SW2(config-vlan)#name Customer
```

在 VTP 服务器（SW2）上，创建 VLAN。

步骤四：添加 VLAN 成员

```
SW2(config)#interface range fastEthernet 0/1-2
SW2(config-if-range)#switchport mode access
SW2(config-if-range)#switchport access vlan 10
SW2(config-if-range)#exit
SW2(config)#interface range fastEthernet 0/3-11
SW2(config-if-range)#switchport mode access
SW2(config-if-range)#switchport access vlan 20
SW2(config-if-range)#exit
SW2(config)#interface range fastEthernet 0/12-20
SW2(config-if-range)#switchport mode access
SW2(config-if-range)#switchport access vlan 30
SW2(config-if-range)#exit
SW2(config)#interface range fastEthernet 0/21-24
SW2(config-if-range)#switchport mode access
SW2(config-if-range)#switchport access vlan 40
```

VTP 客户端 SW1 和 SW3 会通过 VTP 自动学习到 Server 上创建的 4 个 VLAN，端口成员的添加完全同 SW2。因为 3 台交换机的端口规划完全一致。

步骤五：配置 DHCP 服务器为其 VLAN 分配地址

配置各台专用服务器，兼作为 DHCP 服务器，为相应 VLAN 主机自动分配地址。DHCP 服务器地址及分配参数如表 3-9 所示。

表 3-9 DHCP 服务器地址及分配参数

服务器名称	IP 地址/网络前缀	DHCP 池起始地址/子网掩码	自动分配地址数
Server11	192.168.0.250/24	192.168.0.1/24	5
Server21	192.168.1.250/24	192.168.1.1/24	26
Server31	192.168.2.250/24	192.168.2.1/24	26
Server41	192.168.3.250/24	192.168.3.1/24	11

下面以 VLAN 10 的服务器 Server11 为例，学习在 Packet Tracer 下如何配置 DHCP 服务器。单击服务器 Server11，选择"Servives/DHCP"，将 DHCP 的 Service 置于"on"状态，输入 Start IP Address、Subnet Mask、Maximum number 对应的值，然后点击"Save"按钮。如图 3-25 所示。

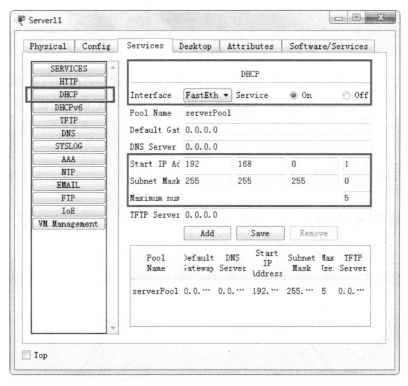

图 3-25 配置 Server11 的 DHCP

接下来，我们需要验证 VLAN 10 中的主机 PC11 和 PC12 是否已通过 DHCP 成功获取 VLAN 10 网段的地址。如图 3-26 所示 PC11 和 PC12 已经成功获取地址。

图 3-26　PC11 和 PC12 DHCP 自动获取地址

经测试 VLAN 20、VLAN 30 以及 VLAN 40 的主机均已成功自动获取 IP 地址，确保了同一部门的主机可以通信，不同部门间主机不能互访。不同 VLAN 间通信需要借助三层设备，我们会在第 4 章学习。

3.5　挑战过关练习

3.5.1　挑战过关练习一

挑战如下三个问题。

① 请仔细看图 3-27 所示网络拓扑回答问题：PC1 和 PC2 能否通信？请说出理由。

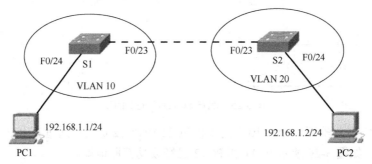

图 3-27　挑战过关练习一（A）

② 如图 3-28 所示，在以太网交换机 Switch1 上创建了 4 个 VLAN，请问该交换机上共有

几个 MAC 地址表？请说出理由。

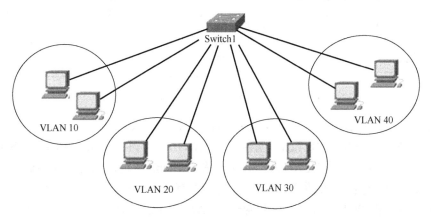

图 3-28　挑战过关练习一（B）

③ 以太网交换机最多可以支持多少个 VLAN？请说明理由。

3.5.2　挑战过关练习二

挑战 HY 同学提出的两个问题：

① 参照图 3-29 所示拓扑，主机 PC1 和 PC2 均属于 VLAN10，其所连交换机 SW1 和 SW2 之间的 Trunk 链路经集线器 Hub 中继，请对图 3-30 所示的测试结果进行分析，说明主机 PC1 和 PC2 都不能与 PC3 通信的理由。

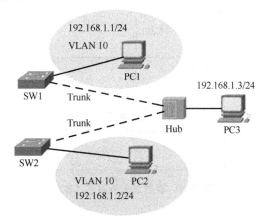

图 3-29　挑战过关练习二（A）

PDU List Window

Fire	Last Status	Source	Destination	Type	Color	Time(sec	Periodic	Num	Edit	Delete
●	Successful	PC1	PC2	ICMP	■	0.000	N	0	(edit)	(delete)
●	Failed	PC1	PC3	ICMP	■	0.000	N	1	(edit)	(delete)
●	Failed	PC2	PC3	ICMP	■	0.000	N	2	(edit)	(delete)

图 3-30　挑战过关练习二（A）连通性测试结果

② 参照图 3-31 所示拓扑，在交换机 Switch1 上创建了 VLAN 10，将端口 Fa0/1 和 Fa0/2 分配给该 VLAN；在交换机 Switch2 创建了 VLAN10 和 VLAN20，并将端口 Fa0/3 分配给 VLAN 10，端口 Fa0/4 分配给 VLAN 20。Switch1 和 Switch2 通过彼此的 Fa0/1 端口实现了互连，Switch2 的 Fa0/1 设置为 trunk 模式，4 台 PC 地址配置成同一网段。对图 3-32 所示测试结果进行分析，请回答主机 PC1 为什么不能和同一 VLAN 的 PC3 通信？又为何能和不同 VLAN 的 PC2 通信？详细说明理由。

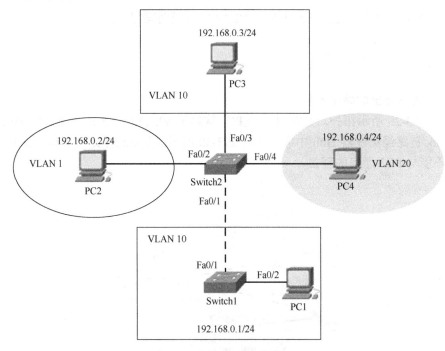

图 3-31　挑战过关练习二（B）

PDU List Window

Fire	Last Status	Source	Destination	Type	Color	Time(sec	Periodic	Num	Edit	Delete
●	Failed	PC1	PC4	ICMP		0.000	N	0	(edit)	(delete)
●	Failed	PC1	PC3	ICMP		0.000	N	1	(edit)	(delete)
●	Successful	PC1	PC2	ICMP		0.000	N	2	(edit)	(delete)

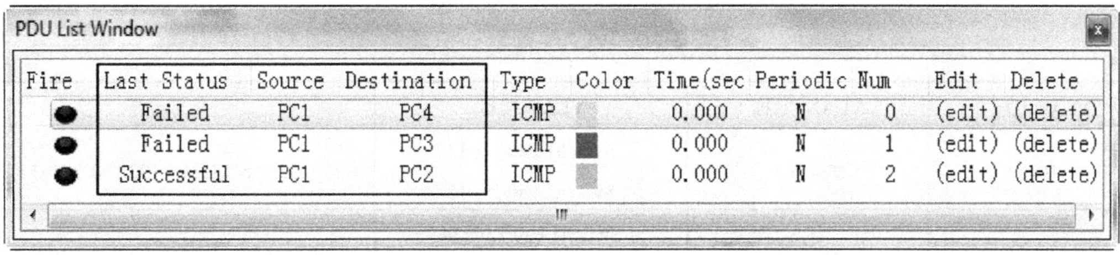

图 3-32　挑战过关练习二（B）连通性测试结果

3.5.3　挑战过关练习三

1. 任务背景

挑战 CFL 老师给大家设计的实验考题。如图 3-33 所示，三台 Cisco 2960 交换机通过千兆接口实现了互连。图中有 Teacher、Student、Worker 和 Staff 四个 VLAN，每个 VLAN 都有各自的专用服务器负责为客户端自动分配地址，目前的网络环境不允许与外界通信，VLAN 间也不可以通信，请用同一网络地址空间 192.168.0.0/24 为各 VLAN 分配相应地址，如表 3-10 所示。

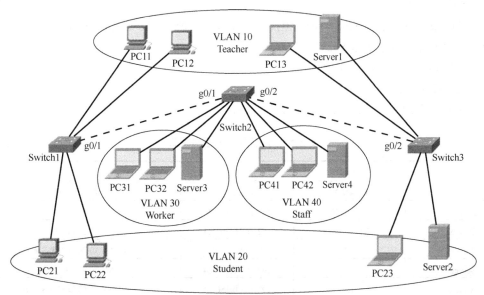

图 3-33　挑战过关练习三

表 3-10 VLAN 规划一览表

VLAN ID	VLAN 名称	VLAN 端口	地址分配
10	Teacher	Switch1 Fa0/1～12 Switch3 Fa0/1～12	192.168.0.1/24～192.168.0.50/24
20	Student	Switch1 Fa0/13～24 Switch3 Fa0/13～24	192.168.0.51/24～192.168.0.100/24
30	Worker	Switch2 Fa0/1～12	192.168.0.101/24～192.168.0.150/24
40	Staff	Switch2 Fa0/13～24	192.168.0.151/24～192.168.0.200/24
99	Admin	—	192.168.0.201/24～192.168.0.203/24

2. 挑战任务

① 搭建网络拓扑。

打开 Cisco Packet Tracer 模拟器，搭建如图 3-33 所示拓扑，端口连接参考表 3-10。

② 配置 VTP 域。

- VTP 域名为 ytvc，VTP 口令为 jsjwl，VTP 模式指定 Switch1 为服务器模式，Swtich3 为客户机模式，Switch2 为透明模式。

③ 配置 VLAN。

- 参照表 3-10，在 VTP Server 上创建 VLAN，并通过 VTP 通告给 VTP Client 端；
- 在 Switch2 创建本地 VLAN；
- 按规划，为各 VLAN 添加端口成员。

④ 配置 IP 地址。

- 参照表 3-11 为各设备分配相应的 IP 地址。

表 3-11 设备固定地址分配表

设备名称	设备接口	地址分配	备注
Switch1	VLAN 99	192.168.0.201/24	管理 VLAN
Switch2	VLAN 99	192.168.0.202/24	管理 VLAN
Switch3	VLAN 99	192.168.0.203/24	管理 VLAN
Server1	NIC	192.168.0.50/24	VLAN 10 的 DHCP Server
Server2	NIC	192.168.0.100/24	VLAN 20 的 DHCP Server
Server3	NIC	192.168.0.150/24	VLAN 30 的 DHCP Server
Server4	NIC	192.168.0.200/24	VLAN 40 的 DHCP Server
PCxx	NIC	DHCP	终端 PC

⑤ 配置 DHCP 服务。
- 配置 DHCP 服务器，使其能为各 VLAN 主机自动分配地址；
- 配置 DHCP 客户端，使其均能通过 DHCP 获取正确的地址。

⑥ VLAN 测试。
- 同一 VLAN 的主机可以通信，不同 VLAN 的主机不能通信。

至此，"学习 VLAN 技术"一章已接近尾声。本章通过前后相互关联的 5 个应用场景，记录了 YH 食品有限公司企业网发展的五段历程。随着公司规模的壮大，我们逐步掌握了单交换机和多交换机 VLAN 的配置、Trunk 的两种封装、Trunk 的静态指定以及 DTP 的动态协商方法，分析了 Trunk 常见的 3 种故障和 VTP 的 3 种类型通告，理解了 Trunk 和 VTP 的工作原理，并能应用 Trunk 和 VTP 技术以及 VTP 修剪功能实现网络的升级和优化。本章最后设计的 3 个挑战过关练习的目的是带领你去发现问题，引领你去积极思考，激发你去探究解决问题的途径。只有经历这样一段学习过程，我们对知识的理解才会是深刻的，才能做到灵活运用所学技术去解决实际问题。

第4章 >>>

学习直连路由

本章要点

- 认识 Cisco 路由设备
- 认识物理接口直连路由
- 认识逻辑接口直连路由
- 分析路由表
- 网络工程师技术面试闯关
- 挑战过关练习

4.1 认识 Cisco 路由设备

第 3 章，我们学习了基于二层交换机用 VLAN 软件技术实现广播域划分的方法，从而实现部门间的安全隔离。本章，"学习直连路由"将帮我们解决 VLAN 间的通信问题。"路由器因爱而生"讲述了互联网上一段传奇的爱情故事，挖掘了路由器伟大、神圣而又美丽的诞生背景。正因为路由器才诞生了网络行业巨头 Cisco 公司，接下来 Cisco 公司为其长远发展收购了生产交换机的 Crescendo 公司，从而产生了 Cisco 交换机。故事的引入让我们不仅学习了知识，又了解了网络的历史以及设备的诞生背景。本章共设计了 9 个教学案例，包括 2 个实验、4 个应用场景和 1 道面试闯关考题，还有 2 个挑战过关练习。每个案例都非常独特，让你耳目一新，这种设计目的在于使读者对本章的网络设备及相关技术充满好奇。

4.1.1 Cisco 路由器因爱而生

1973 年，毕业于美国宾夕法尼亚大学沃顿商学院的莱昂纳德·波萨克（Leonard Bosack）进入 DEC 公司，成为了一名硬件工程师。1979 年，他进入斯坦福大学计算机系，继续深造学习。不久与商学院计算机中心主任桑蒂·勒娜（Sandy Lerner）相爱。次年，波萨克和勒娜完婚。波萨克婚后继续学习，获得计算机硕士学位，并担任了斯坦福大学计算机系的计算机中心主任。成为美国国防部推动"ARPANET"建设的关键人物，而"ARPANET"就是当代互联网 Internet 的鼻祖。

两位计算机主任联姻被传为佳话，更为传奇的是他们设计了一种新型的网络互联设备——世界上第一台路由器。因此互联网上一直流传着一段伟大而又浪漫的爱情故事——路由器因爱而生。

勒娜所在的商学院实验室的计算机网络和波萨克所在计算机系网络间不能互通，因此勒娜无数次对丈夫抱怨，她写给波萨克的情书不能及时传送给丈夫。是爱情的力量让波萨克决心研制一台设备实现夫妇网络间的互通。1984 年，波萨克利用 William Yeager 和 Andy Bechtolsheim 构建的斯坦福大学路由器的源代码，在自家车库里设计和制造出了一台"多协议路由器"，这是波萨克送给勒娜最好的礼物。后来，这台路由器被试用于斯坦福校园网（SUNet），将校园内不兼容的计算机局域网整合在一起，形成了一个统一的网络。斯坦福 16 平方英里的校园顿时欢呼一片，5 000 台计算机奇迹般实现了互连。该路由器标志着联网时代的真正到来。

之后，波萨克夫妇向斯坦福大学请求制造和销售这种多协议路由器，但由于知识产权归属问题没有得到校方支持。于是，他们决定自己创办公司。用什么名字注册呢？因为夫妇二人喜欢旧金山"San Francisco"的读音，更喜欢旧金山代表性的建筑"金门大桥"，所以想用"San Francisco"注册。

美国法律规定，禁止用城市名作为任何产品和公司命名。最终敲定公司名称采用"San Francisco"后五个字母"cisco"，把"金门大桥"作为公司徽标，寓意是通过信号进行连接和沟通。

"思科系统公司"（Cisco Systems, Inc）就此于 1984 年 12 月在硅谷圣何塞（San Jose）正式诞生。

在信息时代，那座闻名于世的金门大桥也因 Cisco 架构的畅通无阻的"桥梁"，而变得更加知名。也正是源于路由器，将世界不同的网络连接在一起，才形成了今天庞大的互联网。而 Cisco 也因路由器改变了整个世界，成为目前引领全球互联网发展的行业巨头。

4.1.2　认识路由器

4.1.1 节，我们已了解全世界第一台路由器诞生的背景。1984 年波萨克夫妇创办 Cisco 公司，1986 年 3 月公司推出第一款路由器产品——AGS（先进网关服务器，如图 4-1 所示）并开始销往海外，从而掀起了一场全球性的通信革命。

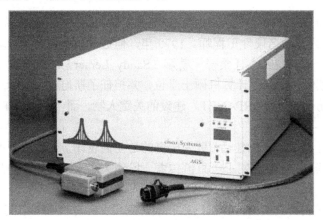

图 4-1　世界上第一台路由器产品（AGS）

1. 路由器的功能

路由器是网络互联设备，它工作在 OSI 网络层，又被称为三层设备。路由器（Router）又称网关设备（Gateway），用于连接多个逻辑上分开的网络。逻辑网络代表一个单独的网络或者一个子网。当数据从一个子网传输到另一个子网时，可通过路由器的路由功能来完成，所谓路由功能就是选择路径功能。路由器具有判断网络地址和选择 IP 路径的功能，支持多种不同的网络协议，因此，它能在多种网络环境中，建立灵活的连接。路由器是互联网络的枢纽，是"交通警察"。

2. 路由器的分类

路由器按不同的标准有不同的分类，常见的分类有以下几种。

（1）按性能档次分

按性能档次分，有高、中、低端路由器。通常将路由器吞吐量大于 40 Gbps 的路由器称为高档路由器，其吞吐量在 25～40 Gbps 的路由器称为中档路由器，而将吞吐量低于 25 Gbps 的路由器看作低档路由器。以 Cisco 公司的路由器为例，12000 系列为高端路由器，7500 以下系列为中低端路由器。

（2）按硬件结构分

按硬件结构分，有模块化路由器和非模块化路由器。模块化结构可以灵活地配置路由器，以适应企业不断增加的业务需求；非模块化结构只能提供固定的端口。通常，中高端路由器为模块化结构，低端路由器为非模块化结构。

（3）按接入功能分

按接入功能分，有骨干级路由器、企业级路由器和接入级路由器

- 骨干级路由器的主功能是实现企业级网络互联，其特点是数据吞吐量较大，非常重要；基本性能要求是高速、高可靠性。为了获得高可靠性，网络系统普遍采用诸如热备份、双电源、双数据通路等传统冗余技术。
- 企业级路由器连接许多终端系统，其特点是连接对象多，系统相对简单，数据流量较小，采用便宜方法尽可能多连接端点，能支持不同的服务质量。
- 接入级路由器主要应用于连接家庭或 ISP 内的小型企业客户群体。

（4）按所处位置分

按所处位置分，有边界路由器和中间节点路由器。

3. Cisco 路由器的型号

Cisco 路由器有很多系列，如 Cisco1800、Cisco1900、Cisco2800、Cisco2900、Cisco3600、Cisco7200 等。Cisco 路由器的命名规则都是以 Cisco 开头，如 Cisco1941。其中，Cisco 表示品牌，前两位数字 19 表示产品系列是 1900，后两位 41 表示 1900 系列中的具体型号。Cisco1941 路由器产品如图 4-2 和图 4-3 所示。

图 4-2　Cisco 1941 路由器的机身　　　　　图 4-3　Cisco 1941 路由器的背板

在 Cisco Packet Tracer 中，我们在选择路由器型号时往往看到的是设备的逻辑图标，通过点击"Physical"标签或用鼠标单击路由器可以进一步了解路由器的物理结构。表 4-1 显示的是典型路由器的逻辑图标和物理结构的对应图。

表 4-1　Packet Tracer 典型路由器举例

路由器	逻辑图标	物理结构
1941		
2901		
Router-PT		

从表 4-1 所列举的三款路由器，我们不难看出路由器的逻辑图标是一样的，但是物理结构却有很大的差异，硬件结构不同，接口也不相同。1941 和 2901 都是生产型设备，即对应着真正的产品型号；而 Router-PT 属于教学模型设备，支持多类型接口（光口、电口；十兆口、百兆口、千兆口；LAN 口、WAN 口），可以方便不同网络类型的实验，建议在生产型设备不能满足实验需求时再选择使用教学模型设备。

4.1.3 Cisco 交换机诞生背景

Cisco 路由器于 1986 年开始销往海外市场，公司得到迅猛发展。1990 年 Cisco 上市，并一度成为美国乃至全球最有影响力的十大 IT 公司之一。

1993 年以前，在路由器市场，没有任何一家公司可以撼动 Cisco 的霸主地位。路由器对连接不同协议的网络有着不可替代的作用，但对同种类型网络的连接实属大材小用。随着 Internet 的发展及局域网的兴起，产生了连接同类型网络的主流技术以太网和 ATM。由于路由器速度慢、价格高的特点，以太网交换机和 ATM 交换机对 Cisco 路由器造成了巨大威胁。一些大客户对 Cisco 路由器的兴趣逐渐开始转向交换机，于是触发了 Cisco 收购克雷森多公司（Crescendo）的念头，Crescendo 公司（Catalyst 交换机产自该公司）是美国知名交换机公司。1993 年秋天，Crescendo 公司成为 Cisco 历史上第一个被收购的公司，1994 年，Cisco 推出第一种面向客户端/服务器模式工作组的智能 Cisco Catalyst 系列交换机，第一批新产品就是来自对 Crescendo 的收购。

Cisco 以雄厚的财力在硅谷和全世界寻觅机会，抓住机遇以扩展其已有的产品系列，不断提出全新发展战略。约翰·钱伯斯（John Chambers）1991 年加入 Cisco，1995 年就任总裁兼首席执行官，在他任职 CEO 的 20 年时间内，带领 Cisco 在全世界范围内掀起了一场并购风暴。先后并购了百余家大大小小的企业，一步步引领着 Cisco 从提供单一路由器产品的小企业走向提供端到端连网方案的行业巨头。Cisco 并购是非常成功的发展战略，为公司不断输送新鲜血液、融合技术、凝聚人才，同时不断推进技术革新，增强公司的软、硬实力，不断巩固其在互联网行业的霸主地位。

4.1.4 认识交换机

交换机从广义上讲可分为广域网交换机（WAN Switches）和局域网交换机（LAN Switches）。广域网交换机主要应用于电信领域，提供通信基础平台；而局域网交换机则应用于局域网，连接终端设备，如 PC、服务器及网络打印机等。本书我们只研究局域网交换机。1993 年，伴随着局域网交换设备的出现，国内掀起了交换网络技术的热潮。交换机以简单、低价、性能高和端口密集等特点深受互联网客户的青睐。

下面我们所介绍交换机都指的是局域网交换机，而最常见的交换机是以太网交换机，本书对交换机的配置都基于以太网交换机。

1. 交换机的功能

交换机将局域网分割为多个冲突域，每个冲突域都有独立的带宽，因此极大地提高了局域网性能。交换机默认工作在 OSI 第二层，默认情况下所有端口属于一个广播域。正如第 3 章所学，我们可以通过二层 VLAN 技术分割广播域。交换机内部的 CPU 会在每个端口成功连接时，将 MAC 地址与源端口的映射关系形成一张 MAC 地址表。交换机就是依据该 MAC 地址表进行数据帧的转发及过滤的。

2. 交换机的分类

以太网交换机从不同的角度可以实现不同的分类，常见的分类有以下几种。

(1) 从传输介质和传输速度分

从传输介质和传输速度分，有以太网交换机、快速以太网交换机、千兆以太网交换机、万兆以太网交换机。

(2) 从规模应用上分

从规模应用上分，有企业级交换机、部门级交换机和工作组级交换机等。一般来讲，企业级交换机都是机架式的，从应用规模来看，当作为骨干交换机时，是支持 500 个信息点以上大型企业应用的交换机；部门级交换机可以是机架式的（插槽数较少），也可以是固定配置式的，是支持 300 个信息点以下中型企业的交换机；工作组级交换机是固定配置式的（功能较为简单），支持 100 个信息点以内的交换机。

(3) 按硬件结构分

按硬件结构分，有固定端口交换机和模块化交换机。

(4) 按工作协议层分

按工作协议层分，有二层交换机和三层交换机。二层交换机基于 MAC 地址转发数据帧，工作在 OSI 数据链路层，是二层设备；三层交换机可以基于 IP 地址转发数据包，工作在 OSI 网络层，是三层设备。

3. Cisco 交换机的型号

Cisco 交换机也有很多系列，如 Catalyst 2950、Catalyst 2960、Catalyst 3560、Catalyst3750、Catalyst4000、Catalyst4500、Catalyst6000、Catalyst6500 和 Catalyst7600 系列等。Cisco 交换机命名都以 WS 开头，如下所述。

(1) WS-C3560-24PS-S 和 WS-C3560-24PS-E

- WS：代表交换设备；
- C：表示带光纤接口；
- 3560：表示是 Catalyst 3560 系列；
- 24：表示端口数量为 24 个；
- P：代表支持 POE（Power Over Ethernet），即支持以太网供电；

- S：代表 SFP（Small Form-factor Pluggable），即扩展接口支持小型可插拔光模块；
- S：代表 Standard Image，即 IOS 是标准版的 SMI 映像；
- E：代表 Enhanced Image，即 IOS 映像是增强版 EMI 映像。

（2）WS-C3750G-48TS-S

- G：代表 Gigabits per second，即表明所有接口都是支持每秒千兆的速率，默认则表明其主要端口均为 10/100 Mbps 或者 100 Mbps；
- 48：表示端口数量为 48 个；
- T：代表 Twisted Pair，即支持介质类型是双绞线，端口是电口，非光纤接口。

一款 48 口的 Cisco Catalyst 3750 系列的交换机如图 4-4 所示。Packet Tracer 典型交换机举例如表 4-2 所示。

图 4-4　Cisco 3750 系列交换机

表 4-2　Packet Tracer 典型交换机举例

交 换 机	逻 辑 图 标	物 理 结 构
3560-24PS		
2960-24TT		
Switch-PT		

对比表 4-2 所列举的三款交换机，我们发现 3560 交换机的图标与其他两款不一样，这是三层交换机的逻辑图标，三层交换机能工作在 OSI 第三层，可以连接不同网段的局域网，实现同类型网络间的互连。2960 和 Switch-PT 都是二层交换机，不能实现网间互连。3560 和 2960 交换机都是生产型设备，而 Switch-PT 属于教学模型设备，支持多类型接口（电口、光口；十兆口、百兆口、千兆口），可以方便教学与实验，建议在生产型设备不能满足实验需求时再选择使用教学模型设备。

4.1.5 路由设备对比

从以上可知，路由器和三层交换机都可以作为路由设备工作在网络层，实现网间互联。但实际上，在路由功能上交换机从来没有真正替代过路由器，因为它们采用的技术不同，核心功能也不同，其在互联网中担任的角色也不尽相同。路由器的主要功能在于协议转换和路径选择，而三层交换机的主要功能在于快速实现数据交换。下面我们将根据两种设备的特点进行详细比对。

1. 路由器

路由器端口类型丰富，既支持广域网连接，如 Serial、ISDN 等，又支持局域网的连接，如 Ethernet、Token Ring 等，还支持多传输介质网络的连接，如光纤、双绞线、电话线和专线等。路由器支持的协议很多，如 IP、IPX、Apple Talk 等三层协议，也支持如 HDLC、Frame Relay、PPP 等二层封装协议。路由器路由能力强，适合大型网络间的互连。路由器主要的功能不是在端口间的快速交换，而是选择最佳路径、负载分担、链路备份，最重要的是与其他网络进行路由信息交换。

随着互联网的发展，对网络设备的集成业务提出了更高的要求。路由器也在不断革新，以适应新时代的需求。目前路由器在提高处理能力的同时，又增加了对多种业务的支持，尤其是出现了大量具有交换功能的路由器，如图 4-5 所示，就是 Cisco 推出的一款 2951 ISR 集成业务路由器。

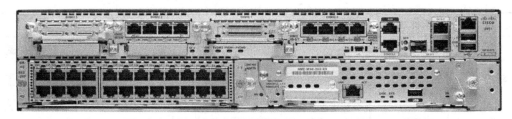

图 4-5 Cisco 2951 集成业务路由器

由图 4-5 可知，路由器可以支持多端口交换模块，集路由与交换为一体，可满足多业务的需求。如在路由器上实现 VLAN 技术，使交换机和路由器的界线也变得愈来愈模糊。为此，本章专门设计了一个有关 ISR 的多业务综合案例。

2. 三层交换机

三层交换机具有部分路由器功能，支持的接口类型少，其最重要目的是加快大型局域网内的数据交换，做到"一次路由，多次转发"。所谓"一次路由，多次转发"是指三层交换机在完成对第一个新 IP 数据包路由后，产生一个 MAC 地址与 IP 地址的映射表，当具有相同目标 IP 地址的数据包再次通过时，它会直接依据此表在二层模块上完成转发，从而提高 IP 数据包的转发效率。IP 数据包路由的过程由三层交换机硬件高速实现，而像路由信息更新、路由表维护、路由计算及路由选择等功能，则由软件实现，从而解决了传统路由器软件路由的速度问题。三层交换技术实质上就是二层交换技术和三层转发技术的融合。传统二层交换技术是基于数据链路层的数据帧交换技术，而三层交换技术则是基于网络层数据包的高速转发技术，采用三层交换机既可实现网络路由功能，又可优化网络性能。

三层交换机是基于 IP 协议设计的，接口类型简单，拥有很强的二层帧处理能力，同时又具有几乎二层交换机的速度，价格相对便宜。因此可以说，三层交换机具有"路由器的功能、交换机的性能"。

综上所述，路由器与三层交换机之间存在着本质区别。在局域网中进行多子网连接，最好选用三层交换机，特别是在不同子网数据交换频繁的环境中；如果子网间的通信不是很频繁，或要与其他类型的网络连接，采用路由器也无可厚非。实际环境中对路由设备的选择还要根据实际需求来定。

4.2 认识物理接口直连路由

因为路由器支持多协议，所以既可以支持同构网络的互连也支持异构网络的互连。例如，路由器可以连接 LAN 和 LAN、WAN 和 WAN、LAN 和 WAN 等。路由器之所以能够连接多类型的网络是因为它支持丰富类型的物理网络接口。所谓物理接口就是路由器的硬件接口。我们这里所说的物理接口是指路由器的网络接口，而不是像 Console 那样的控制端口。

4.2.1 认识路由器物理接口

1. 路由器接口功能

路由器是网间互连设备，工作在 OSI 第三层。路由器对网络的互连功能是通过其接口实现的。路由器的每个物理接口连接一个独立的网络，两个接口的地址不能在一个网段。也就是说路由器的每个物理接口都连接了一个广播域，即路由器通过物理接口实现了广播域的划分。

2. 路由器接口类型

路由器不仅有 LAN 接口，也有 WAN 接口。LAN 接口我们只介绍当前路由器主流的以太网接口，WAN 接口只介绍常用的同步串行接口（Serial 接口）。因为以太网支持多种类型的传输介质，如双绞线、同轴电缆、光纤等，所以以太网有不同类型的网络，接口如 RJ-45、AUI、SC 等。由于同轴电缆属于淘汰的以太网传输介质，因此我们不再介绍 AUI 接口。

以太网按照支持的传输速度可分为 Ethernet（标准以太网）、FastEthernet（快速以太网）、GigabitEthernet（千兆以太网）、10GigabitEthernet（万兆以太网），网速分别为 10 Mbps、100 Mbps、1 000 Mbps、10 000 Mbps。Cisco Packet Tracer 尚不支持万兆网络，因此 Cisco 路由器的以太网接口名称分别为 Ethernet、FastEthernet 和 GigabitEthernet。以太网口常用作连接企业局域网。

3. 路由器接口识别

路由器的接口有固化接口和扩展接口，路由器对物理接口数量的支持是有限的。每台路由器都有固化的以太网口，也可以通过扩展插槽添加相应的接口模块来增加接口数量。像串口这样的广域网接口需要根据实际情况来扩展。我们以 Cisco2811 路由器为例来了解典型模块和接口，如表 4-3 所示。

表 4-3　路由器典型物理接口一览表

模块型号	物理模块	模块接口	接口名称	传输介质	接口类型
WIC-1T			Serial 接口	串行线缆（DTE/DCE）	WAN 口
WIC-2T			Serial 接口	串行线缆（DTE/DCE）	WAN 口
NM-1FE-TX			RJ-45 接口	双绞线	LAN 口
NM-1FE-FX			SC 接口	光纤	LAN 口

4. 路由器模块识别

识别路由器的模块对扩展路由器接口是非常重要的。实际上，并不是每一种模块都能得到所有路由器支持，例如，高端路由器就不支持低端模块，如 Cisco2911 路由器就不支持百兆以太网模块。下面我们以 Packet Tracer 常用典型模块为例来介绍网络模块的含义及所支持的路由器类型，如表 4-4 所示。

表 4-4　路由器典型网络模块一览表

模块型号	模块含义	适用路由器
WIC-1T	WAN 接口卡，提供 1 个 Serial 接口	Cisco1841、2811、2620XM、2621XM
WIC-2T	WAN 接口卡，提供 2 个 Serial 接口	Cisco1841、2811、2620XM、2621XM
HWIC-2T	WAN 接口卡，提供 2 个高速 Serial 接口	Cisco1941、2901、2911、1841、2811
WIC-1EENT	WAN 接口卡，提供 1 个 10 Mbps 以太网接口	Cisco1841、2811
NM-1FE-TX	网络模块，提供 1 个支持双绞线的 100 Mbps 以太网接口	Cisco2811、2620XM、2621XM
NM-1FE-FX	网络模块，提供 1 个支持光纤的 100 Mbps 以太网接口	Cisco2811、2620XM、2621XM
NM-2FE-2W	网络模块，提供 2 个支持双绞线的 100 Mbps 以太网接口和 2 个 WAN 扩展槽	Cisco2811、2620XM、2621XM

若要添加扩展模块，一定要先将路由器电源关闭，否则无法添加成功，因为模块不支持热插拔。添加模块后，再打开路由器电源开关，重启设备，系统会自动检测到新增模块。在 CLI 界面，我们可以通过 **show ip interface brief** 命令查看新增路由器物理接口的名称、数量及状态。确定了接口名称之后，我们可以为其分配 IP 地址和子网掩码，激活接口，再使用 **show ip route [connected]** 命令来查看直连接口的路由（直连路由）。

4.2.2　实验一：配置路由器物理接口

实验描述：请按照图 4-6 规划的网络地址配置路由器 R0 的物理接口，要求接口地址采用其所在网段的最大可用 IP 地址。查看 R0 的路由表，确保其所有物理接口的直连路由均显示在路由表中，若缺少直连路由，请你想办法自行解决。

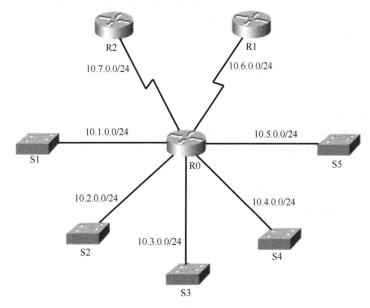

图 4-6　配置路由器物理接口

步骤一：查看路由器 R0 的物理接口

```
R0#show ip interface brief
Interface              IP-Address      OK? Method Status                Protocol
FastEthernet0/0        unassigned      YES manual administratively down  down
GigabitEthernet1/0     unassigned      YES manual administratively down  down
Ethernet2/0            unassigned      YES manual administratively down  down
FastEthernet3/0        unassigned      YES manual administratively down  down
GigabitEthernet4/0     unassigned      YES manual administratively down  down
```

Serial5/0	unassigned	YES manual administratively down	down
Serial6/0	unassigned	YES manual administratively down	down

步骤二：配置路由器 R0 的物理接口

```
R0(config)#interface fastEthernet 0/0
R0(config-if)#no shutdown
R0(config-if)#ip address 10.1.0.254 255.255.255.0
R0(config)#interface gigabitEthernet 1/0
R0(config-if)#no shutdown
R0(config-if)#ip address 10.2.0.254 255.255.255.0
R0(config)#interface ethernet 2/0
R0(config-if)#no shutdown
R0(config-if)#ip address 10.3.0.254 255.255.255.0
R0(config)#interface fastEthernet 3/0
R0(config-if)#no shutdown
R0(config-if)#ip address 10.4.0.254 255.255.255.0
R0(config)#interface gigabitEthernet 4/0
R0(config-if)#no shutdown
R0(config-if)#ip address 10.5.0.254 255.255.255.0
R0(config)#interface serial 5/0
R0(config-if)#no shutdown
R0(config-if)#ip address 10.6.0.254 255.255.255.0
R0(config)#interface serial 6/0
R0(config-if)#no shutdown
R0(config-if)#ip address 10.7.0.254 255.255.255.0
```

步骤三：查看路由器 R0 的直连路由

```
R0#show ip route
Codes: C - connected, S - static, I - IGRP, R - RIP, M - mobile, B - BGP
       D - EIGRP, EX - EIGRP external, O - OSPF, IA - OSPF inter area
       N1 - OSPF NSSA external type 1, N2 - OSPF NSSA external type 2
       E1 - OSPF external type 1, E2 - OSPF external type 2, E - EGP
       i - IS-IS, L1 - IS-IS level-1, L2 - IS-IS level-2, ia - IS-IS   inter area
       * - candidate default, U - per-user static route, o - ODR
       P - periodic downloaded static route

Gateway of last resort is not set
```

```
              10.0.0.0/24 is subnetted, 5 subnets
C             10.1.0.0 is directly connected, FastEthernet0/0
C             10.2.0.0 is directly connected, GigabitEthernet1/0
C             10.3.0.0 is directly connected, Ethernet2/0
C             10.4.0.0 is directly connected, FastEthernet3/0
C             10.5.0.0 is directly connected, GigabitEthernet4/0
```

经查看发现直连路由缺少 10.6.0.0/24 和 10.7.0.0/24。所以需要进一步排错。

步骤四：路由器 R0 直连路由的故障排错

通过命令 show ip interface brief，进一步查看路由器 R0 物理接口的简明配置信息。

```
R0#show ip interface brief
Interface              IP-Address       OK?    Metod     Status      Protocol
FastEthernet0/0        10.1.0.254       YES    manual    up          up
GigabitEthernet1/0     10.2.0.254       YES    manual    up          up
Ethernet2/0            10.3.0.254       YES    manual    up          up
FastEthernet3/0        10.4.0.254       YES    manual    up          up
GigabitEthernet4/0     10.5.0.254       YES    manual    up          up
Serial5/0              10.6.0.254       YES    manual    down        down
Serial6/0              10.7.0.254       YES    manual    down        down
```

以上输出我们发现，接口 Serial5/0 和 Serial6/0 的接口状态是"down down"，其余接口均是"up up"，说明这两个接口的物理层出现故障。检查拓扑发现前 5 个接口连接的是交换机端口，而后面的两个接口连接的是路由器端口。默认情况下，交换机端口的状态是 up 的，而路由器的接口状态是 down。所以当路由器 R0 的所有接口开启时，R0 连接交换机的直连接口的状态均为"up up"，而连接路由器接口另一端依然没有激活，所以状态是"down down"。

接下来查看两条链路对端的 R1 和 R2，激活对应路由器的接口。

```
R1(config)#interface serial 0/0/0
R1(config-if)#no shutdown
%LINK-5-CHANGED: Interface Serial0/0/0, changed state to up
%LINEPROTO-5-UPDOWN: Line protocol on Interface Serial0/0/0, changed state to up


R2(config)#interface serial 0/0/0
R2(config-if)#no shutdown
%LINK-5-CHANGED: Interface Serial0/0/0, changed state to up
%LINEPROTO-5-UPDOWN: Line protocol on Interface Serial0/0/0, changed state to up
R0#
%LINK-5-CHANGED: Interface Serial5/0, changed state to up
%LINEPROTO-5-UPDOWN: Line protocol on Interface Serial5/0, changed state to up
```

```
%LINK-5-CHANGED: Interface Serial6/0, changed state to up
%LINEPROTO-5-UPDOWN: Line protocol on Interface Serial6/0, changed state to up
```
激活路由器 R1 和 R2 的串行接口之后，我们发现 R0 对应的接口状态变成了"up up"。

步骤五：再次查看路由器 R0 的路由表

```
R0#show ip route
Codes: C - connected, S - static, I - IGRP, R - RIP, M - mobile, B - BGP
       D - EIGRP, EX - EIGRP external, O - OSPF, IA - OSPF inter area
       N1 - OSPF NSSA external type 1, N2 - OSPF NSSA external type 2
       E1 - OSPF external type 1, E2 - OSPF external type 2, E - EGP
       i - IS-IS, L1 - IS-IS level-1, L2 - IS-IS level-2, ia - IS-IS inter area
       * - candidate default, U - per-user static route, o - ODR
       P - periodic downloaded static route

Gateway of last resort is not set

     10.0.0.0/24 is subnetted, 7 subnets
C       10.1.0.0 is directly connected, FastEthernet0/0
C       10.2.0.0 is directly connected, GigabitEthernet1/0
C       10.3.0.0 is directly connected, Ethernet2/0
C       10.4.0.0 is directly connected, FastEthernet3/0
C       10.5.0.0 is directly connected, GigabitEthernet4/0
C       10.6.0.0 is directly connected, Serial5/0
C       10.7.0.0 is directly connected, Serial6/0
```

从以上输出我们发现，路由表新增了两条直连路由 10.6.0.0/24 和 10.7.0.0/24。至此配置路由器物理接口的直连路由实验已经顺利完成。

4.2.3　场景一：采用传统路由器实现 VLAN 间路由

🔍**场景一：** 在第 3 章 3.2.2 节场景二中，我们对发展中近四十人规模、分布在两层楼办公、拥有两台交换机的 YH 食品有限公司企业网做了 VLAN 部署。通过 VLAN 技术实现了财务部、销售部和生产部 3 个部门间网络的安全隔离。现在公司需要 3 个部门间能协调工作，我们又该如何升级改造网络呢？请采用 4.2.1 节所学的技术升级网络，并要求各部门的 PC 能动态获取地址，避免地址冲突问题发生，每个 VLAN 的网关是本网段内最小 IP 地址，要求 PC 动态获取的起始地址为 192.168.x.10（x=0,1,2）。

注意：在第 3 章 3.2.2 节场景二 VLAN 部署基础上进行增量实施。

根据网络需求，了解目前网络的基础状况，如 4-7 左图所示。公司 3 个部门处于 3 个 VLAN，属于 3 个网段，划分 VLAN 的实质是把交换机分成多台逻辑上独立的交换机，网间互访需要三层路由设备，所以可用路由器的物理接口实现网间通信，如 4-7 右图所示。按照第 3 章 3.2.2 场景二的 VLAN 规划表，我们做了如图 4-8 所示的连接。

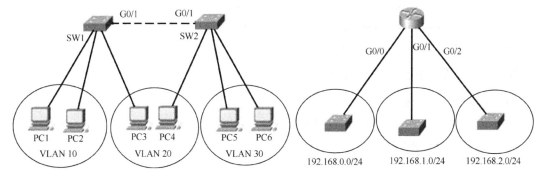

图 4-7　企业网升级前（左图）与升级需求（右图）对比图

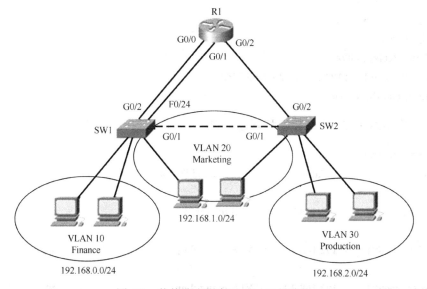

图 4-8　传统路由器实现 VLAN 间路由

企业网络升级具体实施步骤如下所述。

步骤一：将 SW1 的端口添加到相应 VLAN

```
SW1(config)#interface G0/2
SW1(config-if-range)#switchport mode access
SW1(config-if)#switchport access vlan 10
```

```
SW1(config-if-range)#exit
SW1(config)#interface F0/24
SW1(config-if)#switchport mode access
SW1(config-if)#switchport access vlan 20
```

将 SW2 的端口添加到相应 VLAN。

```
SW2(config)#interface G0/2
SW2(config-if)#switchport mode access
SW2(config-if)#switchport access vlan 30
```

步骤二：配置路由器 R1 的物理接口

```
R1(config)#interface G0/0
R1(config-if)#ip address 192.168.0.1 255.255.255.0
R1(config-if)#no shutdown
R1(config-if)#exit
R1(config)#interface G0/1
R1(config-if)#ip address 192.168.1.1 255.255.255.0
R1(config-if)#no shutdown
R1(config-if)#exit
R1(config)#interface G0/2
R1(config-if)#ip address 192.168.2.1 255.255.255.0
R1(config-if)#no shutdown
```

步骤三：查看路由器 R1 的接口和直连路由

```
R1#show ip interface brief
Interface              IP-Address      OK? Method Status                Protocol

GigabitEthernet0/0     192.168.0.1     YES manual up                    up

GigabitEthernet0/1     192.168.1.1     YES manual up                    up

GigabitEthernet0/2     192.168.2.1     YES manual up                    up

Vlan1                  unassigned      YES unset  administratively down down

R1#show ip route connected
    C    192.168.0.0/24   is directly connected, GigabitEthernet0/0
    C    192.168.1.0/24   is directly connected, GigabitEthernet0/1
    C    192.168.2.0/24   is directly connected, GigabitEthernet0/2
```

步骤四：将路由器 R1 配置为 DHCP 服务器

```
R1(config)#ip dhcp excluded-address 192.168.0.1 192.168.0.9
R1(config)#ip dhcp excluded-address 192.168.1.1 192.168.1.9
R1(config)#ip dhcp excluded-address 192.168.2.1 192.168.2.9
R1(config)#ip dhcp pool Finance
R1(dhcp-config)#network 192.168.0.0 255.255.255.0
R1(dhcp-config)#default-router 192.168.0.1
R1(dhcp-config)#exit
R1(config)#ip dhcp pool Marketing
R1(dhcp-config)#network 192.168.1.0 255.255.255.0
R1(dhcp-config)#default-router 192.168.1.1
R1(dhcp-config)#exit
R1(config)#ip dhcp pool Production
R1(dhcp-config)#network 192.168.2.0 255.255.255.0
R1(dhcp-config)#default-router 192.168.2.1
```

步骤五：网络连通性测试

验证 192.168.0.0/24 网段的 PC 能否自动获取 IP 地址。如图 4-9 所示，已经成功分配到地址。

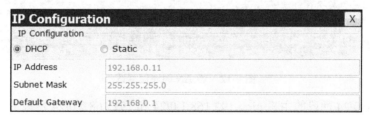

图 4-9　验证 192.168.0.0/24 网段 PC 的 DHCP 获取

验证 192.168.1.0/24 网段的 PC 能否自动获取 IP 地址。如图 4-10 所示，已经成功获取。

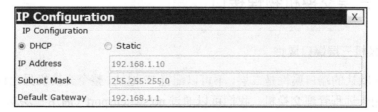

图 4-10　验证 192.168.1.0/24 网段 PC 的 DHCP 获取

验证 192.168.2.0/24 网段的 PC 能否自动获取 IP 地址。如图 4-11 所示，已成功获取。

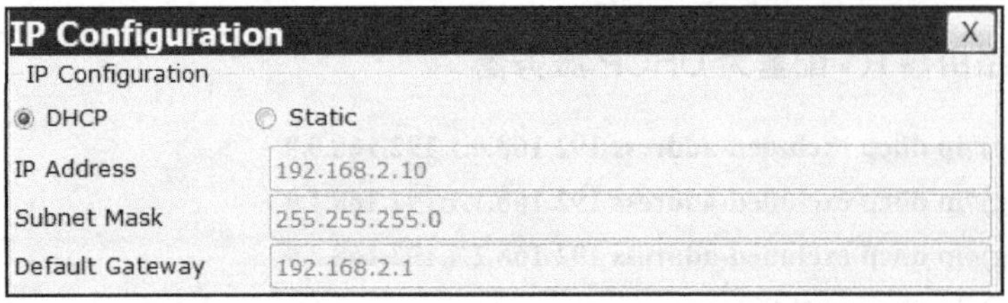

图 4-11 验证 192.168.2.0/24 网段 PC 的 DHCP 获取

下面我们选取 VLAN10（192.168.0.0/24）中的 PC，测试与 VLAN20 和 VLAN30 的连通性，如图 4-12 所示。

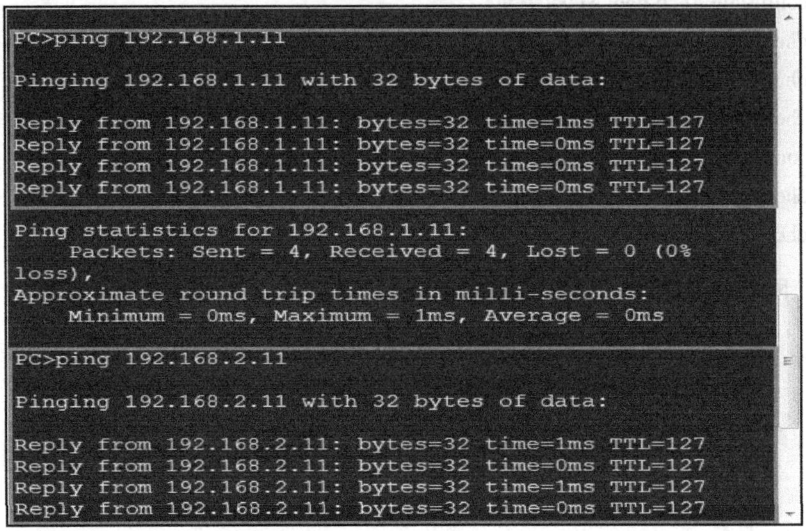

图 4-12 VLAN 间通信测试

经 192.168.0.0/24 网段的主机 ping 192.168.1.0/24 和 192.168.2.0/24 网段的主机，均 ping 通，至此成功实现了 VLAN 间通信的需求。

4.2.4 认识三层交换机物理接口

1. 三层交换机三层端口属性

三层交换机默认的端口属性是二层，不可以配置 IP 地址，整个交换机只允许为管理 VLAN 配置一个 IP，用于远程管理交换机。我们可以通过 no switchsport（启用三层功能）命令，将三层交换机某些端口属性切换到三层，这样，这些端口就可以配置不同网段的 IP 地址，如同路由器的物理接口。

2. 三层交换机二层端口属性

三层交换机的端口允许一些工作在二层,一些工作在三层,即允许二层和三层端口并存,同样,我们也可以通过 switchport(启用二层功能)命令将端口属性从三层切换到二层。

3. 二层和三层端口属性小结

三层交换机上,具有二层属性的端口,可以分配到指定 VLAN 也可以承载多个 VLAN 的数据,即端口模式可以是 Access 模式也可以是 Trunk 模式。但是,三层属性的端口,就没有 Access 和 Trunk 模式,就不支持 VLAN,因为它在物理上对应着一个独立的网段,每个端口对应一个独立的 IP 地址,连接一个广播域。

需要注意的是,三层交换机的同一端口,其二层和三层属性不能同时共存,也就是二层和三层在某一时刻只能启用一个。

4.2.5 实验二:配置三层交换机物理端口

实验描述:以 Cisco Catalyst 3560 系列交换机为例,完成三层交换机物理端口的配置。本实验中,请按照图 4-13 所示网络地址配置三层交换机物理端口 fastethernet0/1~4 为三层端口,IP 地址为 172.16.x.254/24(x=1,2,3,4),剩下的端口为二层端口,同属于 VLAN 1。配置 VLAN 1 的 IP 地址为 172.16.0.254/24。要求所有物理接口的直连路由均出现在路由表中;最后测试 PC 间的连通性,并记录实验测试结果。

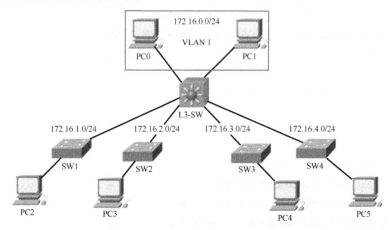

图 4-13 配置三层交换机物理接口

步骤一:配置三层交换机的物理接口

L3-SW(config)#**ip routing**   //三层交换机默认未开启路由功能
L3-SW(config)#**interface fastEthernet 0/1**

```
L3-SW(config-if)#no switchport
L3-SW(config-if)#ip address 172.16.1.254 255.255.255.0    //交换机端口默认已激活
L3-SW(config-if)#interface fastEthernet 0/2
L3-SW(config-if)#no switchport
L3-SW(config-if)#ip address 172.16.2.254 255.255.255.0
L3-SW(config-if)#interface fastEthernet 0/3
L3-SW(config-if)#no switchport
L3-SW(config-if)#ip address 172.16.3.254 255.255.255.0
L3-SW(config-if)#interface fastEthernet 0/4
L3-SW(config-if)#no switchport
L3-SW(config-if)#ip address 172.16.4.254 255.255.255.0
```

步骤二：配置三层交换机的管理 VLAN

```
L3-SW(config-if)#interface vlan 1
L3-SW(config-if)#no shutdown
L3-SW(config-if)#ip address 172.16.0.254 255.255.255.0
```

步骤三：检查三层交换机的直连路由

```
L3-SW#show ip route
Codes: C - connected, S - static, I - IGRP, R - RIP, M - mobile, B - BGP
       D - EIGRP, EX - EIGRP external, O - OSPF, IA - OSPF inter area
       N1 - OSPF NSSA external type 1, N2 - OSPF NSSA external type 2
       E1 - OSPF external type 1, E2 - OSPF external type 2, E - EGP
       i - IS-IS, L1 - IS-IS level-1, L2 - IS-IS level-2, ia - IS-IS inter area
       * - candidate default, U - per-user static route, o - ODR
       P - periodic downloaded static route

Gateway of last resort is not set

     172.16.0.0/24 is subnetted, 5 subnets
C       172.16.0.0 is directly connected, Vlan1
C       172.16.1.0 is directly connected, FastEthernet0/1
C       172.16.2.0 is directly connected, FastEthernet0/2
C       172.16.3.0 is directly connected, FastEthernet0/3
C       172.16.4.0 is directly connected, FastEthernet0/4
```

从以上路由表的输出我们发现，除了三层交换机的 4 个物理接口的直连路由均已出现在路由表中外，VLAN1 的网络地址也写进了路由表，其标记的接口是 Vlan1，显然该接口也是三层接口。Vlan1 到底是一个什么接口呢？为什么会出现在路由表中呢？这个问题我们会在 4.3.4

节中详细讨论。

经测试，我们发现 PC 间彼此都能互相 ping 通，验证了三层交换机确实具有路由功能。

4.2.6 场景二：三层交换机物理接口 VLAN 间路由

> 🔍**场景二**：在本章 4.2.3 节场景一中，我们采用路由器的物理接口实现了对 YH 食品有限公司企业网的升级。LZK 工程师觉得在企业网中用路由器实现 VLAN 间通信实属大材小用，而且不能充分发挥局域网高速转发数据的优势，那么就请 LZK 工程师拿出他的升级方案，我们一起来分享。

图 4-14 是 LZK 工程师展示的升级前后的网络对比图，我们不难看出他采用了三层交换机的物理接口实现部门间的通信，具体网络连接如图 4-15 所示。

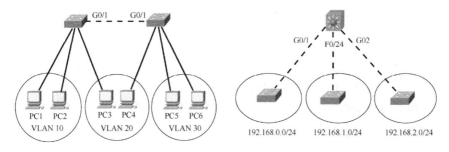

图 4-14　企业网升级前（左图）与升级规划（右图）对比图

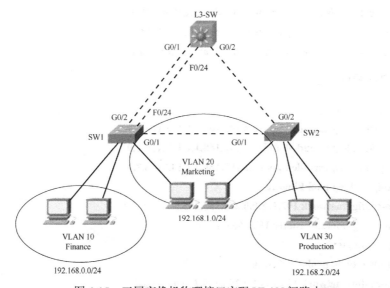

图 4-15　三层交换机物理接口实现 VLAN 间路由

由图 4-15 可知，三层交换机升级方案较路由器升级方案的区别在于设备不一样，但对二层交换机的连接完全一致，所以对 SW1 和 SW2 的实施省略，具体见 4.2.3 场景一的配置。

三层交换机的具体实施步骤如下所述。

步骤一：配置三层交换机物理接口

```
L3-SW(config)#interface G0/1
L3-SW(config-if)#no switchport
L3-SW(config-if)#ip address 192.168.0.1 255.255.255.0
L3-SW(config-if)#no shutdown
L3-SW(config-if)#exit
L3-SW(config)#interface F0/24
L3-SW(config-if)#no switchport
L3-SW(config-if)#ip address 192.168.1.1 255.255.255.0
L3-SW(config-if)#no shutdown
L3-SW(config-if)#exit
L3-SW(config)#interface G0/2
L3-SW(config-if)#no switchport
L3-SW(config-if)#ip address 192.168.2.1 255.255.255.0
L3-SW(config-if)#no shutdown
```

步骤二：查看接口和直连路由

```
L3-SW#show ip route connected
    C    192.168.0.0/24    is directly connected, GigabitEthernet0/1
    C    192.168.1.0/24    is directly connected, FastEthernet0/24
    C    192.168.2.0/24    is directly connected, GigabitEthernet0/2
```

步骤三：自动获取地址

```
L3-SW(config)#ip routing
L3-SW(config)#ip dhcp excluded-address 192.168.0.1 192.168.0.9
L3-SW(config)#ip dhcp excluded-address 192.168.1.1 192.168.1.9
L3-SW(config)#ip dhcp excluded-address 192.168.2.1 192.168.2.9
L3-SW(config)#ip dhcp pool Finance
L3-SW(dhcp-config)#network 192.168.0.0 255.255.255.0
L3-SW(dhcp-config)#default-router 192.168.0.1
L3-SW(dhcp-config)#exit
L3-SW(config)#ip dhcp pool Marketing
L3-SW(dhcp-config)#network 192.168.1.0 255.255.255.0
L3-SW(dhcp-config)#default-router 192.168.1.1
```

L3-SW(dhcp-config)#**exit**
L3-SW(config)#**ip dhcp pool Production**
L3-SW(dhcp-config)#**network 192.168.2.0 255.255.255.0**
L3-SW(dhcp-config)#**default-router 192.168.2.1**

步骤四：网络联通测试

在网段 192.168.0.0/24~192.168.2.0/24 PC 均已成功自动获取地址，经测试，任意两部门的主机均可以 ping 通，即通过三层交换机的物理接口成功实现了 VLAN 间的通信。

4.3 认识逻辑接口直连路由

路由器不仅有丰富的物理硬件接口，同时也有丰富的逻辑软件接口。本节研究的是逻辑接口的直连路由，是基于路由设备（路由器和三层交换机）的虚拟接口的直连路由。路由设备的逻辑接口实质上都是基于三层设备的虚拟三层接口，是软件接口。逻辑接口包括了路由器的子接口、三层交换机的 SVI 接口，以及路由器和三层交换机均支持的环回接口。

4.3.1 认识路由器子接口

1. 子接口定义

子接口（Subinterface）是通过协议和技术将路由器的一个物理接口虚拟出来的多个逻辑接口。子接口属于逻辑三层接口，是基于软件的虚拟接口，每个子接口可以配置不同网段的 IP 地址。划分了子接口的物理接口可以连接多个逻辑网络。

2. 子接口意义

从功能、作用上来讲，子接口与物理接口没有任何区别。子接口打破了路由器对物理接口数量的限制，比如说有的路由器最多支持两个以太网口，如果单台路由器实现三个网络间的互联，物理接口数量不能满足需求。但是，若采用子接口，即使更多网络间的互连，单台路由器也可以实现。从经济角度来讲，子接口可以大大节省成本。

3. 子接口缺点

多个子接口共用一个物理接口，性能比用单个物理接口差，在负载大的情况下容易造成网络瓶颈。即子接口数量越多，性能就越差。由于独立的物理接口无带宽争用现象，与子接口相比，物理接口性能更好。

4. 子接口应用

在拥有多个 VLAN 的交换机上，若要实现 VLAN 间通信，可以通过路由器子接口实现。

路由器通过为子接口封装 IEEE 802.1q 协议，子接口地址被配置为相应 VLAN 的网关，就可以实现 VLAN 间通信，这就是所谓的"单臂路由"。子接口不仅可以应用于 LAN，也可以应用于 WAN，如帧中继网络（Frame-Relay）。

5. 子接口配置

本节我们只介绍子接口在局域网中的应用，也就是"单臂路由"，又名"独臂路由"，即通过路由器的一个物理接口承载多个 VLAN 的数据，最终实现 VLAN 间通信。单臂路由的前提条件是基于 VLAN 的通信，需要划分 VLAN，规划不同网段地址，并且连接路由器的交换机端口必须配置成 Trunk 模式。接下来在路由器上划分子接口，封装 IEEE 802.1q 且允许特定 VLAN 通过，再配置 IP 地址，也就是通过子接口直连路由实现 VLAN 间通信。

子接口具体配置命令如下：

```
interface interface_id.subinterface_id    //在接口模式下，创建子接口（如 interface f0/0.10）
encapsulation dot1q vlan_id               //在子接口模式下，封装 IEEF 802.1q 允许相应 VLAN 通过
ip address ip_address subnet_mask         //配置子接口 IP 地址为对应 VLAN 的网关
举例：
interface fastEthernet0/0
no shutdown                               //激活物理接口，子接口才能启用
no ip address                             //物理接口不该有 IP 地址
interface fastEthernet0/0.10              //子接口编号尽量与 VLAN ID 一致以增加可读性
encapsulation dot1q 10
ip address 192.168.0.254 255.255.255.0
interface fastEthernet0/0.20
encapsulation dot1q 20
ip address 192.168.1.254 255.255.255.0
```

注意：因为路由器子接口是软件接口，子接口默认开启 no shutdown 命令。若物理接口 shutdown，则所有子接口均被禁用；若物理接口 no shutdown 则所有子接口均被启用。划分子接口要确保物理接口没有 IP 地址，所以最好使用 no ip address 命令保证物理接口不配置 IP 地址。若对端设置了 native vlan，则必须在对应子接口配置 native vlan，确保一致性。

4.3.2 场景三：单臂路由器实现 VLAN 间路由

🔍 **场景三**：在第 3 章 3.4.5 节场景五中，我们了解到 YH 食品有限公司迎来了发展中的春天，业务量剧增，公司拥有了财务、市场、生产和客服 4 个部门。部门成员打破了原来集中办公的局面，分散在三层楼上工作，每个部门都配有专用服务器。该阶段的企业网对方便网络管理提出了更高的要求，因此采用了 VTP 技术同步 VLAN 信息，同时让专用服务器兼

作为 DHCP 服务器为各自部门的 PC 动态分配地址。现在公司的需求是在原来部门间进行安全隔离，在同部门成员能通信的基础上，加强部门间的沟通与协作，请采用本章所学技术帮助企业升级网络，确保部门间的互通与交流。要求 VLAN 网关采用最大可用 IP 地址。

注意： 在第 3 章 3.4.5 节场景五中企业网已部署的基础上进行增量实施。

根据目前网络需求，了解网络现状，企业网采用 VLAN 技术实现了 4 个逻辑网络的划分，4 个 VLAN 对应公司的 4 个部门。因为部门间没有互访，因此没有分配网关地址，也没有相应的路由设备，如 4-16 左图所示。现在要实现 4 个部门通信，即 4 个 VLAN 间的通信，就需要 4 个网关。如果用路由器实现显然需要 4 个物理接口，由于物理接口数量的限制，我们考虑可以用上节课的子接口来解决部门间的通信问题，网络规划逻辑图如 4-16 右图所示。按照第 3 章 3.4.5 节场景五的 VLAN 规划表，我们做了如图 4-17 所示的连接。

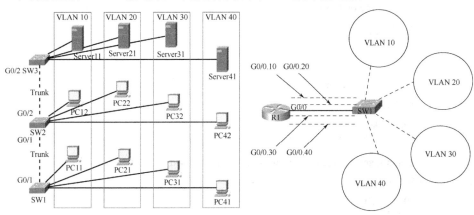

图 4-16　企业网升级前（左图）与网络需求（右图）对比图

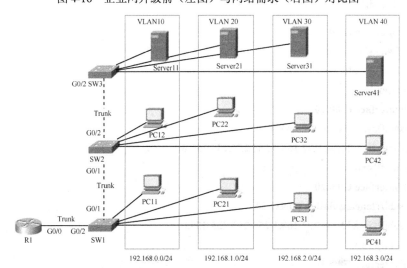

图 4-17　单臂路由器实现 VLAN 间路由

企业网升级的具体实施步骤如下。

步骤一：交换机 SW1 上启 Trunk

SW1(config)#**interface G0/2**
SW1(config-if)#**switchport mode trunk**

通过检查发现，Trunk 已经形成。

SW1#**show interfaces trunk**

Port	Mode	Encapsulation	Status	Native vlan
Gig0/1	on	802.1q	trunking	1
Gig0/2	on	802.1q	trunking	1

Port	Vlans allowed on trunk
Gig0/1	1-1005
Gig0/2	1-1005

Port	Vlans allowed and active in management domain
Gig0/1	1,10,20,30,40
Gig0/2	1,10,20,30,40

Port	Vlans in spanning tree forwarding state and not pruned
Gig0/1	1,10,20,30,40
Gig0/2	1,10,20,30,40

步骤二：配置路由器 R1 的 4 个子接口

R1(config)#**interface G0/0**
R1(config-if)#**no shutdown**
R1(config-if)#**no ip address**
R1(config-if)#**exit**
R1(config)#**interface G0/0.10**
R1(config-subif)#**encapsulation dot1Q 10**
R1(config-subif)#**ip address 192.168.0.254 255.255.255.0**
R1(config-subif)#**exit**
R1(config)#**interface G0/0.20**
R1(config-subif)#**encapsulation dot1Q 20**
R1(config-subif)#**ip address 192.168.1.254 255.255.255.0**
R1(config-subif)#**exit**
R1(config)#**interface G0/0.30**

R1(config-subif)#**encapsulation dot1Q 30**
R1(config-subif)#**ip address 192.168.2.254 255.255.255.0**
R1(config-subif)#**exit**
R1(config)#**interface G0/0.40**
R1(config-subif)#**encapsulation dot1Q 40**
R1(config-subif)#**ip address 192.168.3.254 255.255.255.0**

查看哪些接口的状态是 up，发现 4 个子接口状态均已是 up。

R1#**show ip interface brief | include up up**
GigabitEthernet0/0 unassigned YES unset up up
GigabitEthernet0/0.10 192.168.0.254 YES manual up up
GigabitEthernet0/0.20 192.168.1.254 YES manual up up
GigabitEthernet0/0.30 192.168.2.254 YES manual up up
GigabitEthernet0/0.40 192.168.3.254 YES manual up up

步骤三：查看路由器的直连路由

R1#**show ip route connected**
C 192.168.0.0/24 is directly connected, GigabitEthernet0/0.10
C 192.168.1.0/24 is directly connected, GigabitEthernet0/0.20
C 192.168.2.0/24 is directly connected, GigabitEthernet0/0.30
C 192.168.3.0/24 is directly connected, GigabitEthernet0/0.40

步骤四：部门互通测试

192.168.0.0/24 网段 PC 已成功自动获取地址，如图 4-18 所示。

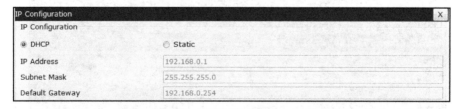

图 4-18　VLAN10 主机 DHCP 获取成功

192.168.1.0/24 网段 PC 已成功自动获取地址，如图 4-19 所示。

图 4-19　VLAN20 主机 DHCP 获取成功

192.168.2.0/24 网段 PC 已成功自动获取地址，如图 4-20 所示。

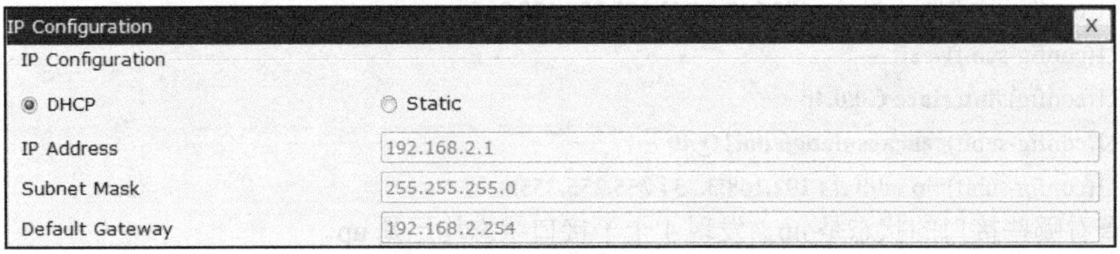

```
IP Configuration                                               X
IP Configuration
◉ DHCP                    ○ Static
IP Address                192.168.2.1
Subnet Mask               255.255.255.0
Default Gateway           192.168.2.254
```

图 4-20　VLAN30 主机 DHCP 获取成功

192.168.3.0/24 网段 PC 已成功自动获取地址，如图 4-21 所示。

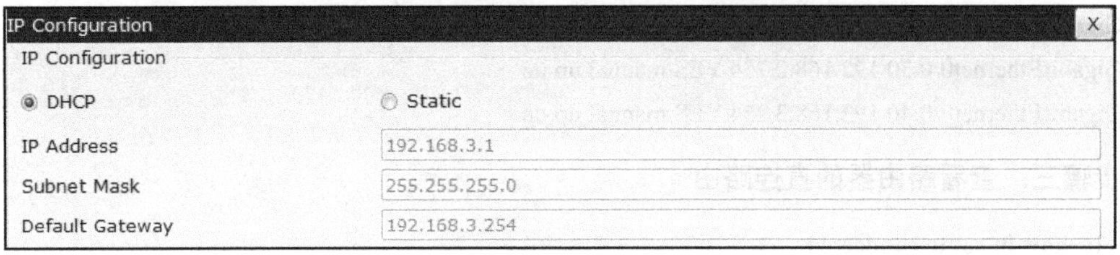

```
IP Configuration                                               X
IP Configuration
◉ DHCP                    ○ Static
IP Address                192.168.3.1
Subnet Mask               255.255.255.0
Default Gateway           192.168.3.254
```

图 4-21　VLAN40 主机 DHCP 获取成功

用 VLAN10 网段的主机 ping 192.168.1.1（VLAN20 主机）和 192.168.2.1（VLAN30），均可以 ping 通，如图 4-22 所示。

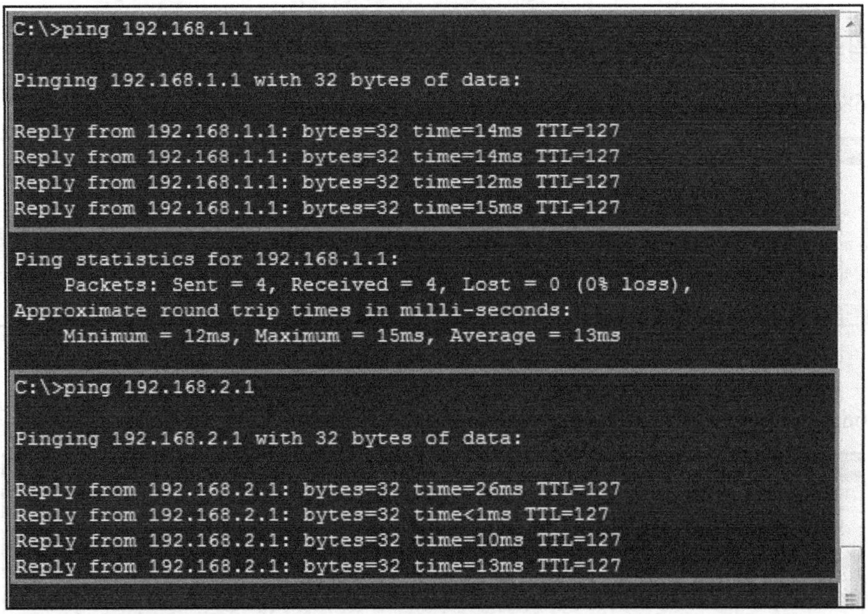

```
C:\>ping 192.168.1.1

Pinging 192.168.1.1 with 32 bytes of data:

Reply from 192.168.1.1: bytes=32 time=14ms TTL=127
Reply from 192.168.1.1: bytes=32 time=14ms TTL=127
Reply from 192.168.1.1: bytes=32 time=12ms TTL=127
Reply from 192.168.1.1: bytes=32 time=15ms TTL=127

Ping statistics for 192.168.1.1:
    Packets: Sent = 4, Received = 4, Lost = 0 (0% loss),
Approximate round trip times in milli-seconds:
    Minimum = 12ms, Maximum = 15ms, Average = 13ms

C:\>ping 192.168.2.1

Pinging 192.168.2.1 with 32 bytes of data:

Reply from 192.168.2.1: bytes=32 time=26ms TTL=127
Reply from 192.168.2.1: bytes=32 time<1ms TTL=127
Reply from 192.168.2.1: bytes=32 time=10ms TTL=127
Reply from 192.168.2.1: bytes=32 time=13ms TTL=127
```

图 4-22　VLAN 10 与 VLAN 20 和 VLAN 30 的主机互通测试

VLAN 10 网段的主机 ping 192.168.3.1（VLAN40 网段主机）也能 ping 通，如图 4-23 所示。

```
C:\>ping 192.168.3.1

Pinging 192.168.3.1 with 32 bytes of data:

Reply from 192.168.3.1: bytes=32 time=15ms TTL=127
Reply from 192.168.3.1: bytes=32 time=36ms TTL=127
Reply from 192.168.3.1: bytes=32 time=15ms TTL=127
Reply from 192.168.3.1: bytes=32 time=16ms TTL=127

Ping statistics for 192.168.3.1:
    Packets: Sent = 4, Received = 4, Lost = 0 (0% loss),
Approximate round trip times in milli-seconds:
    Minimum = 15ms, Maximum = 36ms, Average = 20ms
```

图 4-23　VLAN 10 与 VLAN 40 的主机互通测试

经测试，任意两部门的主机均可以 ping 通，即通过单臂路由器成功实现了 4 个部门间网络的互通。

4.3.3　认识三层交换机 SVI 接口

1. SVI 接口定义

SVI（Switch Virtual Interface）交换虚拟接口，是基于三层交换机创建的 VLAN 接口，是逻辑三层接口，其作用是实现 VLAN 间的路由和桥接功能。一个 SVI 接口对应一个 VLAN，一个 VLAN 仅可以有一个 SVI。

2，SVI 接口用途

SVI 接口用途就是在 VLAN 间提供通信路由。所以，应当为所有 VLAN 配置 SVI 接口，以便可以在 VLAN 间通信。SVI 是联系 VLAN 的 IP 接口，一个 SVI 只能和一个 VLAN 相联系。SVI 实质是 VLAN 虚接口，只不过它是虚拟的，用于连接整个 VLAN。

3. SVI 接口类型

① 主机管理接口：管理员可以利用该接口管理交换机，如默认 VLAN1。
② VLAN 网关接口：用于三层交换机跨 VLAN 间路由，可以用 interface vlan 创建 SVI，然后为其配置 IP 地址，作为各 VLAN 的网关。

3. SVI 接口创建

在全局配置模式下用 **interface** *vlan vlan_id* 命令建立 SVI 接口，也可以用 **no interface vlan** *vlan_id* 删除。不能删除 VLAN 1 的 SVI 接口（VLAN 1），因为 VLAN 1 接口是系统默认创建。

4. SVI 接口状态

SVI 接口状态为 up 的条件：
① SVI 对应的 VLAN 存在，并且在 VLAN 数据库中为激活状态（默认激活，可关闭）。
② VLAN 虚接口 SVI 存在，并且可管理（no shutdown 启用），可以是默认或用户创建。
③ VLAN 中至少存在一个二层端口（access 或 trunk）的链路状态为 up，并且该链路在 VLAN 中的生成树为转发状态（Spanning-tree Forwarding Status）。在后续章节中学习。

5. SVI 接口优势

相比单臂路由的路由器子接口。
① SVI 转发速度快，因经交换机硬件完成路由和交换，比单臂路由快很多。
② 不局限于一条链路，可在三层交换机与二层交换机间采用二层技术实现链路捆绑，以获得更高带宽。
③ SVI 延迟非常低，因为它不需要离开交换机。

4.3.4　场景四：三层交换机 SVI 实现 VLAN 间路由

> 🔍**场景四**：在本章 4.3.2 节场景三中，我们采用单臂路由器实现了对当前阶段 YH 食品有限公司企业网的升级。LZY 工程师觉得在企业网中用路由器虚拟软件接口实现 VLAN 间通信确实能大大节省网络成本，但随着企业网的发展，在部门增加的情况下，多个子接口争用一条物理信道，显然会降低网络性能，无法满足交换式局域网对带宽的需求。那么就请 LZY 工程师分享他的升级方案。

图 4-24 是 LZY 工程师分享的升级前后的网络对比图，我们不难看出他采用三层交换机 SVI 虚拟软件接口实现部门间的通信，具体网络连接如图 4-25 所示。

本场景 SW1 的实施方案与 4.3.2 节场景三单臂路由器实现 VLAN 间路由完全相同，故省略。三层交换机 SVI 的具体实施步骤如下。

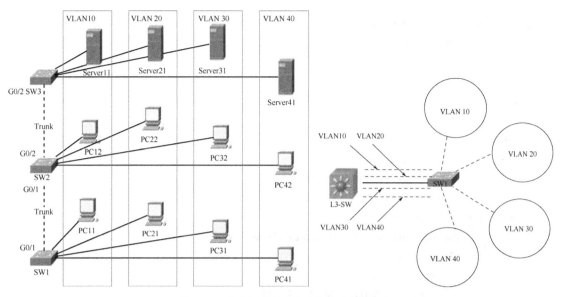

图 4-24　企业网升级前与网络规划对比图

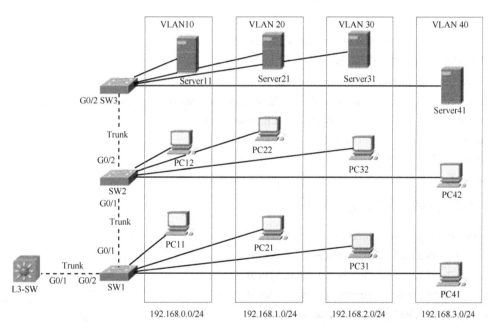

图 4-25　采用三层交换机 SVI 接口实现 VLAN 间路由

步骤一：配置三层交换机的 SVI 接口

Switch>**enable**
Switch#**configure terminal**

```
Switch(config)#hostname L3-SW
L3-SW(config)#ip routing
L3-SW(config)#vlan 10
L3-SW(config-vlan)#vlan 20
L3-SW(config-vlan)#vlan 30
L3-SW(config-vlan)#vlan 40
L3-SW(config-vlan)#exit
L3-SW(config)#interface vlan 10
L3-SW(config-if)#ip address 192.168.0.254 255.255.255.0
L3-SW(config-if)#exit
L3-SW(config)#interface vlan 20
L3-SW(config-if)#ip address 192.168.1.254 255.255.255.0
L3-SW(config-if)#exit
L3-SW(config)#interface vlan 30
L3-SW(config-if)#ip address 192.168.2.254 255.255.255.0
L3-SW(config-if)#exit
L3-SW(config)#interface vlan 40
L3-SW(config-if)#ip address 192.168.3.254 255.255.255.0
```

查看 SVI 接口信息，发现接口状态均为"up"。

```
L3-SW#show ip interface brief | include Vlan
Vlan1              unassigned       YES unset    administratively down down
Vlan10             192.168.0.254    YES manual up                        up
Vlan20             192.168.1.254    YES manual up                        up
Vlan30             192.168.2.254    YES manual up                        up
Vlan40             192.168.3.254    YES manual up                        up
```

步骤二：查看 SVI 接口直连路由

```
L3-SW#show ip route connected
  C   192.168.0.0/24   is directly connected, Vlan10
  C   192.168.1.0/24   is directly connected, Vlan20
  C   192.168.2.0/24   is directly connected, Vlan30
  C   192.168.3.0/24   is directly connected, Vlan40
```

步骤三：部门互通测试

经测试，PC 均已动态获取对应网段的地址，且不同部门间实现了互通，即成功通过三层交换机的 SVI 接口实现了 VLAN 间的互通。

4.3.5 认识设备环回接口

1. 环回接口定义

环回（Loopback）接口是路由器内部的逻辑接口，是一个虚拟的、工作在三层的软件接口。目前很多三层及以上设备均支持环回接口，如路由器、三层交换机、PC、服务器等（本书我们只介绍路由器和三层交换机的环回接口）。

2. 环回接口特点

环回接口不会分配给物理端口，因此永远无法将其连接到任何其他设备上。它被视为是一个软件接口，只要路由器运行正常，该接口就会自动处于 up 状态。环回接口在路由器上与物理接口一样，也是独占一个网段的 IP 地址。设备默认没有任何环回接口，可以手动创建。

3. 环回接口用途

环回接口非常有用，大多数平台都支持使用环回接口来模拟真正的物理接口。环回接口可用于模拟网络、辅助测试以及远程管理设备等。

4. 环回接口意义

（1）具有稳定性

环回接口相比路由器和三层交换机的物理接口，它具有绝对的稳定性，不会因为某一个物理接口 down 而被 down 掉。在追求稳定性的网络协议，如 OSPF 和 BGP 中，环回接口地址会优先选择作为路由器的 Router-ID（后续章节中会详细介绍），这样可以使整个设备的标识稳定可靠。

（2）可节省地址

因为环回接口不会和任何接口公用一个网段，其子网掩码往往使用 255.255.255.255（/32），这样就可以节省大量 IP 地址。环回接口的地址一般和业务地址没有关系，可独立规划。

（3）可用于测试

创建并配置好环回接口之后，它的地址是能被 ping 或 telnet 的，这就可以被用来测试网络的连通性。

5. 环回接口创建

Router(config)#**interface loopback** *number*   //number 的取值范围 <0~2147483647>
Router(config-if)#

4.4 分析路由表

4.4.1 认识路由表

1. 路由表定义

路由表（Routing Table），又称转发数据库，是路由器或者其他网络互联设备上存储的一张路由信息表（RIB），其中存储了直连网络以及远程网络的相关信息。路由器就是依据路由表对数据包做转发决定的。

2. 路由表特点

路由表是保存在内存 RAM 中的数据库文件，一旦路由器关机或重启，路由表中的信息将会全部丢失。只要配置文件存在，路由器就会逐步加载路由表，先加载自身接口的直连路由，然后再加载通过静态以及通过动态路由协议获取到的远程网络信息。

3. 路由表来源

路由表获取路由信息有三种途径：直连路由、静态路由、动态路由。

- **直连路由**：路由器直连的软、硬件接口的网络地址或接口地址，属于本地路由（路由代码为 C、L）。
- **静态路由**：通过手工配置添加到路由表中的远程网络或主机地址（路由代码为 S）。
- **动态路由**：通过路由协议动态获知的路由条目（路由代码为 R、O、D、B 等）。

4. 路由表项类型

- **网络路由**：即主网路由，是指子网掩码等于有类网络的子网掩码的路由。如 110.0.0.0/8、190.22.0.0116、200.5.19.0/24 都是主网路由，因为它们的子网掩码分别等于 A 类的 8，B 类的子网掩码 16 以及 C 类的子网掩码 24。
- **子网路由**：是指子网掩码大于有类网络的子网掩码的路由。如 10.1.0.0/24 是子网路由，因为其子网掩码 24 大于其有类的子网掩码 8。
- **超网路由**：是指子网掩码小于有类掩码的网络地址。如超网路由 172.16.0.0/12，子网掩码 12 小于 B 类子网掩码 16；超网路由 192.168.0.0/22，子网掩码 22 小于 C 类的 24。
- **主机路由**：是一种特殊的子网掩码为/32 的路由。如 192.168.11.22/32，是主机地址。
- **默认路由**：是指目标地址为 0.0.0.0/0 的特殊路由，目标网络地址为 0.0.0.0，子网掩码也为 0.0.0.0。

本书我们所研究的路由表，都是基于 IP 路由协议的路由表，路由器依据路由表来转发相应的 IP 数据包。

4.4.2　认识直连路由

配置路由器的接口并使用 no shutdown 命令将其激活（有的接口默认激活）后，该接口必须在收到来自其他设备（如路由器、交换机、集线器、PC）的载波信号后其状态才能视为"up"（开启）。一旦接口为"up"（开启）状态，该接口所在的网络就会作为直接相连网络而加入路由表。在三层设备接口（物理接口、逻辑接口）的网络地址中，直接被写进路由表，我们称之为直连路由（Connected Route）。

若路由器仅仅配置了接口 IP 地址和子网掩码,此时路由表里只有直连路由,可以使用 show ip route 命令查看直连路由；若配置了其他协议，直连路由可以通过命令 show ip route connected 单独查看，如下所示：

```
R1#show ip route connected
C    10.0.12.0/24    is directly connected, Serial0/0/0
C    10.0.13.0/24    is directly connected, Serial0/0/1
C    10.0.14.0/24    is directly connected, Serial0/1/0
C    10.0.15.0/24    is directly connected, Serial0/1/1
C    10.0.16.0/24    is directly connected, Serial0/2/0
C    10.0.17.0/24    is directly connected, FastEthernet0/0
C    217.2.22.3/32   is directly connected, Loopback0
```

从路由器 R1 的路由输出可知，该路由器共有 7 条直连路由，其中有 5 条是广域网接口的直连路由，1 条是以太网接口的直连路由，还有 1 条环回接口的直连路由。这 7 条直连路由有 6 条是物理硬件接口的直连路由，1 条是逻辑软件接口的直连路由。

在路由器上配置静态或动态路由之前，路由器只知道与自己直接相连的网络（直连路由）。这些直连路由是在配置静态或动态路由之前唯一显示在路由表中的路由，它对于路由器进行路由决定起着重要的作用。如果路由器没有直连路由，也就不会有静态和动态路由的存在。如果路由器接口未配置 IP 地址和子网掩码，路由器就不能从该接口将数据包发送出去。正如 PC 的以太网卡在没有配置 IP 地址和子网掩码的情况下，PC 也不能将 IP 数据包从该接口发送出去。

直连路由是路由表添加其他类型路由的基础，在确保直连路由没有问题的情况下方可进行后续其他类型路由的相关配置。这也是我们对路由进行排错的第一步，即确保直连路由配置没有任何问题。

4.4.3　分析路由表结构

路由表结构与电话簿非常相似。电话簿中包含大量的姓名和电话号码，这些信息都按照姓氏的字母顺序排序。姓名类似路由表中的目标地址，电话号码类似路由表的送出接口或下一跳地址。查找姓名时，电话簿中的姓名（路由条目）越少查找速度越快。例如，搜索一本含有10页1 000个条目的电话薄，远比搜索本100页10 000个条目的电话薄轻松。即路由表中的路由条目越少，查找速度越快。为了使路由表保持较小体积，路由表通常用网络地址加子网掩码来取代单一主机IP地址。

下面我们通过一台路由器RouterX的路由表输出，研究一下路由表的结构。

```
RouterX#show ip route
Codes: C - connected, S - static, I - IGRP, R - RIP, M - mobile, B - BGP
       D - EIGRP, EX - EIGRP external, O - OSPF, IA - OSPF inter area
       N1 - OSPF NSSA external type 1, N2 - OSPF NSSA external type 2
       E1 - OSPF external type 1, E2 - OSPF external type 2, E - EGP
       i - IS-IS, L1 - IS-IS level-1, L2 - IS-IS level-2, ia - IS-IS inter area
       * - candidate default, U - per-user static route, o - ODR
       P - periodic downloaded static route

Gateway of last resort is 0.0.0.0 to network 0.0.0.0

     10.0.0.0/24 is subnetted, 21 subnets
C       10.0.12.0 is directly connected, Serial0/0/0
C       10.0.13.0 is directly connected, Serial0/0/1
C       10.0.14.0 is directly connected, Serial0/1/0
C       10.0.15.0 is directly connected, Serial0/1/1
C       10.0.16.0 is directly connected, Serial0/2/0
C       10.0.17.0 is directly connected, FastEthernet0/0
B       10.2.1.0 [20/0] via 10.0.12.2, 00:20:03
B       10.2.2.0 [20/0] via 10.0.12.2, 00:20:03
B       10.2.3.0 [20/0] via 10.0.12.2, 00:20:03
B       10.2.4.0 [20/0] via 10.0.12.2, 00:20:03
B       10.2.5.0 [20/0] via 10.0.12.2, 00:20:03
R       10.4.1.0 [120/1] via 10.0.14.4, 00:00:12, Serial0/1/0
R       10.4.2.0 [120/1] via 10.0.14.4, 00:00:12, Serial0/1/0
O       10.5.1.0 [110/65] via 10.0.15.5, 00:01:27, Serial0/1/1
O       10.5.2.0 [110/65] via 10.0.15.5, 00:00:56, Serial0/1/1
O       10.5.3.0 [110/65] via 10.0.15.5, 00:00:46, Serial0/1/1
D       10.6.1.0 [90/2297856] via 10.0.16.6, 00:19:55, Serial0/2/0
D       10.6.2.0 [90/2297856] via 10.0.16.6, 00:19:55, Serial0/2/0
D       10.6.3.0 [90/2297856] via 10.0.16.6, 00:19:55, Serial0/2/0
```

D   10.6.4.0 [90/2297856] via 10.0.16.6, 00:19:55, Serial0/2/0
S   10.7.1.0 [1/0] via 10.0.17.7
           217.2.22.0/32 is subnetted, 1 subnets
C   217.2.22.3 is directly connected, Loopback0
S*   0.0.0.0/0 is directly connected, Serial0/0/1

路由表实质是一张简化的二维表，它的行是记录，代表的是路由条目；它的列包含了很多字段，下面我们通过以上输出将这张二维关系表展示出来，如表 4-5 所示。

表 4-5 RouterX 的二维路由表

路由类型	路由代码	目标地址	子网掩码	送出接口	下一跳	管理距离/度量
直连路由 (Connected Route)	C	10.0.12.0	/24	Serial0/0/0	—	[0/0]
		10.0.13.0	/24	Serial0/0/1	—	
		10.0.14.0	/24	Serial0/1/0	—	
		10.0.15.0	/24	Serial0/1/1	—	
		10.0.16.0	/24	Serial0/2/0	—	
		10.0.17.0	/24	Fa0/0	—	
		217.2.22.3	/24	Loopback0	—	
静态路由 (Static Route)	S	10.7.1.0	/24	—	10.0.17.7	[1/0]
	S*	0.0.0.0	/24	Serial0/0/1		
动态路由 (Dynamic Route)	R	10.4.1.0	/24	Serial0/1/0	10.0.14.4	[120/1]
		10.4.2.0	/24	Serial0/1/0	10.0.14.4	[120/1]
	O	10.5.1.0	/24	Serial0/1/1	10.0.15.5	[110/65]
		10.5.2.0	/24	Serial0/1/1	10.0.15.5	[110/65]
		10.5.3.0	/24	Serial0/1/1	10.0.15.5	[110/65]
	D	10.6.1.0	/24	Serial0/2/0	10.0.16.6	[90/2297856]
		10.6.2.0	/24	Serial0/2/0	10.0.16.6	[90/2297856]
		10.6.3.0	/24	Serial0/2/0	10.0.16.6	[90/2297856]
		10.6.4.0	/24	Serial0/2/0	10.0.16.6	[90/2297856]
	B	10.2.1.0	/24	—	10.0.12.2	[20/0]
		10.2.2.0	/24	—	10.0.12.2	[20/0]
		10.2.3.0	/24	—	10.0.12.2	[20/0]
		10.2.4.0	/24	—	10.0.12.2	[20/0]
		10.2.5.0	/24	—	10.0.12.2	[20/0]

有关路由代码的具体含义，我们通过表 4-6 来进一步学习。

从 4-6 表中，我们看到还有一个"L"的路由代码。在 Cisco 路由设备 IPv6 的路由表里面都有"L"代码，只要有接口状态是 up 且配置了正确地址，除了产生直连路由外，还会有一个 /32 的接口地址，标记了"L"。目前，Packet Tracer 中较新版本的 IOS 其 IPv4 的路由表均支持"L"代码。下面我们看一下 RouterY 的路由表输出，查找带"L"的路由代码。

表 4-6 路由来源代码解析表

路由代码	代表含义	路由来源代码含义
C	Connected	标识直连网络
S	Static	标识手动创建的通往特定网络的静态路由
R	RIP	标识使用 RIP 路由协议从相邻路由器动态获取的网络
O	OSPF	标识使用 OSPF 路由协议从相邻路由器动态获取的网络
D	EIGRP（DUAL 算法）	标识使用 EIGRP 路由协议从相邻路由器动态获知的网络
B	BGP	标识使用 BGP 路由协议从相邻路由器动态获取的网络
L	Local	标识路由器接口分配的具体地址，使路由器能优先确定（/32，最长匹配）属于其接口的数据包并直接接收

```
RouterY#show ip route
Codes: L - local, C - connected, S - static, R - RIP, M - mobile, B - BGP
       D - EIGRP, EX - EIGRP external, O - OSPF, IA - OSPF inter area
       N1 - OSPF NSSA external type 1, N2 - OSPF NSSA external type 2
       E1 - OSPF external type 1, E2 - OSPF external type 2, E - EGP
       i - IS-IS, L1 - IS-IS level-1, L2 - IS-IS level-2, ia - IS-IS inter area
       * - candidate default, U - per-user static route, o - ODR
       P - periodic downloaded static route

Gateway of last resort is 0.0.0.0 to network 0.0.0.0

     172.16.0.0/16 is variably subnetted, 4 subnets, 3 masks
C       172.16.1.160/28 is directly connected, Loopback1
L       172.16.1.167/32 is directly connected, Loopback1
C       172.16.1.176/30 is directly connected, Port-channel 1
L       172.16.1.178/32 is directly connected, Port-channel 1
     200.1.1.0/24 is variably subnetted, 2 subnets, 2 masks
C       200.1.1.0/30 is directly connected, Serial0/3/0
L       200.1.1.1/32 is directly connected, Serial0/3/0
S*   0.0.0.0/0 is directly connected, Serial0/3/0
```

从以上输出我们可以看到路由器的三个直连网络，分别产生了一条标记为"L"，子网掩码为/32 的路由条目，我们把掩码为/32 的路由称为"主机路由"。

4.4.4 剖析路由表原理

路由器的主要工作就是为经过路由器的每个数据包寻找一条最佳的传输路径，并将该数据有效地送达目的站，这就是所谓的路由器最佳路径选择。由此可见，选择最佳路径的策略即路由算法是路由器选路的关键所在。为完成这项工作，路由器中专门有一个路由表，保存着各种传输路径的相关数据，供路由选择时使用，表中包含的信息决定了数据转发的策略。打个比方，路由表就像我们平时使用的地图一样，标识出各种路线。路由表中保存着目标网络的标识信息、到达目标的距离和下一站路由器的地址等信息。路由表可以由管理员手工配置好，也可以配置路由协议让路由器自动获取邻居的信息，动态生成路由表。

当路由器收到一个 IP 数据包时，它会使用路由表中的子网掩码去匹配数据包，匹配时使用"最长匹配原则"，即匹配掩码最长的那条路由，它认为这样匹配的结果更加精确。

例如，路由表里有两条路由，其目的地址相同，都是 172.16.0.0，但掩码不同，分别为/22 和/23，两条路由对应着不同的出口 Serial1/1 和 Serial1/0。如下所示：

```
Router#show ip route 172.16.0.0
    D     172.16.0.0/22 [90/2172416] via 172.16.2.138, 00:03:56, Serial1/1
    D     172.16.0.0/23 [90/2172416] via 172.16.2.134, 00:03:56, Serial1/0
```

此时，如果我们在终端 PC 上发布 ping 命令，ping 172.16.1.1，则路由器就会收到一个目的地址为 172.16.1.1 的数据包。接下来，路由器就会查找路由表，并依据最长匹配原则，选择最佳路径，最终选出来的结果如下：

```
    D     172.16.0.0/23   [90/2172416] via 172.16.2.134, 00:03:56, Serial1/0
```

因为 172.16.0.0/23 这条路由能匹配 23 位，比 172.16.0.0/22 匹配的 22 位要长，匹配更精确。最长匹配原则是 Cisco IOS 路由器默认的路由查找方式。当路由器收到一个 IP 数据包时，它会将数据包的目的 IP 地址与自己本地路由表中的表项进行逐位查找，直到找到匹配度最长的条目，这就是最长匹配原则。

IP 数据包从源主机到达目标主机，中途可能经过若干台路由器的接力转发，每台路由器都会依据自己的路由表做转发决定。默认采用最长匹配原则，匹配，则转发；不匹配，则找默认路由；默认路由不存在，则丢弃。路由器的转发是逐跳进行的，到达目标经过的中途路由器都必须有目标网络的路由。数据是双向的，路由器只关心数据包去的路径，不负责回包的路径。

我们现将路由表原理总结如下：

- 每台路由器依据自己的路由表做转发决定；
- 一台路由器路由表中包含的某些信息并不代表其他路由器也包含相同信息；
- 不同网络主机间的通信，路由器的路由表只负责为其提供转发路径，不负责提供接收路径，即路由器的路由表不提供往返路径。

4.4.5 路由表的层次化

思科 IP 路由表并不是一个平面数据库。路由表实际上是一个分层结构，在查找路由并转发数据包时，这样的分层结构可以加快查找进程。在此结构中包含若干个层级，本节只讨论只有两个层级的路由表，即第一级路由和第二级路由。

- 第一级路由：网络路由、超网路由、默认路由；
- 第二级路由：子网路由。

不管是第一级路由还是第二级路由，最终对转发起决定性作用的都是最终路由。什么是最终路由呢？最终路由就是包含下一跳 IP 地址或送出接口的路由表条目，它可以是第一级路由也可以是第二级路由。下面我们通过路由器 RouterZ 的路由输出来深入学习路由表的层级并对第一级路由、第二级路由以及最终路由做一下归类。

```
RouterZ#show ip route
================Output Omitted================
Gateway of last resort is 0.0.0.0 to network 0.0.0.0

C       192.168.0.0/24 is directly connected, Loopback0
C       192.168.1.0/24 is directly connected, Loopback1
        192.168.2.0/26 is subnetted, 4 subnets
C       192.168.2.0 is directly connected, Loopback2
C       192.168.2.64 is directly connected, Loopback3
C       192.168.2.128 is directly connected, Loopback4
C       192.168.2.192 is directly connected, Loopback5
        192.168.3.0/24 is variably subnetted, 4 subnets, 3 masks
C       192.168.3.0/25 is directly connected, Loopback6
C       192.168.3.128/26 is directly connected, Loopback7
C       192.168.3.192/27 is directly connected, Loopback8
C       192.168.3.224/27 is directly connected, Loopback9
C       192.168.4.0/22 is directly connected, Loopback10
S       192.168.255.0/24 [1/0] via 192.168.3.9
S*      0.0.0.0/0 is directly connected, Loopback9
```

通过以上路由表的输出，我们统计出共有 15 个路由条目。有路由来源代码的是 13 条，没有路由来源代码的是 2 条。这 13 条路由都包含送出接口或下一跳地址，因此都是最终路由。很明显路由表中的记录出现了分层，外层（第一级路由）和内层（第二级路由），下面我们通过表 4-7 将记录进行如下归类。

表 4-7 路由表记录归类

路由类别	对应路由表中的路由条目		记录数	备注
第一级路由	192.168.0.0/24 192.168.1.0/24 192.168.2.0/26 192.168.3.0/24	192.168.4.0/22 192.168.255.0/24 0.0.0.0/0	7	网络路由 超网路由 默认路由
第二级路由	192.168.2.0/26 192.168.2.64/26 192.168.2.128/26 192.168.2.192/26	192.168.3.0/25 192.168.3.128/26 192.168.3.192/27 192.168.3.224/27	8	子网路由

从表 4-7 中我们发现，第一级路由 192.168.2.0/26 既不是网络路由又不是超网路由和默认路由，为什么它会是第一级路由呢？这就是我们下面要介绍的另外一种特殊类型的第一级路由——父路由。

所谓的父路由就是有子网的路由。因为子网路由是第二级路由，所以父路由自然就是第一级路由。路由表中 192.168.2.0/26 有 4 个子网，所以它是第一级路由；192.168.3.0/24 也有子网，也是第一级路由，又称第一级父路由。但两个父路由存储形式不同，父路由 192.168.2.0 的子网掩码为/26，是下面子网的掩码，在子网掩码等长的情况下，子路由掩码就显示在父路由上，如图 4-26 所示，子网都是一级子网，且掩码都为/26。

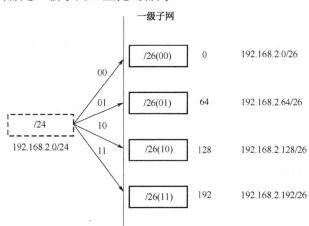

图 4-26 等长子网掩码的路由

父路由 192.168.3.0/24 则显示正常，因为它的 4 个子网掩码不同，因此网络被分成三级子网，如图 4-27 所示，掩码分别为/25、/26 和/27。因为子网的掩码不同，父路由显示父路由的掩码，子路由显示子路由的掩码。

第一级父路由相当于其子路由的一个标题，不包含下一跳或送出接口。只要向路由表中增加一个子网，就会在表中自动创建第一级父路由。父路由是伴随着子路由生成而自动生成的，也会伴着所有子路由消失而自动消失。路由器在查询时，首先匹配第一级父路由，只有父路由的有类地址与数据包的目的 IP 地址匹配，才会检查子路由，这样会大大加快查找速度。

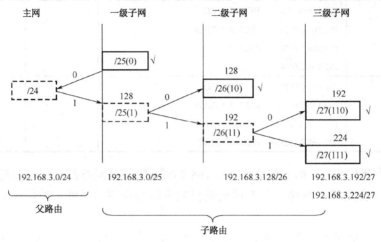

图 4-27 不等长掩码的路由

4.5 网络工程师技术面试闯关

1. 任务背景

YH 食品有限公司遇到了前所未有的发展机遇，目前互联网业务已占据公司销售总额的近 70%。公司在 LWJ 销售经理的再三请求下，终于获批新增一个 IT 部门，负责公司企业网的日常运维工作。现在 IT 部门急需招聘两名具有 CCNA 认证的网络工程师，LWJ 请求 CFL 老师出面帮忙面试，为公司出一道面试用的专业技术考题，并给出参考答案。

2. 试题要求

① 实验考题，采用模拟器；
② 考试时间为 60 分钟；
③ 难度适中；
④ 尽量能考查出应聘人员实际解决问题的能力，熟悉思科常用设备。

以下是 CFL 老师为 YH 的 IT 部门出的面试考题，内容如下。

已知 X 是一台网络设备，请根据 X 的路由表推断网络拓扑，采用 Packet Tracer 7.0 模拟器进行设备选型、拓扑搭建、配置实施，产生如下完全一致的路由表，还要实现全网互通。要求在 X 设备上配置 DHCP 服务，使终端能动态获取地址。

请根据提供的 6 个任务来完成相应的实验步骤，在任务四中输出你产生的路由表项。

要求：为设备统一命名，路由器命名为 R1、R2……；交换机则命名为 SW1、SW2……。
X 设备的路由表输出如下：

```
X#show ip route
Codes: C - connected, S - static, I - IGRP, R - RIP, M - mobile, B - BGP
       D - EIGRP, EX - EIGRP external, O - OSPF, IA - OSPF inter area
       N1 - OSPF NSSA external type 1, N2 - OSPF NSSA external type 2
       E1 - OSPF external type 1, E2 - OSPF external type 2, E - EGP
       i - IS-IS, L1 - IS-IS level-1, L2 - IS-IS level-2, ia - IS-IS inter area
       * - candidate default, U - per-user static route, o - ODR
       P - periodic downloaded static route

Gateway of last resort is not set

     10.0.0.0/24 is subnetted, 4 subnets
C       10.1.0.0 is directly connected, Loopback0
C       10.1.1.0 is directly connected, Loopback1
C       10.1.2.0 is directly connected, Loopback2
C       10.1.3.0 is directly connected, Loopback3
     172.16.0.0/24 is subnetted, 2 subnets
C       172.16.0.0 is directly connected, Vlan10
C       172.16.1.0 is directly connected, Vlan20
C    192.168.0.0/24 is directly connected, FastEthernet0/0
C    192.168.1.0/24 is directly connected, FastEthernet0/1.10
C    192.168.2.0/24 is directly connected, FastEthernet0/1.20
```

CFL 老师给出面试题目的参考答案如下（拓扑结构可以不唯一，但输出结果唯一）。

任务一：推断物理设备

因为设备"X"有路由表输出，所以它可能是一台路由器也可能是一台三层交换机。因为路由表中存在子接口，所以推断 X 是一台路由器。又因为存在有两个 SVI 接口，因此它又是一台三层交换机，进一步推断 X 是一台集成多业务路由器（ISR），即有三层交换模块的路由器。

经分析发现该设备的物理接口为 FastEthernet，因此是百兆端口的路由器，排除了 1900 系列、2900 系列等相对高端的路由器。进一步查看支持三层交换模块的路由器。

任务二：搭建网络拓扑

推断网络拓扑如图 4-28 所示。

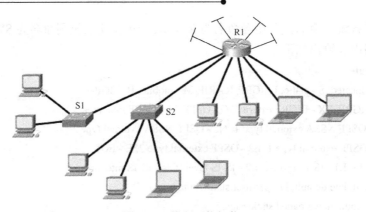

图 4-28　推断网络拓扑

任务三：规划网络地址

网络地址规划如图 4-29 所示。

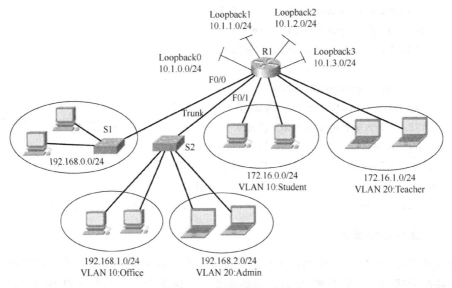

图 4-29　网络地址规划

任务四：配置实施网络

1. 配置当前路由器的直连路由

（1）配置路由器环回接口

```
R1(config)#interface loopback 0
R1(config-if)#ip address 10.1.0.1 255.255.255.0
```

```
R1(config-if)#interface loopback 1
R1(config-if)#ip address 10.1.1.1 255.255.255.0
R1(config-if)#interface loopback 2
R1(config-if)#ip address 10.1.2.1 255.255.255.0
R1(config-if)#interface loopback 3
R1(config-if)#ip address 10.1.3.1 255.255.255.0
```

（2）配置三层交换模块

```
R1#vlan database
% Warning: It is recommended to configure VLAN from config mode,
    as VLAN database mode is being deprecated. Please consult user
    documentation for configuring VTP/VLAN in config mode.
R1(vlan)#vlan 10 name Student
VLAN 10 added:
    Name: Student
R1(vlan)#vlan 20 name Teacher
VLAN 20 modified:
    Name: Teacher

R1(config)#interface range fastEthernet 0/0/0-3
R1(config-if-range)#switchport mode access
R1(config-if-range)#switchport access vlan 10
R1(config-if-range)#exit
R1(config)#interface range fastEthernet 0/1/0-3
R1(config-if-range)#switchport mode access
R1(config-if-range)#switchport access vlan 20
R1(config)#interface vlan 10
R1(config-if)#ip address 172.16.0.254 255.255.255.0
R1(config)#interface vlan 20
R1(config)#ip address 172.16.1.254 255.255.255.0
```

（3）配置路由器物理接口

```
R1(config)#interface F0/0
R1(config-if)#ip address 192.168.0.254 255.255.255.0
R1(config-if)#no shutdown
```

（4）配置路由器子接口

```
R1(config)#interface F0/1
R1(config-if)#no ip address        //为了强调，物理接口不允许有 IP 地址
R1(config-if)#no shutdown          //激活物理接口，其下子接口均被激活
R1(config-if)#interface F0/1.10
R1(config-subif)#encapsulation dot1Q 10
R1(config-subif)#ip address 192.168.1.254 255.255.255.0
R1(config-subif)#interface F0/1.20
R1(config-subif)#encapsulation dot1Q 20
R1(config-subif)#ip address 192.168.2.254 255.255.255.0
```

（5）查看接口和直连路由

```
R1#show ip interface brief | include                up            up
FastEthernet0/0         192.168.0.254    YES     manual up     up
FastEthernet0/1         unassigned       YES     unset   up    up
FastEthernet0/1.10      192.168.1.254    YES     manual up     up
FastEthernet0/1.20      192.168.2.254    YES     manual up     up
FastEthernet0/0/0       unassigned       YES     unset   up    up
FastEthernet0/0/1       unassigned       YES     unset   up    up
FastEthernet0/1/0       unassigned       YES     unset   up    up
FastEthernet0/1/1       unassigned       YES     unset   up    up
Loopback0               10.1.0.1         YES     manual up     up
Loopback1               10.1.1.1         YES     manual up     up
Loopback2               10.1.2.1         YES     manual up     up
Loopback3               10.1.3.1         YES     manual up     up
Vlan10                  172.16.0.254     YES     manual up     up
Vlan20                  172.16.1.254     YES     manual up     up

R1#show ip route
Codes: C - connected, S - static, I - IGRP, R - RIP, M - mobile, B - BGP
       D - EIGRP, EX - EIGRP external, O - OSPF, IA - OSPF inter area
       N1 - OSPF NSSA external type 1, N2 - OSPF NSSA external type 2
       E1 - OSPF external type 1, E2 - OSPF external type 2, E - EGP
       i - IS-IS, L1 - IS-IS level-1, L2 - IS-IS level-2, ia - IS-IS inter area
       * - candidate default, U - per-user static route, o - ODR
       P - periodic downloaded static route
```

Gateway of last resort is not set

    10.0.0.0/24 is subnetted, 4 subnets
C   10.1.0.0 is directly connected, Loopback0
C   10.1.1.0 is directly connected, Loopback1
C   10.1.2.0 is directly connected, Loopback2
C   10.1.3.0 is directly connected, Loopback3
    172.16.0.0/24 is subnetted, 2 subnets
C   172.16.0.0 is directly connected, Vlan10
C   172.16.1.0 is directly connected, Vlan20
C   192.168.0.0/24 is directly connected, FastEthernet0/0
C   192.168.1.0/24 is directly connected, FastEthernet0/1.10
C   192.168.2.0/24 is directly connected, FastEthernet0/1.20

2. 配置路由器为 DHCP 服务器

R1(config)#**ip dhcp excluded-address 192.168.0.254**
R1(config)#**ip dhcp excluded-address 192.168.1.254**
R1(config)#**ip dhcp excluded-address 192.168.2.254**
R1(config)#**ip dhcp excluded-address 172.16.0.254**
R1(config)#**ip dhcp excluded-address 172.16.1.254**
R1(config)#**ip dhcp pool LAN1**
R1(dhcp-config)#**network 192.168.0.0 255.255.255.0**
R1(dhcp-config)#**default-router 192.168.0.254**
R1(dhcp-config)#**ip dhcp pool OFFICE**
R1(dhcp-config)#**network 192.168.1.0 255.255.255.0**
R1(dhcp-config)#**default-router 192.168.1.254**
R1(dhcp-config)#**ip dhcp pool ADMIN**
R1(dhcp-config)#**network 192.168.2.0 255.255.255.0**
R1(dhcp-config)#**default-router 192.168.2.254**
R1(dhcp-config)#**ip dhcp pool STUDENT**
R1(dhcp-config)#**network 172.16.0.0 255.255.255.0**
R1(dhcp-config)#**default-router 172.16.0.254**
R1(dhcp-config)#**ip dhcp pool TEACHER**
R1(dhcp-config)#**network 172.16.1.0 255.255.255.0**
R1(dhcp-config)#**default-router 172.16.1.254**

3. 配置交换机 VLAN 及 Trunk

```
S2(config)#interface fastEthernet 0/24
S2(config-if)#switchport mode trunk
```

```
S2#show interfaces trunk
Port        Mode        Encapsulation   Status      Native vlan
Fa0/24      on          802.1q          trunking    1

Port        Vlans allowed on trunk
Fa0/24      1-1005

Port        Vlans allowed and active in management domain
Fa0/24      1,10,20

Port        Vlans in spanning tree forwarding state and not pruned
Fa0/24      1,10,20
```

任务五：自动获取地址

192.168.0.0/24 网段 PC 已成功自动获取地址，如图 4-30 所示。

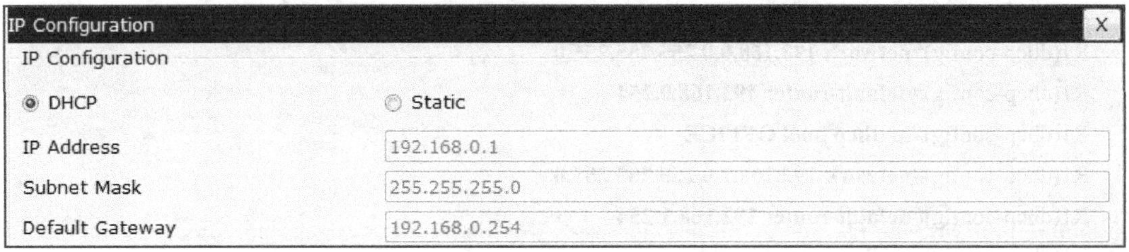

```
IP Configuration                                                    X
  IP Configuration
    ◉ DHCP              ○ Static
    IP Address          192.168.0.1
    Subnet Mask         255.255.255.0
    Default Gateway     192.168.0.254
```

图 4-30　验证 192.168.0.0/24 网段 PC 的 DHCP 获取

192.168.1.0/24 网段 PC 已成功自动获取地址，如图 4-31 所示。

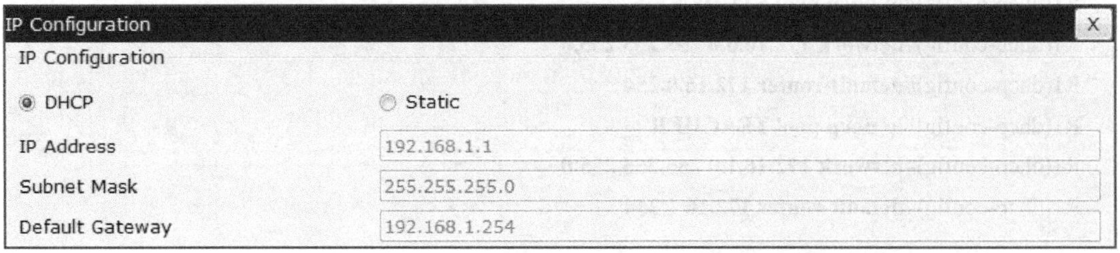

```
IP Configuration                                                    X
  IP Configuration
    ◉ DHCP              ○ Static
    IP Address          192.168.1.1
    Subnet Mask         255.255.255.0
    Default Gateway     192.168.1.254
```

图 4-31　验证 192.168.1.0/24 网段 PC 的 DHCP 获取

192.168.2.0/24 网段 PC 已成功自动获取地址，如图 4-32 所示。

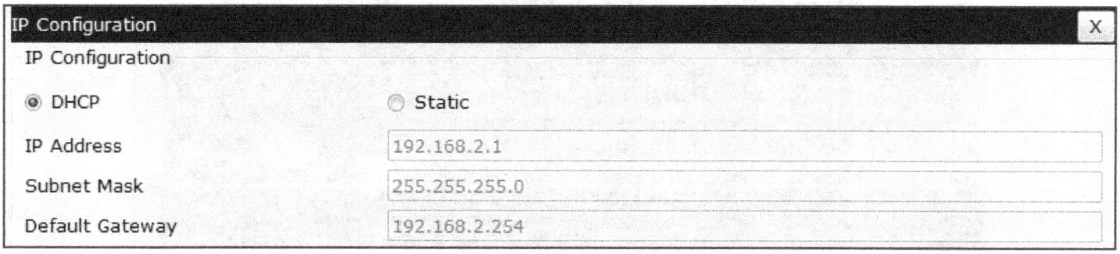

```
IP Configuration                                                X
  IP Configuration
    ⦿ DHCP              ○ Static
    IP Address          192.168.2.1
    Subnet Mask         255.255.255.0
    Default Gateway     192.168.2.254
```

图 4-32　验证 192.168.2.0/24 网段 PC 的 DHCP 获取

172.16.0.0/24 网段 PC 已成功自动获取地址，如图 4-33 所示。

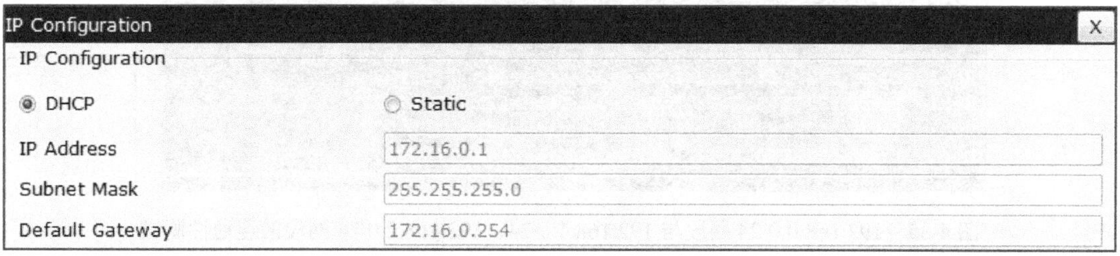

```
IP Configuration                                                X
  IP Configuration
    ⦿ DHCP              ○ Static
    IP Address          172.16.0.1
    Subnet Mask         255.255.255.0
    Default Gateway     172.16.0.254
```

图 4-33　验证 172.16.0.0/24 网段 PC 的 DHCP 获取

172.16.1.0/24 网段 PC 已成功自动获取地址，如图 4-34 所示。

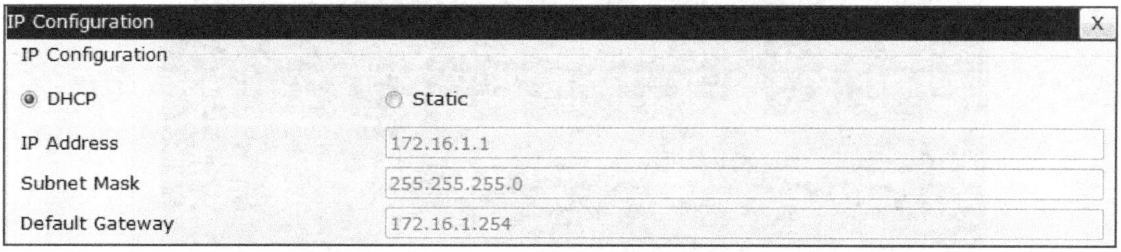

```
IP Configuration                                                X
  IP Configuration
    ⦿ DHCP              ○ Static
    IP Address          172.16.1.1
    Subnet Mask         255.255.255.0
    Default Gateway     172.16.1.254
```

图 4-34　验证 172.16.1.0/24 网段 PC 的 DHCP 获取

任务六：完成网络连通测试

用 192.168.0.0/24 网段的主机 ping 192.168.1.0/24 和 192.168.2.0/24 网段的主机，均可 ping 通，如图 4-35 所示。

用 192.168.0.0/24 网段的主机 ping 172.16.0.0/24 和 172.16.1.0/24 网段的主机，均可 ping 通，如图 4-36 所示。

```
C:\>ping 192.168.1.1

Pinging 192.168.1.1 with 32 bytes of data:

Reply from 192.168.1.1: bytes=32 time=13ms TTL=127
Reply from 192.168.1.1: bytes=32 time=15ms TTL=127
Reply from 192.168.1.1: bytes=32 time=13ms TTL=127
Reply from 192.168.1.1: bytes=32 time=11ms TTL=127

Ping statistics for 192.168.1.1:
    Packets: Sent = 4, Received = 4, Lost = 0 (0% loss),
Approximate round trip times in milli-seconds:
    Minimum = 11ms, Maximum = 15ms, Average = 13ms

C:\>ping 192.168.2.1

Pinging 192.168.2.1 with 32 bytes of data:

Reply from 192.168.2.1: bytes=32 time=1ms TTL=127
Reply from 192.168.2.1: bytes=32 time=12ms TTL=127
Reply from 192.168.2.1: bytes=32 time=14ms TTL=127
Reply from 192.168.2.1: bytes=32 time<1ms TTL=127
```

图 4-35　192.168.0.0/24 网段与 192.168.1.0/24、192.168.2.0/24 网段的连通性测试

```
C:\>ping 172.16.0.2

Pinging 172.16.0.2 with 32 bytes of data:

Reply from 172.16.0.2: bytes=32 time=1ms TTL=127
Reply from 172.16.0.2: bytes=32 time=13ms TTL=127
Reply from 172.16.0.2: bytes=32 time=10ms TTL=127
Reply from 172.16.0.2: bytes=32 time<1ms TTL=127

Ping statistics for 172.16.0.2:
    Packets: Sent = 4, Received = 4, Lost = 0 (0% loss),
Approximate round trip times in milli-seconds:
    Minimum = 0ms, Maximum = 13ms, Average = 6ms

C:\>ping 172.16.1.2

Pinging 172.16.1.2 with 32 bytes of data:

Reply from 172.16.1.2: bytes=32 time=13ms TTL=127
Reply from 172.16.1.2: bytes=32 time=10ms TTL=127
Reply from 172.16.1.2: bytes=32 time=13ms TTL=127
Reply from 172.16.1.2: bytes=32 time<1ms TTL=127
```

图 4-36　192.168.0.0/24 网段与 172.16.0.0/24、172.16.1.0/24 网段的连通性测试

经反复测试，结果是任意网段的两台主机均可以 ping 通，即网络成功实现了全网互通。

4.6 挑战过关练习

4.6.1 挑战过关练习一

1. 任务背景

图 4-37 所示网络拓扑中有 3 台交换机：S1、S2、S3，它们之间通过千兆接口实现了互连。图中共有 4 个 VLAN：VLAN10、VLAN20、VLAN30 和 VLAN40，其端口分配见图中标注。VLAN 间通信通过三层交换机实现。请参照图 4-37 完成挑战任务。

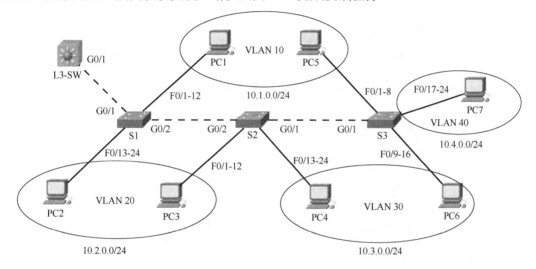

图 4-37 挑战过关练习一

2. 挑战任务

- 配置 4 台网络设备的主机名；
- 在 3 台交换机上创建相应 VLAN，分配对应的端口；
- 在 3 台交换机上创建管理 VLAN，VLAN ID 为 99，管理 IP 分别是 10.0.0.1/24、10.0.0.2/24、10.0.0.3/24；
- 配置 Trunk，关闭 DTP 协商，配置本征 VLAN 为 VLAN 99；
- 配置三层交换机 SVI，通过 SVI 实现 VLAN 间通信；
- 配置三层交换机为 DHCP 服务器，为 VLAN 分配地址，地址池名为 VLAN 10、VLAN 20、VLAN 30、VLAN 40；
- 配置 PC 通过 DHCP 自动获取地址；
- 连通性测试，各 PC 间均可以通信。

4.6.2 挑战过关练习二

1. 任务背景

请打开 Cisco Packet Tracer 模拟器，在设备选区选择一台 2811 型号的路由器并将其拖至工作区。关闭 2811 路由器的电源，添加 3 个交换模块，1 块 16 口交换模块 NM-ESW-161，两块 4 口 HWIC-4ESW 模块，如图 4-38 所示。

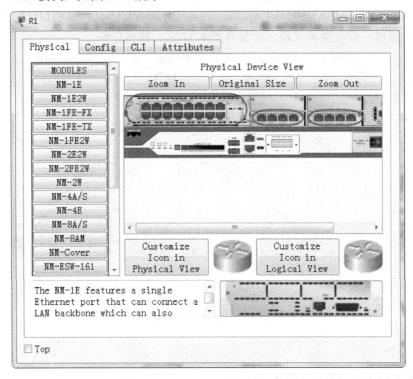

图 4-38　为 2811 路由器添加模块

再从设备选区中选择两台型号为 2950-24 的交换机，拖至工作区，如图 4-39 所示。选取 13 台 PC 和 1 台服务器，并按照图 4-39 更改 Display Name。

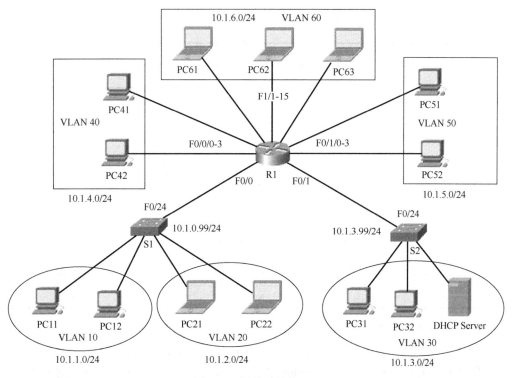

图 4-39 挑战过关练习二

2. 挑战任务

- 请参照表 4-8 所示，在相应设备上配置主机名并完成 VLAN 的配置。

表 4-8 挑战练习 VLAN 规划一览表

VLAN ID	VLAN 名称	端口成员	地址规划
10	VLAN10	S1:F0/1-12	10.1.1.0/24
20	VLAN20	S1:F0/13-23	10.1.2.0/24
30	VLAN30	S2:F0/1-24,G0/1-2	10.1.3.0/24
40	VLAN40	R1:F0/0/0-3	10.1.4.0/24
50	VLAN50	R1: F0/1/0-3	10.1.5.0/24
60	VLAN60	R1:F1/1-15	10.1.6.0/24

- 请参照表 4-9 所示，在相应设备上完成相应接口的地址配置。

表 4-9　设备接口地址规划一览表

设备名称	接口名称	地址分配
R1	F0/0.10	10.1.1.254/24
	F0/0.20	10.1.2.254/24
	F0/0.99	10.1.0.254/24
	F0/1	10.1.3.254/24
	VLAN 40	10.1.4.254/24
	VLAN 50	10.1.5.254/24
	VLAN 60	10.1.6.254/24
S1	VLAN 99	10.1.0.99/24
S2	VLAN 30	10.1.3.99/24
DHCP Server	NIC	10.1.3.250/24
PCx	NIC	DHCP

- 配置 DHCP 服务器，使其能为所有 VLAN 自动分配地址。
- 注意：跨网段 DHCP 地址分配，需要做 DHCP 中继，即在 VLAN 所在网关接口开启 DHCP 中继，命令为 ip helper-address *ip address*（地址为 DHCP 服务器的 IP 地址）
- 配置 PC 自动获取地址。
- 全网连通性测试，任意两台 PC 均可以相互通信。

"学习直连路由"一章到此即将结束。本章通过设置的 2 个实验、4 个应用场景、1 个面试闯关考题和 2 个挑战过关练习，将本章所学技术进行了"有机整合"。其中 4 个应用场景紧密围绕 YH 食品有限公司发展的需求，带动我们不断去升级网络，更新技术。设计的 YH 食品有限公司 IT 部门技术面试闯关试题，实际上是对第 3、4 章所学技术的一个大整合。技术涵盖了 VLAN 技术、Trunk 技术、DHCP、VLAN 间路由的 4 种实现方式。题目采用逆向思维，从一张路由表展开，教你如何分析表项、推断设备、画出拓扑、规划地址、实施网络、输出与要求一致的路由表。面试题目的设计旨在提高你的综合分析能力、逻辑推理能力以及对所学技术的综合应用能力。最后的 2 个挑战闯关练习进一步开阔了你的视野，打开你的思路，带你走上一条探索学习的道路，其过程充满了趣味性和挑战性。

第5章

学习静态路由

本章要点

- 认识静态路由
- 配置标准静态路由
- 配置特殊静态路由
- 企业网综合配置案例
- 拓展思维案例

第 4 章,"学习直连路由",是路由器自己学习本地的路由,而本章,"学习静态路由"是用户或管理员通过手工方式来获取远程的路由,从而实现远程连接。本章设计了 10 个应用场景,1 个企业网综合配置案例和 4 个课外拓展案例来贯穿整个教学内容,教学案例丰富。其中包含下一跳和出接口两种标准静态路由,以及默认路由、主机路由、汇总路由、黑洞路由、浮动路由、等价路由和递归路由共 7 种特殊静态路由。每种技术都有其典型的应用场景,可使你身临其境,进行轻松有趣的学习。

5.1 认识静态路由

静态路由是指由用户或网络管理员手工配置的路由。当网络拓扑结构或链路状态发生变化时,需要用户或网络管理员手工去修改路由表中的相关信息。静态路由信息在默认情况下是私有的,不会传递给其他路由器。当然,网管员也可通过对路由器进行相应设置使之与其他路由器共享。

5.1.1 静态路由的特点

静态路由不会随着网络拓扑的改变而自动调整,路由器之间不会发送通告消息,这样,可以节省网络带宽和路由器的运算资源。静态路由是单向的,适合小型网络或结构比较稳定的网络。表 5-1 列出了动态路由和静态路由功能的比较,通过对比不难看出一种方式的优点恰恰是另一种方式的缺点。

表 5-1 动态路由和静态路由的对比

	动 态 路 由	静 态 路 由
对资源的占用率	占用 CPU、RAM 和链路带宽	不占额外资源
配置的复杂程度	不受网络规模的限制	网络规模增大配置趋向复杂
拓扑改变的影响	能自动适应网络拓扑的改变	需要管理员维护变化的路由
网络安全性影响	不安全	更安全
网络拓扑的适应	简单和复杂的网络拓扑均可	简单网络拓扑

5.1.2 静态路由的应用

静态路由一般适用于比较简单的网络环境,这样的网络环境可使网管员能清楚地了解网络拓扑结构,便于配置正确的路由信息,其主要应用总结如表 5-2 所示。

表 5-2 静态路由的应用场合

应 用 场 合	场 景 描 述
小型网络	网络中仅含几台路由设备，且不会显著增长
末节网络	只能通过单条路径访问的网络，路由器只有一个邻居
通过单 ISP 接入 Internet 的网络	企业边界路由器接入 ISP 的网络环境
集中星型拓扑结构的大型网络	由一个中央节点向多个分支呈放射状连接的网络

5.1.3 静态路由的类型

静态路由主要有以下 4 种类型。
- 标准静态路由：普通的、常规的通往目的网络的路由。
- 默认静态路由：将 0.0.0.0/0 作为目的网络地址的路由，可以匹配所有数据包。
- 汇总静态路由：将多条静态路由汇总成一条静态路由，可减少路由条目，优化路由表。
- 浮动静态路由：为一条路由提供备份的静态路由，当链路出现故障时选择走备用链路。

5.1.4 静态路由的语法

静态路由的配置语法：
Router(config)#ip route {destination-address} {mask} {next-hop-address |exit- interface} [distance]
命令参数解释如下。
- destination-address：目的地址，采用点分十进制表示，可以是具体的网络地址，也可以是一个具体 IP 地址、甚至可以是一个汇聚的地址；
- mask：可以是子网掩码，也可以是主机掩码；
- next-hop-address：下一跳地址，即通往目的地址使用的邻居路由器入口 IP 地址；
- exit- interface：本地送出接口，即通往目的地址使用的本地路由设备的送出接口；
- distance：静态路由条目的管理距离，默认值为 1，其取值范围为 1~255。

5.2 配置标准静态路由

5.2.1 场景一：配置下一跳静态路由

> 场景一：今天网络课，CFL 老师要给大家讲静态路由的配置。老师给同学们布置了

课前预习任务,大家完成得都不错。课前 WMC 同学向老师反映:"静态路由"一章太简单,让老师直接跳过,讲下一章的"动态路由"。老师笑着说:"我觉得这章不简单,反而有点难啊!"。同学们惊诧地望着老师。接下来,老师要求同学们打开 Packet Tracer,做老师布置的课堂作业,一并检验大家的预习效果。作业要求如下所述。

A、B、C 公司因业务往来日趋频繁,现需要将三家公司通过 WAN 接口实现互连。A、B、C 公司的企业网边界路由器分别为 R1、R2 和 R3。考虑到各公司内部网络结构简单,要求采用下一跳配置静态路由实现公司网络间的互连,互连拓扑如图 5-1 所示。

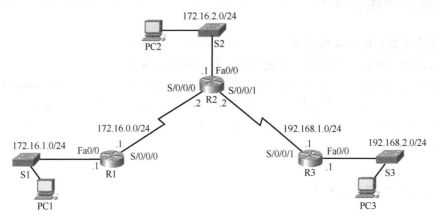

图 5-1　配置下一跳静态路由拓扑

同学们很快完成了老师布置的任务,接下来老师分享了她自己整理的实验步骤,并要求同学们看完回答后面的问题,实验过程如下。

步骤一:路由器基本配置

首先,为 3 台路由器接口配置正确的 IP 地址,确保路由表中显示所有直连路由,否则,需进一步排错。

R1#show ip interface brief | include manual			
FastEthernet0/0	172.16.1.1	YES manual up	up
Serial0/0/0	172.16.0.1	YES manual up	up

R2#show ip interface brief | include manual			
FastEthernet0/0	172.16.2.1	YES manual up	up
Serial0/0/0	172.16.0.2	YES manual up	up
Serial0/0/1	192.168.1.2	YES manual up	up

R3#show ip interface brief | include manual

FastEthernet0/0	192.168.2.1	YES manual up	up
Serial0/0/1	192.168.1.1	YES manual up	up

步骤二：配置下一跳静态路由

R1(config)#**ip route 172.16.2.0 255.255.255.0 172.16.0.2**
R1(config)#**ip route 192.168.2.0 255.255.255.0 172.16.0.2**

R2(config)#**ip route 172.16.1.0 255.255.255.0 172.16.0.1**
R2(config)#**ip route 192.168.2.0 255.255.255.0 192.168.1.1**

R3(config)#**ip route 172.16.1.0 255.255.255.0 192.168.1.2**
R3(config)#**ip route 172.16.2.0 255.255.255.0 192.168.1.2**

步骤三：检查静态路由

R1>**show ip route**
Codes: C - connected, S - static, I - IGRP, R - RIP, M - mobile, B - BGP
Gateway of last resort is not set

    172.16.0.0/24 is subnetted, 3 subnets
C   172.16.0.0 is directly connected, Serial0/0/0
C   172.16.1.0 is directly connected, FastEthernet0/0
S   172.16.2.0 [1/0] via 172.16.0.2
S   192.168.2.0/24 [1/0] via 172.16.0.2

R1 的路由表，静态路由添加成功。

R2>**show ip route**
Codes: C - connected, S - static, I - IGRP, R - RIP, M - mobile, B – BGP
Gateway of last resort is not set

    172.16.0.0/24 is subnetted, 3 subnets
C   172.16.0.0 is directly connected, Serial0/0/0
S   172.16.1.0 [1/0] via 172.16.0.1
C   172.16.2.0 is directly connected, FastEthernet0/0
C   192.168.1.0/24 is directly connected, Serial0/0/1
S   192.168.2.0/24 [1/0] via 192.168.1.1

R2 的路由表，静态路由添加成功。

R3>**show ip route**
Codes: C - connected, S - static, I - IGRP, R - RIP, M - mobile, B - BGP

Gateway of last resort is not set

     172.16.0.0/24 is subnetted, 2 subnets
S   172.16.1.0 [1/0] via 192.168.1.2
S   172.16.2.0 [1/0] via 192.168.1.2
C   192.168.1.0/24 is directly connected, Serial0/0/1
C   192.168.2.0/24 is directly connected, FastEthernet0/0

R3 的路由表，静态路由添加成功。

步骤四：网络连通性测试

配置终端 PC 的 IP 地址参数，在主机 PC1 上用 ping 命令测试与 PC2 的连通性。

```
PC>ping 172.16.2.2
Pinging 172.16.2.2 with 32 bytes of data:
Reply from 172.16.2.2: bytes=32 time=2ms TTL=126
Reply from 172.16.2.2: bytes=32 time=2ms TTL=126
Reply from 172.16.2.2: bytes=32 time=1ms TTL=126
Reply from 172.16.2.2: bytes=32 time=7ms TTL=126
//ping 的 Reply 包经过了路由器 R2 和 R1 到达源主机，  TTL=128-2=126
Ping statistics for 172.16.2.2:
    Packets: Sent = 4, Received = 4, Lost = 0 (0% loss),
Approximate round trip times in milli-seconds:
    Minimum = 1ms, Maximum = 7ms, Average = 3ms
```

以上测试结果表明 A 公司和 B 公司实现了互通，我们继续测试 PC1 和 PC3 的连通性。

```
PC>ping 192.168.2.2
Pinging 192.168.2.2 with 32 bytes of data:
Reply from 192.168.2.2: bytes=32 time=3ms TTL=125
Reply from 192.168.2.2: bytes=32 time=3ms TTL=125
Reply from 192.168.2.2: bytes=32 time=2ms TTL=125
Reply from 192.168.2.2: bytes=32 time=2ms TTL=125
//ping 的 Reply 包经过了路由器 R3、R2 和 R1 到达源主机，  TTL=128-3=125
Ping statistics for 192.168.2.2:
    Packets: Sent = 4, Received = 4, Lost = 0 (0% loss),
Approximate round trip times in milli-seconds:
    Minimum = 2ms, Maximum = 3ms, Average = 2ms
```

测试结果表明 A 公司和 C 公司也实现了互连互通。

我们不仅可以在终端 PC 上测试网络的连通性，也可在中间网络设备上进行测试。

例如，在路由器 R2 上测试与 A 公司和 C 公司的连通性。

R2>**ping 172.16.1.1**

Type escape sequence to abort.

Sending 5, 100-byte ICMP Echos to 172.16.1.1, timeout is 2 seconds:

!!!!!

Success rate is 100 percent (5/5), round-trip min/avg/max = 1/4/8 ms

"!!!!!"表示源主机已经成功收到 5 个由目的主机发来的 ping 的回包。

R2>**ping 192.168.2.1**

Type escape sequence to abort.

Sending 5, 100-byte ICMP Echos to 192.168.2.1, timeout is 2 seconds:

!!!!!

Success rate is 100 percent (5/5), round-trip min/avg/max = 1/3/6 ms

以上输出表明，R2 代表的 B 公司实现了与 A 公司和 C 公司的互通。

在进行网络的连通性测试时，我们还可以借助 Packet Tracer 右面工具栏的 ping 工具进行直观测试，如图 5-2 所示。

Fire	Last Status	Source	Destination	Type	Color	Time(sec)	Periodic	Num	Edit	Delete
●	Successful	PC1	PC2	ICMP		0.000	N	0	(edit)	
●	Successful	PC1	PC3	ICMP		0.000	N	1	(edit)	
●	Successful	PC2	PC3	ICMP		0.000	N	2	(edit)	

图 5-2　网络连通测试结果

问题 1：以路由器 R3 的路由表为例，请解释路由条目"S　172.16.1.0 [1/0] via 192.168.1.2"各项的含义。

问题 2：路由器如何使用路由表中的静态路由转发数据包？若路由器 R3 收到一个目的 IP 地址为 172.16.1.1 的数据包，它是怎样查询路由表做转发决定的？数据包一旦匹配了带下一跳 IP 地址的静态路由，能直接被转发吗？请说出理由。

问题 3：将静态路由写进路由表，需要什么条件？

同学们讨论异常激烈，踊跃回答老师提出的问题，本堂课 ZHP、ZHN 和 ZHT 三位同学表现极为突出，他们对三个问题的回答如下：

ZHP 同学回答问题 1：

- S——代表 Static，表示路由来源，即静态获取路由信息；
- 172.16.1.0——表示目的网络地址；
- [1/0]——1 和 0 表分别表示静态路由的管理距离和度量值；
- via——代表的意思是经由；
- 192.168.1.2——表示下一跳 IP 地址，即邻居路由器入口的 IP 地址。

ZHN 同学回答问题 2：

首先，R3 通过查询路由表定位到路由条目"S　**172.16.1.0 [1/0] via 192.168.1.2**"上，从而确定下一跳 IP 地址为 192.168.1.2，这是第一步查询；然后 R3 需要进一步完成二次查询找到对应送出接口，本次需要查找去目的地址 192.168.1.2 的路由条目。本例中，192.168.1.2 的地址恰巧与直连网络 192.168.1.0124 的路由条目（**C 192.168.1.0124 is directly connected, Serial0/0/1**）相匹配，即通过二次查询找到出接口 Serial0/0/1，这就是所谓的**递归查询**。由此我们可以推断出路由器最终的转发是基于接口的，所以，数据包一旦匹配了带下一跳 IP 地址的静态路由，是不能被直接转发的，势必会引发递归查询，即通过下一跳 IP 地址解析成出接口。

ZHT 同学回答问题 3：

在保证命令正确配置的前提下，确保下一跳在路由表中能解析成功，即下一跳 IP 地址在路由表中能找到对应的直连网络。

老师对本节课做了补充，利用带下一跳的静态路由转发数据包，必经递归查询，降低查询效率。如何提高查询效率，我们将会通过 5.2.2 节配置出接口静态路由来优化网络。

5.2.2　场景二：配置出接口静态路由

> 🔎**场景二**：CFL 老师要同学们把"场景一"提供的拓扑文件备份一份，按照图 5-3 所示拓扑，做相应修改：每个局域网增加一台 PC，R1 与 R2 间的串行链路换成 LAN 链路，其他不变。接下来完成老师布置的第二道作业，具体要求如下所述。

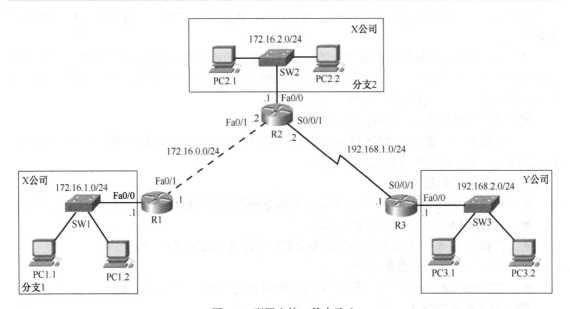

图 5-3　配置出接口静态路由

X 公司的两个分支位于一个工业园区内,其合作伙伴 Y 公司在另一城市,目前因业务关系要求实现 X 与 Y 公司网络间的互连。其中 X 公司的两个分支用局域网口互连,X 公司与 Y 公司通过 WAN 口互连。X 公司分支路由器分别为 R1 和 R2,Y 公司的路由器为 R3。考虑到网络结构相对简单,现要求采用送出接口完成静态路由的配置,并依据场景一来整理实验步骤。

下面,让我们一起来分享 ZHB 同学整理的实验步骤,具体如下所述。

步骤一:路由器的基本配置

配置路由器接口的 IP 地址,确保直连路由均能显示在路由表中。

```
R1>show ip route connected
  C    172.16.0.0/24     is directly connected, FastEthernet0/1
  C    172.16.1.0/24     is directly connected, FastEthernet0/0

R2>show ip route connected
  C    172.16.0.0/24     is directly connected, FastEthernet0/1
  C    172.16.2.0/24     is directly connected, FastEthernet0/0
  C    192.168.1.0/24    is directly connected, Serial0/0/1

R3>show ip route connected
  C    192.168.1.0/24    is directly connected, Serial0/0/1
  C    192.168.2.0/24    is directly connected, FastEthernet0/0
```

步骤二:配置出接口静态路由

```
R1(config)#ip route 172.16.2.0 255.255.255.0 f0/1
R1(config)#ip route 192.168.2.0 255.255.255.0 f0/1

R2(config)#ip route 172.16.1.0 255.255.255.0 f0/1
R2(config)#ip route 192.168.2.0 255.255.255.0 s0/0/1

R3(config)#ip route 172.16.1.0 255.255.255.0 s0/0/1
R3(config)#ip route 172.16.2.0 255.255.255.0 s0/0/1
```

步骤三:检查静态路由

```
R1>show ip route
Codes: C - connected, S - static, I - IGRP, R - RIP, M - mobile, B – BGP

Gateway of last resort is not set
```

```
        172.16.0.0/24 is subnetted, 3 subnets
C       172.16.0.0 is directly connected, FastEthernet0/1
C       172.16.1.0 is directly connected, FastEthernet0/0
S       172.16.2.0 is directly connected, FastEthernet0/1
S    192.168.2.0/24 is directly connected, FastEthernet0/1
```

R1 的路由表添加了两条静态路由。

```
R2>show ip route
Codes: C - connected, S - static, I - IGRP, R - RIP, M - mobile, B - BGP

Gateway of last resort is not set

        172.16.0.0/24 is subnetted, 3 subnets
C       172.16.0.0 is directly connected, FastEthernet0/1
S       172.16.1.0 is directly connected, FastEthernet0/1
C       172.16.2.0 is directly connected, FastEthernet0/0
C    192.168.1.0/24 is directly connected, Serial0/0/1
S    192.168.2.0/24 is directly connected, Serial0/0/1
```

R2 的路由表也成功地添加了两条静态路由。

```
R3>show ip route
Codes: C - connected, S - static, I - IGRP, R - RIP, M - mobile, B - BGP

Gateway of last resort is not set

        172.16.0.0/24 is subnetted, 2 subnets
S       172.16.1.0 is directly connected, Serial0/0/1
S       172.16.2.0 is directly connected, Serial0/0/1
C    192.168.1.0/24 is directly connected, Serial0/0/1
C    192.168.2.0/24 is directly connected, FastEthernet0/0
```

以上输出表明路由器 R3 的路由条目，包括 2 条直连路由和 2 条静态路由，配置正确。请注意场景二中出接口的静态路由显示的是直连（**Directly Connected**），而场景一中下一跳地址的静态路由显示的是"**[1/0]**"，这一点两者不同。

步骤四：网络连通性测试

全网实现了互通，测试结果如图 5-4 所示。

```
Fire  Last Status  Source  Destination  Type  Color  Time(sec)  Periodic  Num  Edit    Delete
      Successful   PC1.1   PC1.2        ICMP         0.000      N         0    (edit)
      Successful   PC2.1   PC2.2        ICMP         0.000      N         1    (edit)
      Successful   PC3.1   PC3.2        ICMP         0.000      N         2    (edit)
      Successful   PC1.1   PC2.1        ICMP         0.000      N         3    (edit)
      Successful   PC1.2   PC2.2        ICMP         0.000      N         4    (edit)
      Successful   PC1.1   PC3.1        ICMP         0.000      N         5    (edit)
      Successful   PC1.1   PC3.2        ICMP         0.000      N         6    (edit)
      Successful   PC2.1   PC3.1        ICMP         0.000      N         7    (edit)
      Successful   PC2.2   PC3.2        ICMP         0.000      N         8    (edit)
```

图 5-4　网络连通测试结果

5.2.3　场景三：下一跳与出接口对 ARP 表的影响

> **场景三**：基于场景二"配置出接口静态路由"，老师要求大家分析出接口静态路由存在的问题。有同学说场景二分明是对场景一"配置下一跳静态路由"的优化，还能有什么问题；有同学反映他们在做连通性测试的时候，发现跨网段通信延迟较长，初始 ping 操作都出现了丢包情况，这算什么优化，反而让网络延迟更大；还有同学说出接口静态路由变得和直连路由一样了，连管理距离及度量[1/0]都没有了。CFL 老师很高兴同学们能反映出这样的问题，看来大家真的是用心在学习。

接下来，老师让大家和她一起在路由器 R1 上按照表 5-3 做 ping 测试，并一起验证表中预测的结果。

表 5-3　场景二中路由器 R1 的 ping 测试

顺序编号	目的主机名或接口名	目的 IP 地址	Ping 结果预测	加入 ARP 表预测
1	PC1.1	172.16.1.2	√	√
2	PC1.2	172.16.1.3	√	√
3	R2 的 F0/1	172.16.0.2	√	√
4	R2 的 F0/0	172.16.2.1	√	√
5	PC2.1	172.16.2.2	√	√
6	PC2.2	172.16.2.3	√	√
7	R2 的 S0/0/1	192.168.1.2	×	×
8	R3 的 S0/0/1	192.168.1.1	×	×
9	R3 的 Fa0/0	192.168.2.1	×	√
10	PC3.1	192.168.2.2	×	√
11	PC3.2	192.168.2.3	×	√

经测试后，同学们一致反映 ping 的结果和表 5-3 预测的完全一致。老师又让同学们在路由

器 R1 上发布命令 show ip arp 或 show arp 来查看结果是否和表 5-3 一致。结果显示了 11 条记录，如下所示。

R1#show ip arp

Protocol	Address	Age (min)	Hardware Addr	Type	Interface
Internet	172.16.0.1	-	0060.2FA8.375C	ARPA	FastEthernet0/1
Internet	172.16.0.2	4	0090.0C82.4089	ARPA	FastEthernet0/1
Internet	172.16.1.1	-	0060.3EA6.63B4	ARPA	FastEthernet0/0
Internet	172.16.1.2	4	0005.5EB2.D143	ARPA	FastEthernet0/0
Internet	172.16.1.3	4	000C.CF03.1497	ARPA	FastEthernet0/0
Internet	172.16.2.1	4	0090.0C82.4089	ARPA	FastEthernet0/1
Internet	172.16.2.2	4	0090.0C82.4089	ARPA	FastEthernet0/1
Internet	172.16.2.3	4	0090.0C82.4089	ARPA	FastEthernet0/1
Internet	192.168.2.1	3	0090.0C82.4089	ARPA	FastEthernet0/1
Internet	192.168.2.2	3	0090.0C82.4089	ARPA	FastEthernet0/1
Internet	192.168.2.3	3	0090.0C82.4089	ARPA	FastEthernet0/1

以上输出比表 5-3 预测的 9 条多了两条，HY 同学反映预测的结果是正确的，多的两条是路由器 R1 自身接口 IP 地址与其 MAC 地址的映射，无须 ping 就能自动学习到。老师对 HY 同学的分析非常满意。

接下来，老师让同学们删除路由器 R1 上的两条配置了出接口的静态路由，采用下一跳重新配置，确保静态路由安装在路由表中，并清空 R1 的 ARP 缓存，再按照表 5-3 在 R1 上重新测试一遍，最后显示 ARP 表。

R1#show arp

Protocol	Address	Age (min)	Hardware Addr	Type	Interface
Internet	172.16.0.1	-	0060.2FA8.375C	ARPA	FastEthernet0/1
Internet	172.16.0.2	1	0090.0C82.4089	ARPA	FastEthernet0/1
Internet	172.16.1.1	-	0060.3EA6.63B4	ARPA	FastEthernet0/0
Internet	172.16.1.2	2	000C.CF03.1497	ARPA	FastEthernet0/0
Internet	172.16.1.3	1	0005.5EB2.D143	ARPA	FastEthernet0/0

这次 ARP 表的显示结果让全班同学惊呆了。老师解释到比较前后两张 ARP 表，我们发现一个问题：R1 前面的 ARP 表，MAC 地址 0090.0C82.4089 对应了 7 个 IP 地址，产生了 7 条映射，是多对一映射（即多个 IP 对应一个 MAC 地址）；而 R1 后面的 ARP 表，MAC 地址 0090.0C82.4089 只对应一个 IP 地址 172.16.0.2（下一跳），即一对一映射。前面 7 条映射，只有一条是真映射，其余是"伪映射"。MAC 地址为 0090.0C82.4089 的 F0/1 接口在跨网段通信中起到了 ARP 代理作用（5.2.4 节详细介绍）。前者的 R1 每次跨网段通信都需要借助 ARP 代理回复 MAC 地址，而后者只需要下一跳 MAC 地址，后者一条映射解决所有跨网段通信问题。同学反应的丢包延迟现象，原因是与不同网段主机通信都会引发 R1 的 ARP 请求，丢包现象是

ARP 请求超时导致的。

以太网通信采用二层 MAC 封装，MAC 地址是 LAN 中的地址；而串行口的 WAN 通信是点到点通信，采用协议指定地址（HDLC 或 PPP 协议，在后续 WAN 中介绍），因此在 WAN 中，用出接口配置静态路由，可以避免递归解析，提高转发效率；但在 LAN 中使用出接口配置静态路由，虽然避免递归解析，但给出口封装带来新的问题，所以 LAN 中最好不用出接口配置静态路由，应使用下一跳，也可以将送出接口和下一跳关联配置，但 Packet Tracer 目前尚不支持。

另外，有的厂商的路由产品，在 LAN 中，配置出接口静态路由会被认为是错误配置。Cisco 路由产品可以关闭或开启 ARP 代理，因此支持出接口静态路由，ARP 代理功能默认是开启的。

至于有的同学反映出接口静态路由和直连路由一样，没有管理距离和度量，CFL 老师给出了如下解释。

出接口静态路由管理距离和度量验证如下：

```
R3#show ip route 172.16.1.0
Routing entry for 172.16.1.0/24
Known via "static", distance 1, metric 0 (connected)    //管理距离是 1，度量是 0
  Routing Descriptor Blocks:
  * directly connected, via Serial0/0/1                 //出接口静态路由
      Route metric is 0, traffic share count is 1
```

下一跳静态路由管理距离和度量验证如下：

```
R3#show ip route 172.16.1.0
Routing entry for 172.16.1.0/24
Known via "static", distance 1, metric 0              //管理距离是 1，度量是 0
  Routing Descriptor Blocks:
  * 192.168.1.2                                        //下一跳静态路由
      Route metric is 0, traffic share count is 1
```

以上输出表明，出接口静态路由管理距离依然是 1，度量是 0，不过是隐形存在的而已。

CFL 老师刚解释完上述问题，XL 同学已经迫不及待地举起手来，她让老师解释一下表 5-3 阴影部分，XL 提出如下两个问题：

问题 1：为什么全网主机间均能互相 ping 通，而路由器 R1 却只能 ping 通前 6 个地址，ping 不通后 5 个呢？

问题 2：既然后 5 个地址都 ping 不通，那为什么 192.168.2.1、192.168.2.2 以及 192.168.2.3 能加入 ARP 缓存表，而 192.168.1.1 和 192.168.1.2 却没有加入呢？这如何解释？

全班同学顿时为 XL 同学能提出这么尖锐的问题而感到震惊，看得出来 CFL 老师对 XL 的问题非常赞赏，她问全班同学谁能为 XL 解答，班上鸦雀无声，老师高兴地对大家说这堂课 XL 同学可以拿到两分的加分。看来这个问题只能抛给老师来解答了。

CFL 老师对问题 1 的解答：全网 PC 均可以相互 ping 通，但路由器 R1 却 ping 不通部分地

址,为什么?路由器都有 PC 所在网段的路由,这没有问题吧!因此这些网段的 PC 间可以互相 ping 通。用路由器 R1 ping 表里的 11 个地址,请问路由器 R1 的源地址是谁? R1 有两个地址 172.16.1.1 和 172.16.0.1,到底哪个作为源地址呢?这个必须搞清楚,答案是出口地址,即 172.16.0.1,那该地址所在的网段 172.16.0.0/24,其他两台路由器有该路由吗?很显然,R2 有,且是直连路由,但 R3 没有,所以与 R3 相关联的 4 个地址 192.168.1.1、192.168.2.1、192.168.2.2 以及 192.168.2.3 是不能与 R1 通信的;R1 出口与 R2 直连,那为什么 R1 却 ping 不通 R2 的接口地址 192.168.1.2 呢?我们知道通信是双向的,R2 有返回 R1 的路由,但 R1 有去 R2 接口 192.168.1.2 所在网段的路由吗?显然没有,所以 R1 ping 不通 192.168.1.2 是正常的。这"5 个不通",大家都明白了吗?同学们茅塞顿开,有的感叹,静态路由问题好多呀!接下来,老师开始解释那"6 个通",这 6 个地址,有 3 个属于 R1 直连网络,当然可以 ping 通,另三个目的地址属于 172.16.2.0/24 网段,R1 有其静态路由,而源地址 172.16.0.1 属于 R2 的直连网络,互相可达,当然可以 ping 通。至此,XL"6 个通,5 个不通"的问题已解答完毕。

CFL 老师对问题 2 的解答:既然 5 个不通,那为什么 3 个地址能写进 ARP 表?2 个却写不进呢?显然 3 个地址的 ARP 请求都得到了回复,所以才会写进 ARP 缓存;2 个地址的 ARP 没有回复,为什么?既然都是跨网段通信,对 ARP 的请求处理结果是一样的,换句话来说,只要发出 ARP 请求,都会得到回复。那为什么离 R1 远的网段 192.168.2.0/24 的 3 个 ARP 请求回复了,而离 R1 近的网段 192.168.1.0/24 的请求却得不到回复呢?这似乎很奇怪,唯一的一种可能性:R1 根本没有发出对 192.168.1.0/24 网段的 2 个地址的 ARP 请求,没有回复那就顺理成章了。就剩最后一步了,为什么 192.168.2.0/24 能发出 3 个 ARP 请求,而 192.168.1.0/24 却发不出那 2 个 ARP 请求呢?已经没有任何悬念了,最后的落脚点在路由上!因为 R1 有去 192.168.2.0/24 的静态路由,却没有去 192.168.1.0/24 的路由,因此去 192.168.1.0/24 两个地址的 ping 包被路由器 R1 丢弃,根本没有机会发出 ARP 请求,解答完毕。

CFL 老师再一次当全班同学的面表扬了 XL 同学,同时也侧面点拨了其他同学,老师的表格为什么设置阴影,其目的就在于想引起同学们的注意,达到抛砖引玉的效果,然后带领大家一起深入学习,是 XL 同学引爆了这个问题。

5.2.4 认识 ARP 代理

上节课,我们已经涉猎了 ARP(地址解析协议)协议。那 ARP 的具体功能是什么?就是在局域网中实现逻辑 IP 地址到物理 MAC 地址的映射,是局域网的专用协议,依据 IP 地址查 MAC 地址的寻址协议。当源主机在网络层将 IP 数据包封装好向下传送给数据链路层时,需封装二层 MAC 地址,而 MAC 地址包括源 MAC 地址和目标 MAC 地址。如何获得目标 MAC 地址呢?这就是 ARP 协议要解决的问题。

那 ARP 代理又是什么呢?ARP 代理又名代理 ARP(Proxy ARP),它应用于局域网,让路由器"假装"目标主机来应答 ARP 请求。ARP 代理就是让一台主机代替另一台主机做 ARP 应答。下面我们分三种情况讨论 ARP 的请求与应答：

情况一：同一网段内 ARP 请求与应答，如图 5-5 所示。

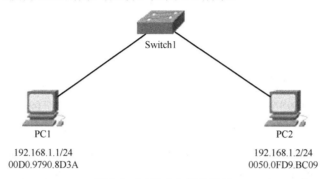

图 5-5　LAN 内主机的通信

同一网段主机 PC1 和 PC2 间的通信，如 PC1 ping PC2，PC1 会发送 ARP 广播请求 PC2 的 MAC 地址，因为是同一广播域，PC2 会将自己的 MAC 地址以单播形式回复给 PC1，PC1 收到应答也会将 PC2 的 IP 地址与其 MAC 地址的映射添加到 PC1 的 ARP 缓存，如下所示：

```
C:\>arp -a

  Internet Address      Physical Address      Type
  192.168.1.2           0050.0fd9.bc09        dynamic
```

情况二：跨网段 ARP 请求与应答（无须 ARP 代理），如图 5-6 所示。

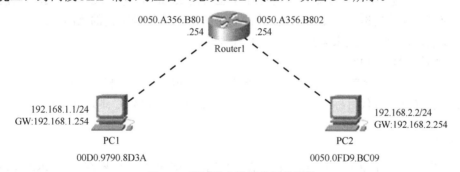

图 5-6　跨网段主机间的正常通信

PC1 跨网段 ping PC2，如果两台主机均配置了正确的网关，ARP 不可能跨网段请求主机 PC2 的 MAC 地址。PC1 通过比较目的 IP 地址发现这是属于不同网段的通信，要借助于网关 192.168.1.254，所以 PC1 请求的是网关 192.168.1.254 的 MAC 地址，并将其 ARP 回复填入到自己的 ARP 缓存表，如下所示：

```
C:\>arp -a

  Internet Address      Physical Address      Type
  192.168.1.254         0050.a356.b801        dynamic
```

情况三：跨网段 ARP 请求与应答（需 ARP 代理），如图 5-7 所示。

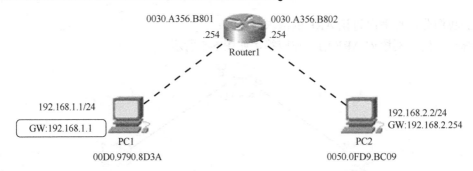

图 5-7 跨网段主机间的 ARP 代理通信

PC1 和 PC2 依然跨网段通信，但此时 PC1 的网关设置成自己的 IP 地址 192.168.1.1（可用真机完成实验，目前 Packet Tracer 尚不支持 PC1 的这种设置），PC1 的这种设置类似于路由器配置了出接口的静态路由，都指向了自己的出接口。此时 Router1 连接 PC1 的物理接口默认使用了 **IP proxy-arp** 命令开启 ARP 代理功能。PC1 ping PC2 依然可以成功，但 PC1 的 ARP 表会有如下记录：

C:\>arp –a		
Internet Address	Physical Address	Type
192.168.2.2	0030.a356.b801	dynamic

此时我们会发现 PC1 的 ARP 缓存表里记录的是 PC2 的 IP 地址，MAC 地址却是 Router1 物理接口的 MAC 地址，这种映射是"假映射"，也就是 Router1 代替 PC2 对无法跨网段的 ARP 请求做了应答，应答的地址就是自己接口的 MAC 地址，这就是所谓的 ARP 代理。

下面我们通过问答形式来总结一下 ARP 代理。

① 什么是 ARP 代理？

答：当路由器收到 ARP 请求时，若请求的目的 IP 地址与源主机不在同一网段，路由器会扮演 ARP 代理的角色，代目标主机回答,告诉源主机它所要的 MAC 地址就是自己接口的 MAC 地址。

② 为什么要有 ARP 代理？

答：路由器的一个重要功能是隔绝广播，即把广播限制在一个网络内。ARP 请求的对象若在同一个局域网内，目标主机就能应答。但如果不在同一局域网呢？为解决此问题，路由器就提供 ARP 代理服务。

③ 除了 ARP 代理，还有什么方法解决跨局域网的 ARP 请求？

答：主机配置默认网关。这是最常用的方法，请求默认网关的 MAC 地址，实现跨网段通信。推荐使用本方法。

④ ARP 代理的主要缺点是什么？

答：能增加某一网段 ARP 的流量，主机需更大的 ARP 缓存，存在安全问题，如 ARP 欺骗（ARP Spoofing）等。

⑤ ARP 代理是否建议使用？

答：不建议使用 ARP 代理，因为会对网络安全造成一定的威胁，有的路由器以太网口默认开启 ARP 代理，建议使用 no ip proxy-arp 关闭 ARP 代理功能。

为加深对 ARP 代理的理解，CFL 老师拿出场景二 X 公司的企业网分支，如图 5-8 所示来做如下解析：

我们知道在以太网通信中要封装以太帧，封装二层 MAC 地址。例如，在路由器 R1 上 ping 172.16.2.2，若采用带下一跳地址配置静态路由 172.16.2.0/24，路由器依据下一跳 172.16.0.2，通过 ARP 请求其 MAC 地址，作为二层封装地址。但场景二配置静态路由 172.16.2.0/24 采用的是出接口 Fa0/1，所以只能请求目的主机 172.16.2.2（与请求端 172.16.0.1 不在一个网段）的 MAC 地址。R1 发出 ARP 广播请求，R2 收到消息并发现请求的是其他网段主机的 MAC 地址，因无法跨网段转发，考虑到路由器 R2 的接口 Fa0/1 已开启 ARP 代理功能，故将 Fa0/1 的 MAC 地址回复给 R1，使得 R1 成功封装成帧。假如关闭路由器 R2 Fa0/1 口的 ARP 代理功能，再清空路由器 R1 缓存的记录，打开 debug 观察 ping 的结果，封装失败的消息将会显示出来，具体操作如下。

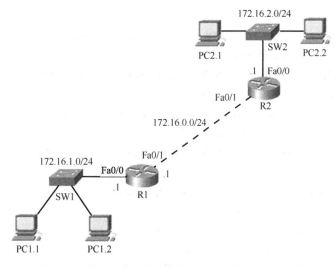

图 5-8 场景二 X 公司的企业网

```
R2(config)#interface fastEthernet 0/1
R2(config-if)#no ip proxy-arp        //关闭 ARP 代理
```

```
R1(config)#interface f0/0
R1(config-if)#shutdown               //清空 Fa0/0 口的 ARP 缓存
R1(config-if)#no shutdown
R1(config-if)#interface f0/1         //清空 Fa0/1 口的 ARP 缓存
R1(config-if)#shutdown
```

R1(config-if)#**no shutdown**

清除 ARP 表的命令 clear arp 或 clear arp-cache 在 Packet Tracer 下没有效果，因此我们采用 down 掉接口再重新激活的办法实现。

R1#**debug ip packet**　　　　　　//打开 debug 功能

Packet debugging is on

此时，R2 不再为路由器 R1 做 ARP 代理，我们来做如下测试：

R1#**ping 172.16.2.2**

Type escape sequence to abort.

Sending 5, 100-byte ICMP Echos to 172.16.2.2, timeout is 2 seconds:

IP: tableid=0, s=172.16.0.1 (local), d=172.16.2.2 (FastEthernet0/1), routed via RIB

IP: s=172.16.0.1 (local), d=172.16.2.2 (FastEthernet0/1), len 128, sending

IP: s=172.16.0.1 (local), d=172.16.2.2 (FastEthernet0/1), len 128, encapsulation failed

　　　　　　//第 1 个 ping 包封装失败

IP: tableid=0, s=172.16.0.1 (local), d=172.16.2.2 (FastEthernet0/1), routed via RIB

IP: s=172.16.0.1 (local), d=172.16.2.2 (FastEthernet0/1), len 128, sending

IP: s=172.16.0.1 (local), d=172.16.2.2 (FastEthernet0/1), len 128, encapsulation failed

　　　　　　//第 2 个 ping 包封装失败

IP: tableid=0, s=172.16.0.1 (local), d=172.16.2.2 (FastEthernet0/1), routed via RIB

IP: s=172.16.0.1 (local), d=172.16.2.2 (FastEthernet0/1), len 128, sending

IP: s=172.16.0.1 (local), d=172.16.2.2 (FastEthernet0/1), len 128, encapsulation failed

　　　　　　//第 3 个 ping 包封装失败

IP: tableid=0, s=172.16.0.1 (local), d=172.16.2.2 (FastEthernet0/1), routed via RIB

IP: s=172.16.0.1 (local), d=172.16.2.2 (FastEthernet0/1), len 128, sending

IP: s=172.16.0.1 (local), d=172.16.2.2 (FastEthernet0/1), len 128, encapsulation failed

　　　　　　//第 4 个 ping 包封装失败

IP: tableid=0, s=172.16.0.1 (local), d=172.16.2.2 (FastEthernet0/1), routed via RIB

IP: s=172.16.0.1 (local), d=172.16.2.2 (FastEthernet0/1), len 128, sending

IP: s=172.16.0.1 (local), d=172.16.2.2 (FastEthernet0/1), len 128, encapsulation failed

　　　　　　//第 5 个 ping 包封装失败

Success rate is 0 percent (0/5)

//源地址为 172.16.0.1，去往目的地址 172.16.2.2 的 ping 包在转发时数据封装失败 "encapsulation failed"

由此可知，只有 R2 的 ARP 代理功能开启，R1 采用出接口静态路由可以实现全网互通；R2 关闭 ARP 代理，出接口静态路由会影响网络通信。

鉴于此，我们对配置标准静态路由总结如下：

- 在 WAN 中（串口连接）配置静态路由：最好采用出接口配置，提高转发效率；
- 在 LAN 中（以太口连接）配置静态路由：建议采用下一跳，避免 ARP 代理通信。

5.3 配置特殊静态路由

5.3.1 场景四：配置默认路由

> **场景四**：今天的网络课 CFL 老师将带领大家一起学习第一种特殊静态路由"默认路由"的配置方法，同学们提前做了预习。讲课之前老师让大家说出预习过程中遇到的问题。LXF 同学问老师什么是末节路由器，并让老师以场景一中的三台路由器为例进行说明；LYH 同学问是否可以用默认路由代替场景一或场景二的标准静态路由；ZHB 同学说两台路由器各有一个局域网，通过串口实现互连，两台路由器是否互为末节路由器，都支持默认路由；YX 同学问什么是边缘路由器；LX 同学问默认路由用的是否广泛；还有两位同学问老师默认路由是否就是默认网关；默认路由是否有静态默认路由与动态默认路由之分，怎么区分，等等。同学们竟然提出了这么多问题，这是 CFL 老师未曾预料到的，她让 XL 同学对这些问题做了汇总，并给出了如表 5-4 的解答。

表 5-4　默认路由问题汇总

问题提出	具 体 问 题	问 题 简 答
LXF	末节路由器	只有一个上游邻居路由器的路由器；如场景一的 R1 和 R3
LYH	场景一默认路由	可以在 R1 和 R3 上配置默认路由，但 R2 不可以
ZHB	两台路由器是否互为末节路由器	不是，不可在两台路由器上配置默认路由，会引起环路
YX	边缘路由器	连接 ISP 运营商的路由器，又被称为企业边界路由器
LX	默认路由的应用是否广泛	非常广泛，可以简化路由表
WXN	默认路由是否是默认网关	默认路由就是默认网关，或被称为最后求助网关（Gateway of Last Resort）
WN	默认路由的分类	分静态默认路由和动态默认路由两种；静态默认路由由用户手工添加；动态默认路由由路由器通过路由协议动态学习

　　默认路由通常被称为"全零路由"，网络地址是 0.0.0.0，子网掩码也是 0.0.0.0，路由表的存在形式为 0.0.0.0/0。按照路由表最长匹配原则，默认路由是无须匹配的路由，因此在路由表中查询的优先级最低。默认路由的作用是让路由器转发目的地址不在路由表中的数据包，它可以匹配所有数据包，其应用场合主要在边缘路由器或末节路由器上。

　　如图 5-9（a）所示的网络拓扑，老师已经给出了具体地址规划表和详细的实验步骤，请根据图 5-9（a）的具体实施步骤完成 5-9（b）的配置，注意两种拓扑的区别。

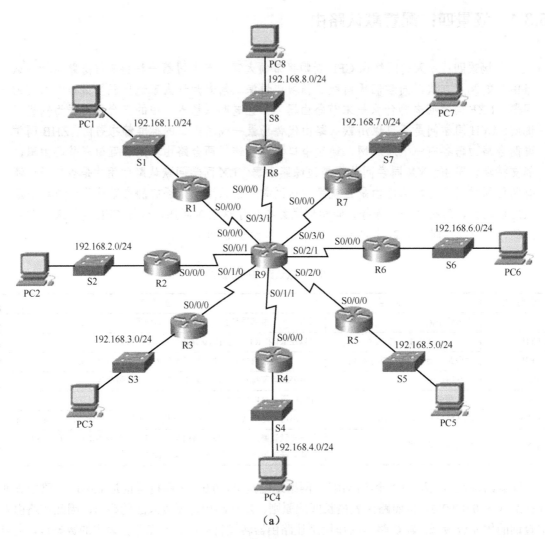

图 5-9 配置静态默认路由

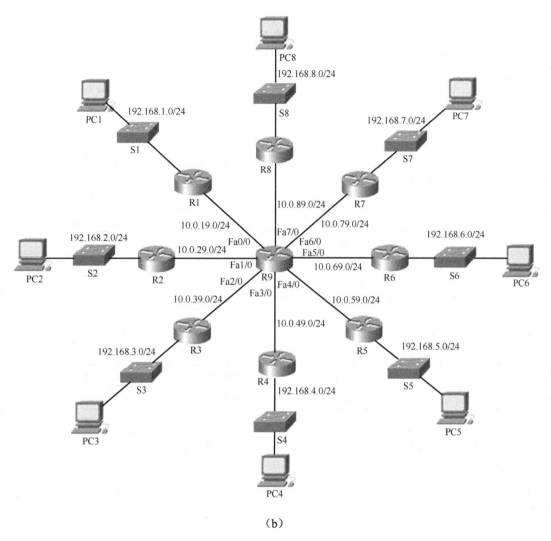

(b)

图 5-9 配置静态默认路由

图 5-9（a）网络拓扑的具体配置步骤如下。

步骤一：路由器基本配置

按照表 5-5 所示的地址规划表，为路由器 R1~R9 配置物理接口，确保各自的直连路由都出现在路由表中。

表 5-5　地址规划表

设 备 名 称	接　口	IP 地址	子 网 掩 码	描　述
R1	S0/0/0	10.0.19.1	255.255.255.0	Link to R9 S0/0/0
	Fa0/0	192.168.1.1	255.255.255.0	Link to S1
R2	S0/0/0	10.0.29.2	255.255.255.0	Link to R9 S0/0/1
	Fa0/0	192.168.2.1	255.255.255.0	Link to S2
R3	S0/0/0	10.0.39.3	255.255.255.0	Link to R9 S0/1/0
	Fa0/0	192.168.3.1	255.255.255.0	Link to S3
R4	S0/0/0	10.0.49.4	255.255.255.0	Link to R9 S0/1/1
	Fa0/0	192.168.4.1	255.255.255.0	Link to S4
R5	S0/0/0	10.0.59.5	255.255.255.0	Link to R9 S0/2/0
	Fa0/0	192.168.5.1	255.255.255.0	Link to S5
R6	S0/0/0	10.0.69.6	255.255.255.0	Link to R9 S0/2/1
	Fa0/0	192.168.6.1	255.255.255.0	Link to S6
R7	S0/0/0	10.0.79.7	255.255.255.0	Link to R9 S0/3/0
	Fa0/0	192.168.7.1	255.255.255.0	Link to S7
R8	S0/0/0	10.0.89.8	255.255.255.0	Link to R9 S0/3/1
	Fa0/0	192.168.8.1	255.255.255.0	Link to S8
R9	S0/0/0	10.0.19.9	255.255.255.0	Link to R1 S0/0/0
	S0/0/1	10.0.29.9	255.255.255.0	Link to R2 S0/0/0
	S0/1/0	10.0.39.9	255.255.255.0	Link to R3 S0/0/0
	S0/1/1	10.0.49.9	255.255.255.0	Link to R4 S0/0/0
	S0/2/0	10.0.59.9	255.255.255.0	Link to R5 S0/0/0
	S0/2/1	10.0.69.9	255.255.255.0	Link to R6 S0/0/0
	S0/3/0	10.0.79.9	255.255.255.0	Link to R7 S0/0/0
	S0/3/1	10.0.89.9	255.255.255.0	Link to R8 S0/0/0

R9#show ip interface brief | include　　　　　　　up　　　　　　　　　up
Serial0/0/0　　　　10.0.19.9　　　　YES manual up　　　　　　　up
Serial0/0/1　　　　10.0.29.9　　　　YES manual up　　　　　　　up
Serial0/1/0　　　　10.0.39.9　　　　YES manual up　　　　　　　up
Serial0/1/1　　　　10.0.49.9　　　　YES manual up　　　　　　　up
Serial0/2/0　　　　10.0.59.9　　　　YES manual up　　　　　　　up

Serial0/2/1	10.0.69.9	YES manual up	up
Serial0/3/0	10.0.79.9	YES manual up	up
Serial0/3/1	10.0.89.9	YES manual up	up

步骤二：配置核心路由器到各分支路由器的静态路由

```
R9(config)#ip route 192.168.1.0 255.255.255.0 Serial0/0/0
R9(config)#ip route 192.168.2.0 255.255.255.0 Serial0/0/1
R9(config)#ip route 192.168.3.0 255.255.255.0 Serial0/1/0
R9(config)#ip route 192.168.4.0 255.255.255.0 Serial0/1/1
R9(config)#ip route 192.168.5.0 255.255.255.0 Serial0/2/0
R9(config)#ip route 192.168.6.0 255.255.255.0 Serial0/2/1
R9(config)#ip route 192.168.7.0 255.255.255.0 Serial0/3/0
R9(config)#ip route 192.168.8.0 255.255.255.0 Serial0/3/1
```

查看核心路由器 R9 的路由表

```
R9#show ip route
Codes: C - connected, S - static, I - IGRP, R - RIP, M - mobile, B - BGP


     10.0.0.0/24 is subnetted, 8 subnets
C       10.0.19.0 is directly connected, Serial0/0/0
C       10.0.29.0 is directly connected, Serial0/0/1
C       10.0.39.0 is directly connected, Serial0/1/0
C       10.0.49.0 is directly connected, Serial0/1/1
C       10.0.59.0 is directly connected, Serial0/2/0
C       10.0.69.0 is directly connected, Serial0/2/1
C       10.0.79.0 is directly connected, Serial0/3/0
C       10.0.89.0 is directly connected, Serial0/3/1
S    192.168.1.0/24 is directly connected, Serial0/0/0
S    192.168.2.0/24 is directly connected, Serial0/0/1
S    192.168.3.0/24 is directly connected, Serial0/1/0
S    192.168.4.0/24 is directly connected, Serial0/1/1
S    192.168.5.0/24 is directly connected, Serial0/2/0
S    192.168.6.0/24 is directly connected, Serial0/2/1
S    192.168.7.0/24 is directly connected, Serial0/3/0
S    192.168.8.0/24 is directly connected, Serial0/3/1
```

步骤三：配置分支路由器到核心路由器的静态默认路由

```
R1(config)#ip route 0.0.0.0 0.0.0.0 Serial0/0/0    //配置出接口的静态路由
```

R2(config)#**ip route 0.0.0.0 0.0.0.0 Serial0/0/0**
R3(config)#**ip route 0.0.0.0 0.0.0.0 Serial0/0/0**
R4(config)#**ip route 0.0.0.0 0.0.0.0 Serial0/0/0**
R5(config)#**ip route 0.0.0.0 0.0.0.0 Serial0/0/0**
R6(config)#**ip route 0.0.0.0 0.0.0.0 Serial0/0/0**
R7(config)#**ip route 0.0.0.0 0.0.0.0 Serial0/0/0**
R8(config)#**ip route 0.0.0.0 0.0.0.0 Serial0/0/0**

查看分支路由器的路由表

R1#**show ip route**
Codes: C - connected, S - static, I - IGRP, R - RIP, M - mobile, B - BGP
Gateway of last resort is 0.0.0.0 to network 0.0.0.0

  10.0.0.0/24 is subnetted, 1 subnets
C 10.0.19.0 is directly connected, Serial0/0/0
C 192.168.1.0/24 is directly connected, FastEthernet0/0
S* 0.0.0.0/0 is directly connected, Serial0/0/0

R6#**show ip route**
Codes: C - connected, S - static, I - IGRP, R - RIP, M - mobile, B - BGP

Gateway of last resort is 0.0.0.0 to network 0.0.0.0

  10.0.0.0/24 is subnetted, 1 subnets
C 10.0.69.0 is directly connected, Serial0/0/0
C 192.168.6.0/24 is directly connected, FastEthernet0/0
S* 0.0.0.0/0 is directly connected, Serial0/0/0

步骤四：测试网络连通性

配置终端 PC 地址，经测试，全网实现了互通。

图 5-9（b）网络拓扑的具体配置步骤如下。

步骤一：路由器基本配置

确保所有路由器均有直连路由。

R9#**show ip interface brief | include**		**up**	**up**
FastEthernet0/0	10.0.19.9	YES manual up	up
FastEthernet1/0	10.0.29.9	YES manual up	up
FastEthernet2/0	10.0.39.9	YES manual up	up

FastEthernet3/0	10.0.49.9	YES manual up	up
FastEthernet4/0	10.0.59.9	YES manual up	up
FastEthernet5/0	10.0.69.9	YES manual up	up
FastEthernet6/0	10.0.79.9	YES manual up	up
FastEthernet7/0	10.0.89.9	YES manual up	up

步骤二：配置核心路由器到各分支路由器的静态路由

```
R9(config)#ip route 192.168.1.0 255.255.255.0 10.0.19.1
R9(config)#ip route 192.168.2.0 255.255.255.0 10.0.29.2
R9(config)#ip route 192.168.3.0 255.255.255.0 10.0.39.3
R9(config)#ip route 192.168.4.0 255.255.255.0 10.0.49.4
R9(config)#ip route 192.168.5.0 255.255.255.0 10.0.59.5
R9(config)#ip route 192.168.6.0 255.255.255.0 10.0.69.6
R9(config)#ip route 192.168.7.0 255.255.255.0 10.0.79.7
R9(config)#ip route 192.168.8.0 255.255.255.0 10.0.89.8
```

查看核心路由器 R9 的静态路由

```
R9#show ip route
Codes: C - connected, S - static, I - IGRP, R - RIP, M - mobile, B - BGP

     10.0.0.0/24 is subnetted, 8 subnets
C       10.0.19.0 is directly connected, FastEthernet0/0
C       10.0.29.0 is directly connected, FastEthernet1/0
C       10.0.39.0 is directly connected, FastEthernet2/0
C       10.0.49.0 is directly connected, FastEthernet3/0
C       10.0.59.0 is directly connected, FastEthernet4/0
C       10.0.69.0 is directly connected, FastEthernet5/0
C       10.0.79.0 is directly connected, FastEthernet6/0
C       10.0.89.0 is directly connected, FastEthernet7/0
S    192.168.1.0/24 [1/0] via 10.0.19.1
S    192.168.2.0/24 [1/0] via 10.0.29.2
S    192.168.3.0/24 [1/0] via 10.0.39.3
S    192.168.4.0/24 [1/0] via 10.0.49.4
S    192.168.5.0/24 [1/0] via 10.0.59.5
S    192.168.6.0/24 [1/0] via 10.0.69.6
S    192.168.7.0/24 [1/0] via 10.0.79.7
S    192.168.8.0/24 [1/0] via 10.0.89.8
```

步骤三：配置分支路由器到核心路由器的静态默认路由

```
R1(config)#ip route 0.0.0.0 0.0.0.0 10.0.19.9    //配置下一跳的静态路由
R2(config)#ip route 0.0.0.0 0.0.0.0 10.0.29.9
R3(config)#ip route 0.0.0.0 0.0.0.0 10.0.39.9
R4(config)#ip route 0.0.0.0 0.0.0.0 10.0.49.9
R5(config)#ip route 0.0.0.0 0.0.0.0 10.0.59.9
R6(config)#ip route 0.0.0.0 0.0.0.0 10.0.69.9
R7(config)#ip route 0.0.0.0 0.0.0.0 10.0.79.9
R8(config)#ip route 0.0.0.0 0.0.0.0 10.0.89.9
```

查看分支路由器 R1 的默认路由。

```
R1#show ip route
Codes: C - connected, S - static, I - IGRP, R - RIP, M - mobile, B - BGP

Gateway of last resort is 10.0.19.9 to network 0.0.0.0

     10.0.0.0/24 is subnetted, 1 subnets
C       10.0.19.0 is directly connected, FastEthernet1/0
C    192.168.1.0/24 is directly connected, FastEthernet0/0
S*   0.0.0.0/0 [1/0] via 10.0.19.9
```

查看分支路由器 R5 的默认路由。

```
R5#show ip route
Codes: C - connected, S - static, I - IGRP, R - RIP, M - mobile, B - BGP

Gateway of last resort is 10.0.59.9 to network 0.0.0.0

     10.0.0.0/24 is subnetted, 1 subnets
C       10.0.59.0 is directly connected, FastEthernet1/0
C    192.168.5.0/24 is directly connected, FastEthernet0/0
S*   0.0.0.0/0 [1/0] via 10.0.59.9
```

步骤四：测试网络连通性

配置终端 PC 地址，经测试，全网实现了互通，测试结果如图 5-10 所示。

Fire	Last Status	Source	Destination	Type	Color	Time(sec)	Periodic	Num	Edit	Delete
●	Successful	PC1	PC2	ICMP		0.000	N	0	(edit)	
●	Successful	PC2	PC3	ICMP		0.000	N	1	(edit)	
●	Successful	PC3	PC4	ICMP		0.000	N	2	(edit)	
●	Successful	PC4	PC5	ICMP		0.000	N	3	(edit)	
●	Successful	PC5	PC6	ICMP		0.000	N	4	(edit)	
●	Successful	PC6	PC7	ICMP		0.000	N	5	(edit)	
●	Successful	PC7	PC8	ICMP		0.000	N	6	(edit)	
●	Successful	PC8	PC1	ICMP		0.000	N	7	(edit)	

图 5-10　默认路由环境网络连通性测试结果

5.3.2　场景五：配置主机路由

> **场景五**：CFL 老师将带领大家学习第二种特殊静态路由"主机路由"。通常"主机路由"是指特定主机的路由，在路由表中它的优先级最高。按照路由表的最长匹配原则，主机路由的掩码最长，需要 32 位精确匹配，掩码为 255.255.255.255，我们把这种掩码叫作主机掩码，在路由表中显示为/32。下面让我们打开 Packet Tracer，根据要求完成以下任务。

- 任务一：按图 5-11 所示拓扑，配置静态路由，让路由器 R1 只转发去 192.168.3.1 和 192.168.3.7 主机的数据包。
- 任务二：按图 5-13 所示拓扑，配置静态路由，让路由器 R1 转发去 192.168.3.4～192.168.3.7 主机的数据包。

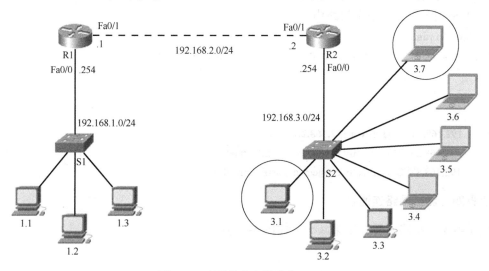

图 5-11　配置特定主机路由（一）

任务一：配置到特定主机的静态路由

步骤一：路由器基本配置

确保两台路由器均有直连路由。

步骤二：配置静态路由

R1(config)#**ip route 192.168.3.1 255.255.255.255 192.168.2.2**
R1(config)#**ip route 192.168.3.7 255.255.255.255 192.168.2.2**

在路由器 R1 上配置到特定主机的静态路由。

R2(config)#**ip route 192.168.1.0 255.255.255.0 192.168.2.1**

步骤三：查看路由表

R1#**show ip route**
Codes: C - connected, S - static, I - IGRP, R - RIP, M - mobile, B - BGP

Gateway of last resort is not set

C 192.168.1.0/24 is directly connected, FastEthernet0/0
C 192.168.2.0/24 is directly connected, FastEthernet0/1
  192.168.3.0/32 is subnetted, 2 subnets
S 192.168.3.1 [1/0] via 192.168.2.2
S 192.168.3.7 [1/0] via 192.168.2.2

R1 的路由表中有两条主机路由。

R2#**show ip route**
Codes: C - connected, S - static, I - IGRP, R - RIP, M - mobile, B - BGP

Gateway of last resort is not set

S 192.168.1.0/24 [1/0] via 192.168.2.1
C 192.168.2.0/24 is directly connected, FastEthernet0/1
C 192.168.3.0/24 is directly connected, FastEthernet0/0

步骤四：测试网络连通性

测试特定主机路由转发结果（一）如图 5-12 所示。

Fire	Last Status	Source	Destination	Type	Color	Time(sec)	Periodic	Num	Edit	Delete
●	Successful	1.1	3.1	ICMP		0.000	N	0	(edit)	
●	Failed	1.1	3.2	ICMP		0.000	N	1	(edit)	
●	Failed	1.1	3.3	ICMP		0.000	N	2	(edit)	
●	Failed	1.1	3.4	ICMP		0.000	N	3	(edit)	
●	Failed	1.1	3.5	ICMP		0.000	N	4	(edit)	
●	Failed	1.1	3.6	ICMP		0.000	N	5	(edit)	
●	Successful	1.1	3.7	ICMP		0.000	N	6	(edit)	

图 5-12 测试特定主机路由转发结果（一）

从以上测试可知，路由器 R1 仅转发了去特定主机 192.168.3.1 和 192.168.3.7 的数据包。

任务二：配置到特组主机的静态路由

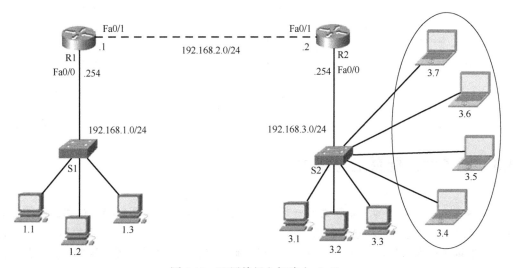

图 5-13 配置特组主机路由（二）

步骤一：路由器基本配置

确保两台路由器均有直连路由。

步骤二：配置静态路由

R1(config)#**ip route 192.168.3.4 255.255.255.252 192.168.2.2** //将 4 台主机地址进行了合并

R2(config)#**ip route 192.168.1.0 255.255.255.0 192.168.2.1**

步骤三：查看 R1 的路由表

R1#**show ip route**
Codes: C - connected, S - static, I - IGRP, R - RIP, M - mobile, B - BGP

```
Gateway of last resort is not set

C    192.168.1.0/24 is directly connected, FastEthernet0/0
C    192.168.2.0/24 is directly connected, FastEthernet0/1
     192.168.3.0/30 is subnetted, 1 subnets
S    192.168.3.4 [1/0] via 192.168.2.2
```

步骤四：测试网络连通性

测试特定主机路由转发结果（二）如图 5-14 所示。

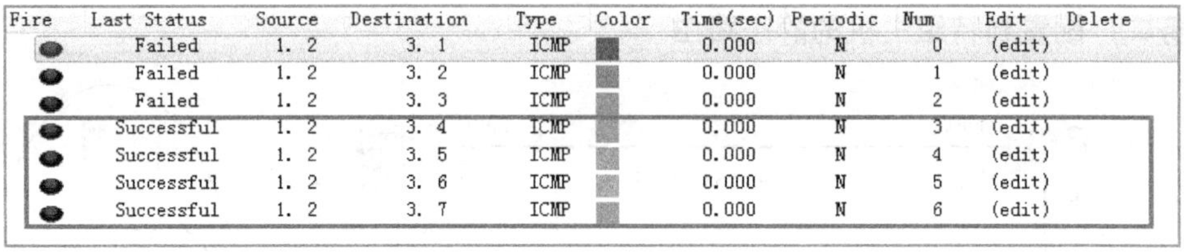

Fire	Last Status	Source	Destination	Type	Color	Time(sec)	Periodic	Num	Edit	Delete
●	Failed	1. 2	3. 1	ICMP		0.000	N	0	(edit)	
●	Failed	1. 2	3. 2	ICMP		0.000	N	1	(edit)	
●	Failed	1. 2	3. 3	ICMP		0.000	N	2	(edit)	
●	Successful	1. 2	3. 4	ICMP		0.000	N	3	(edit)	
●	Successful	1. 2	3. 5	ICMP		0.000	N	4	(edit)	
●	Successful	1. 2	3. 6	ICMP		0.000	N	5	(edit)	
●	Successful	1. 2	3. 7	ICMP		0.000	N	6	(edit)	

图 5-14　测试特定主机路由转发结果（二）

从以上测试可知，路由器 R1 仅转发了去特组主机 192.168.3.4～192.168.3.7 的数据包。

5.3.3　场景六：配置汇总路由

> **场景六**：CFL 老师将带领大家一起学习"汇总路由"，这是我们接触的第三种类型的特殊静态路由，汇总路由和前两种路由一样都是通过设置子网掩码来匹配目的地址的。汇总路由就是将多个网络地址汇总成一个，从而简化路由表，提高查询速度，降低对内存的占用率，减少 CPU 开销。汇总的前提条件是网络地址必须连续，且通往这些地址的路由采用相同出口或下一跳。关于汇总问题，老师提前给大家布置了汇总 B 类私有地址和汇总 C 类私有地址的作业。上课后，老师公布了两个汇总结果 172.16.0.0/12 和 192.168.0.0/16。为检验同学们对汇总路由的理解，老师进行了本节课的测试，如下所述。

如图 5-15 所示，M 公司有 4 个分支，分别经路由器 M-BRx（x=1,2,3,4）连接到公司核心路由器 M-HQ 上；N 公司有两个 LAN，现将 N 公司与 M 公司的核心路由器实现互连。请用静态路由实现全网的互连互通，即 M 公司内部各分支间可以互访，M 与 N 公司间互访。要求采用"最优方案"配置静态路由以简化路由表。

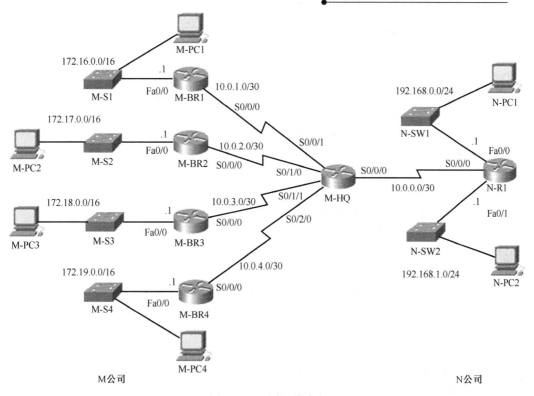

图 5-15 配置汇总路由

以下是 ZHF 同学上交的实验文档,我们一起分享一下她整理的实验步骤。

步骤一:路由器基本配置

确保路由器均有直连路由。部分路由器实验结果如下。

M-HQ#show ip interface brief | include			up	up
Serial0/0/0	10.0.0.1	YES manual up		up
Serial0/0/1	10.0.1.1	YES manual up		up
Serial0/1/0	10.0.2.1	YES manual up		up
Serial0/1/1	10.0.3.1	YES manual up		up
Serial0/2/0	10.0.4.1	YES manual up		up
M-BR1#show ip interface brief | include			up	up
FastEthernet0/0	172.16.0.1	YES manual up		up
Serial0/0/0	10.0.1.2	YES manual up		up
N-R1#show ip interface brief | include			up	up
FastEthernet0/0	192.168.0.1	YES manual up		up
FastEthernet0/1	192.168.1.1	YES manual up		up

Serial0/0/0	10.0.0.2	YES manual up	up

```
M-HQ#show ip route connected
    C    10.0.0.0/30    is directly connected, Serial0/0/0
    C    10.0.1.0/30    is directly connected, Serial0/0/1
    C    10.0.2.0/30    is directly connected, Serial0/1/0
    C    10.0.3.0/30    is directly connected, Serial0/1/1
    C    10.0.4.0/30    is directly connected, Serial0/2/0
```

步骤二：配置分支路由器的静态默认路由

```
M-BR1(config)#ip route 0.0.0.0 0.0.0.0 s0/0/0
M-BR2(config)#ip route 0.0.0.0 0.0.0.0 s0/0/0
M-BR3(config)#ip route 0.0.0.0 0.0.0.0 s0/0/0
M-BR4(config)#ip route 0.0.0.0 0.0.0.0 s0/0/0
```

查看各分支路由器的路由表，以确保静态默认路由已经存在，分支代表 M-BR1 的显示如下。

```
M-BR1#show ip route
Codes: C - connected, S - static, I - IGRP, R - RIP, M - mobile, B - BGP


Gateway of last resort is 0.0.0.0 to network 0.0.0.0

       10.0.0.0/30 is subnetted, 1 subnets
C         10.0.1.0 is directly connected, Serial0/0/0
C      172.16.0.0/16 is directly connected, FastEthernet0/0
S*     0.0.0.0/0 is directly connected, Serial0/0/0
```

步骤三：配置核心路由器到各个分支的静态路由

```
M-HQ(config)#ip route 172.16.0.0 255.255.0.0 s0/0/1
M-HQ(config)#ip route 172.17.0.0 255.255.0.0 s0/1/0
M-HQ(config)#ip route 172.18.0.0 255.255.0.0 s0/1/1
M-HQ(config)#ip route 172.19.0.0 255.255.0.0 s0/2/0
```

步骤四：配置 M 和 N 公司互访的静态汇总路由

配置 M 公司访问 N 公司的静态汇总路由。

```
M-HQ(config)#ip route 192.168.0.0 255.255.254.0 s0/0/0
```

配置 N 公司访问 M 公司的静态汇总路由。

```
N-R1(config)#ip route 172.16.0.0 255.252.0.0 s0/0/0
```

查看 M 公司核心路由器的路由表。

```
M-HQ#show ip route
Codes: C - connected, S - static, I - IGRP, R - RIP, M - mobile, B - BGP
```

```
Gateway of last resort is not set

     10.0.0.0/30 is subnetted, 5 subnets
C       10.0.0.0 is directly connected, Serial0/0/0
C       10.0.1.0 is directly connected, Serial0/0/1
C       10.0.2.0 is directly connected, Serial0/1/0
C       10.0.3.0 is directly connected, Serial0/1/1
C       10.0.4.0 is directly connected, Serial0/2/0
S       172.16.0.0/16 is directly connected, Serial0/0/1
S       172.17.0.0/16 is directly connected, Serial0/1/0
S       172.18.0.0/16 is directly connected, Serial0/1/1
S       172.19.0.0/16 is directly connected, Serial0/2/0
S       192.168.0.0/23 is directly connected, Serial0/0/0        //N 公司的汇总路由
```

查看 N 公司路由器的路由表。

```
N-R1#show ip route
Codes: C - connected, S - static, I - IGRP, R - RIP, M - mobile, B - BGP

Gateway of last resort is not set

     10.0.0.0/30 is subnetted, 1 subnets
C       10.0.0.0 is directly connected, Serial0/0/0
S       172.16.0.0/14 is directly connected, Serial0/0/0         //M 公司的汇总路由
C       192.168.0.0/24 is directly connected, FastEthernet0/0
C       192.168.1.0/24 is directly connected, FastEthernet0/1
```

步骤五：全网连通性测试

全网实现了互连互通，测试结果如图 5-16 所示。

Fire	Last Status	Source	Destination	Type	Color	Time(sec)	Periodic	Num	Edit	Delete
●	Successful	M-PC1	M-PC2	ICMP		0.000	N	0	(edit)	
●	Successful	M-PC2	M-PC3	ICMP		0.000	N	1	(edit)	
●	Successful	M-PC3	M-PC4	ICMP		0.000	N	2	(edit)	
●	Successful	M-PC4	M-PC1	ICMP		0.000	N	3	(edit)	
●	Successful	N-PC1	N-PC2	ICMP		0.000	N	4	(edit)	
●	Successful	M-PC1	N-PC1	ICMP		0.000	N	5	(edit)	
●	Successful	M-PC2	N-PC2	ICMP		0.000	N	6	(edit)	
●	Successful	M-PC3	N-PC1	ICMP		0.000	N	7	(edit)	
●	Successful	M-PC4	N-PC2	ICMP		0.000	N	8	(edit)	

图 5-16　汇总路由网络环境连通性测试

从以上实验我们不难看出，ZHF 同学在 M 公司的核心路由器上采用了带出接口的标准静态路由和汇总静态路由，在 M 公司的分支路由器上采用了静态默认路由，在 N 公司路由器上采用了汇总静态路由实现了全网互通。在 N 公司将 M 公司的分支实现了汇总，同样在 M 公司对 N 公司的网络实现了汇总，从而优化了路由表。

5.3.4 场景七：配置黑洞路由

> 🔑 **场景七**：本场景中，CFL 老师将带领大家学习第四种特殊静态路由黑洞路由。黑洞路由实质上是一条指向 Null 0 接口的特殊路由，命令是 "ip route *destination-address mask* Null0"，它是一种"防环机制"。Null0 接口是一个虚拟三层接口，特点是永不 down 掉且无法配置 IP 地址。在路由表中，一旦匹配了黑洞路由，IP 数据包都将被默默丢弃，而不指明原因，即丢弃 IP 数据包也不会通知源主机。黑洞路由就是为不精确路由汇总服务的，即第四种特殊路由黑洞路由是为第三种特殊路由汇总路由服务的。如果精确汇总，就不会出现黑洞路由。下面 CFL 老师将通过图 5-17 所示拓扑，带领大家一起学习如何配置黑洞路由。

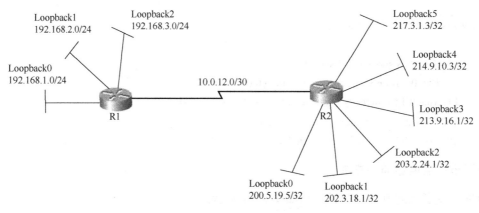

图 5-17　配置黑洞路由

两台路由器 R1 和 R2 通过各自的 S0/0/0 接口实现了互连。R1 和 R2 分别用多个环回接口模拟各自的网段，现在需要配置静态路由实现各网段间的互通。要求尽量简化路由表，提高查询速度。

具体配置步骤如下。

步骤一：路由器基本配置

确保路由器均有直连路由。

```
R1#show ip interface brief | include manual
Serial0/0/0          10.0.12.1        YES manual up                    up
Loopback0            192.168.1.254    YES manual up                    up
Loopback1            192.168.2.254    YES manual up                    up
Loopback2            192.168.3.254    YES manual up                    up
```

路由器 R1 和 R2 的接口均配置了正确的 IP 地址，且状态都为 up。

```
R2#show ip interface brief | include manual
Serial0/0/0          10.0.12.2        YES manual up                    up
Loopback0            200.5.19.5       YES manual up                    up
Loopback1            202.3.18.1       YES manual up                    up
Loopback2            203.2.24.1       YES manual up                    up
Loopback3            213.9.16.1       YES manual up                    up
Loopback4            214.9.10.3       YES manual up                    up
Loopback5            217.3.1.3        YES manual up                    up
```

步骤二：配置静态路由

在路由器 R1 上配置静态默认路由。

R1(config)#**ip route 0.0.0.0 0.0.0.0 s0/0/0**

在路由器 R2 配置静态汇总路由。

R2(config)#**ip route 192.168.0.0 255.255.252.0 s0/0/0**

查看路由器 R1 的路由表。

```
R1#show ip route
Codes: C - connected, S - static, I - IGRP, R - RIP, M - mobile, B - BGP
Gateway of last resort is 0.0.0.0 to network 0.0.0.0

     10.0.0.0/30 is subnetted, 1 subnets
C       10.0.12.0 is directly connected, Serial0/0/0
C    192.168.1.0/24 is directly connected, Loopback0
C    192.168.2.0/24 is directly connected, Loopback1
C    192.168.3.0/24 is directly connected, Loopback2
S*   0.0.0.0/0 is directly connected, Serial0/0/0
```

查看路由器 R2 的路由表。

```
R2#show ip route
Codes: C - connected, S - static, I - IGRP, R - RIP, M - mobile, B - BGP

Gateway of last resort is not set
```

```
         10.0.0.0/30 is subnetted, 1 subnets
C           10.0.12.0 is directly connected, Serial0/0/0
S        192.168.0.0/22 is directly connected, Serial0/0/0
         200.5.19.0/32 is subnetted, 1 subnets
C           200.5.19.5 is directly connected, Loopback0
         202.3.18.0/32 is subnetted, 1 subnets
C           202.3.18.1 is directly connected, Loopback1
         203.2.24.0/32 is subnetted, 1 subnets
C           203.2.24.1 is directly connected, Loopback2
         213.9.16.0/32 is subnetted, 1 subnets
C           213.9.16.1 is directly connected, Loopback3
         214.9.10.0/32 is subnetted, 1 subnets
C           214.9.10.3 is directly connected, Loopback4
         217.3.1.0/32 is subnetted, 1 subnets
C           217.3.1.3 is directly connected, Loopback5
```

通过以上输出，我们发现静态路由已正确填入路由表。

步骤三：测试网络连通性

通过扩展 ping 命令，分别测试网间的连通性，例如，测试 192.168.1.254 所在网段与 217.3.1.3 的连通性，我们可以通过如下命令测试：

```
R1#ping
Protocol [ip]:
Target IP address: 217.3.1.3
Repeat count [5]:
Datagram size [100]:
Timeout in seconds [2]:
Extended commands [n]: y
Source address or interface: 192.168.1.254
Type of service [0]:
Set DF bit in IP header? [no]:
Validate reply data? [no]:
Data pattern [0xABCD]:
Loose, Strict, Record, Timestamp, Verbose[none]:
Sweep range of sizes [n]:
Type escape sequence to abort.
Sending 5, 100-byte ICMP Echos to 217.3.1.3, timeout is 2 seconds:
Packet sent with a source address of 192.168.1.254
```

!!!!!
Success rate is 100 percent (5/5), round-trip min/avg/max = 1/3/8 ms

通过更改源地址和目标地址，反复进行扩展 ping 测试，成功实现全网互通。看似没有任何问题的网络，实际上却存在着很大隐患，在路由器 R2 上将网段 192.168.1.0/24～192.168.3.0/24 汇总成 192.168.0.0/22，实质上该汇总是不精确汇总，地址 192.168.0.0/24 也被汇总在内。在 R1 上做如下 ping 测试：

R1>**ping 192.168.0.1**
Type escape sequence to abort.
Sending 5, 100-byte ICMP Echos to 192.168.0.1, timeout is 2 seconds:
.....   //ping 包超时
Success rate is 0 percent (0/5)

我们发现去往目标地址 192.168.0.1 的 ping 包出现了超时。实际上会引起环路，R1 通过默认路由 0.0.0.0/0 将 IP 包转发给路由器 R2，因为 192.168.0.1 的目标地址匹配 192.168.0.0/22 的路由，因此被路由器 R2 转发给 R1，这样去往目的 192.168.0.1 的 IP 包就会在路由器 R1 和 R2 之间来回传递，还好 IP 报头有一个 TTL 域，每经过一个路由器会减 1，直到 TTL=0，数据包才会被丢弃。在这期间数据包循环传递一定会增加网络开销，对网络性能造成一定的影响。下面我们来探讨如何解决这个问题。

步骤四：配置静态丢弃路由

R2(config)#**ip route 192.168.0.0 255.255.255.0 null0**
查看路由器 R2 的静态路由
R2#**show ip route static**
S   192.168.0.0/22 is directly connected, Serial0/0/0
S   **192.168.0.0/24 is directly connected, Null0**

有了黑洞路由，去往 192.168.0.0/24 网段的数据包就会被送往 Null0 口丢弃，有效地控制了路由环路。

5.3.5 场景八：配置浮动路由

🔍 **场景八**：网络课堂上，CFL 老师向全班同学提出了一个问题：
"我们已经学习了几种特殊静态路由？分别是什么？哪几种有共性？其共性是什么？"
DWC 同学回答道："学习了 4 种，分别是默认路由、主机路由、汇总路由和黑洞路由，其中前 3 种有共性，其共性在于它们都是在目的地址和子网掩码两个参数上做文章"。
CFL 老师满意地点点头，她让 DWC 同学再说详细点。DWC 说，默认路由无须匹配，掩码为/0，网络地址是全零；主机路由是 32 位全部匹配，掩码为/32，目的地址是 IP 地址；

> 汇总路由匹配的位数取决于几个连续网段的相同网络地址位数,假设 X 位相同,则掩码为/X。老师表扬 DWC 同学总结的非常到位,她又补充道:第四种黑洞路由是在出接口上做了文章,即采用特殊的 Null0 接口来丢弃数据包。CFL 老师接着问同学们,我们今天学习的第五种特殊静态路由——浮动静态路由,有哪位同学知道它是在哪个参数上做文章呢?还是 DWC 同学第一个回答:在管理距离 AD 上做文章,老师表扬 DWC 回答得太棒了,可以加两分。

浮动静态路由是一种特殊的静态路由,通过配置一个比主路由管理距离更大的静态路由,保证网络在主路由失效情况下,提供备份路由。在存在主路由的情况下,浮动路由不会出现在路由表中,当主路由失效时,它才会浮出水面出现在路由表中,确保网络不中断,即实现冗余备份。浮动路由应用于对网络可靠性要求较高的场合。我们知道路由表中管理距离越低,路由的优先级越高管理距离为 0~255 的整数值,管理路离为 0 的路由是路由器最可信赖的路由,管理距离为 255 的路由是最不可信的路由,不会写进路由表。直连路由最可信,Cisco 规定直连路由的管理距离为 0,静态路由的管理距离为 1。下面我们将通过图 5-18 学习浮动路由的配置,具体说明如下。

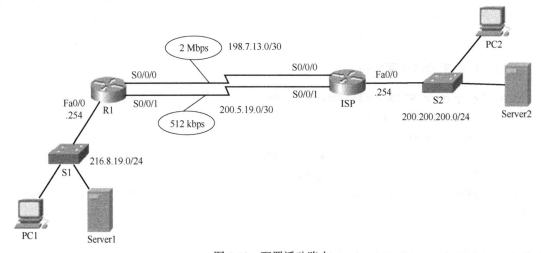

图 5-18 配置浮动路由

ABC 公司是一家小型的 IT 公司,公司通过边界路由器 R1 连接到 ISP。因公司所有业务全部基于 Internet,所以对网络可靠性要求较高,特向 ISP 申请了两条专线,带宽分别为 2 Mbps 和 512 kbps。现要求将带宽为 2 Mbps 的链路设置为主链路,带宽为 512 kbps 的链路设置为备份链路。请配置静态路由实现全网互通。

具体配置如下。

步骤一:路由器的基本配置

确保两台路由器均有直连路由。

第 5 章 学习静态路由

查看路由器 R1 的接口信息：

```
R1#show ip interface brief | include manual
FastEthernet0/0      216.8.19.254     YES manual up                    up
Serial0/0/0          198.7.13.1       YES manual up                    up
Serial0/0/1          200.5.19.1       YES manual up                    up
```

查看路由器 ISP 的接口信息：

```
ISP#show ip interface brief | include manual
FastEthernet0/0      200.200.200.254  YES manual up                    up
Serial0/0/0          198.7.13.2       YES manual up                    up
Serial0/0/1          200.5.19.2       YES manual up                    up
```

步骤二：配置静态路由

```
R1(config)#ip route 0.0.0.0 0.0.0.0 s0/0/0                //主链路管理距离为1
R1(config)#ip route 0.0.0.0 0.0.0.0 s0/0/1 217            //备用链路管理距离为217

ISP(config)#ip route 216.8.19.0 255.255.255.0 s0/0/0      //主链路管理距离为1
ISP(config)#ip route 216.8.19.0 255.255.255.0 s0/0/1 220  //备用链路管理距离为220
```

管理距离具有本地意义，与其他路由器的管理距离没有关系。

步骤三：测试主链路转发数据包

```
R1#show ip route | begin    Gateway
Gateway of last resort is 0.0.0.0 to network 0.0.0.0

     198.7.13.0/30 is subnetted, 1 subnets
C       198.7.13.0 is directly connected, Serial0/0/0
     200.5.19.0/30 is subnetted, 1 subnets
C       200.5.19.0 is directly connected, Serial0/0/1
C    216.8.19.0/24 is directly connected, FastEthernet0/0
S*   0.0.0.0/0 is directly connected, Serial0/0/0
```

从 PC1 发送到 PC2 的跟踪包，如图 5-19 所示。

```
C:\>tracert 200.200.200.1

Tracing route to 200.200.200.1 over a maximum of 30 hops:

  1    0 ms      0 ms      0 ms     216.8.19.254
  2    0 ms      1 ms      0 ms     198.7.13.2
  3   10 ms     10 ms     11 ms     200.200.200.1

Trace complete.
```

图 5-19　浮动路由网络环境主路径测试

图 5-19 的输出显示 PC1 到 PC2 的数据包经 S0/0/0 口的主链路转发。

步骤四：测试备份链路转发数据包

Down 掉主链路：

R1(config)#**interface s0/0/0**
R1(config-if)#**shutdown**

%LINK-5-CHANGED: Interface Serial0/0/0, changed state to administratively down
%LINEPROTO-5-UPDOWN: Line protocol on Interface Serial0/0/0, changed state to down

查看主链路 down 掉后 R1 的路由表：

R1#**show ip route | begin Gateway**
Gateway of last resort is 0.0.0.0 to network 0.0.0.0

   200.5.19.0/30 is subnetted, 1 subnets
C 200.5.19.0 is directly connected, Serial0/0/1
C 216.8.19.0/24 is directly connected, FastEthernet0/0
S* **0.0.0.0/0 is directly connected, Serial0/0/1**

继续从 PC1 发送到 PC2 的跟踪包，如图 5-20 所示。

```
C:\>tracert 200.200.200.1

Tracing route to 200.200.200.1 over a maximum of 30 hops:

  1    1 ms      0 ms      0 ms      216.8.19.254
  2    0 ms      0 ms      0 ms      200.5.19.2
  3    8 ms     11 ms      0 ms      200.200.200.1

Trace complete.
```

图 5-20 浮动路由网络环境备份路径测试

从以上跟踪结果来看，主链路出现故障之后，数据包自动切换到了备用链路。当主链路恢复时，备用链路的路由就会隐藏，主路由出现在路由表。

5.3.6 场景九：配置等价路由

> 🔍 **场景九**：第六种特殊静态路由"等价路由"与第五种特殊静态路由"浮动路由"密切相关。浮动路由和主路由是交替出现在路由表中的，正常情况下主路由在路由表中，当主路径出现故障时，浮动路由才会出现在路由表中。我们配置浮动路由，实质上就是修改其管理距离，使其比"主路由"的管理距离大。配置等价路由实质上是两条通往相同目的地址的路由，管理距离相同但路径不同，如图 5-21 所示。

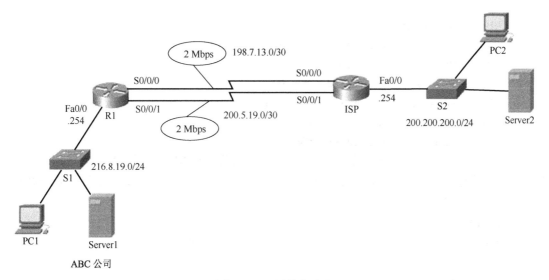

图 5-21　配置等价路由

假设 ABC 公司向 ISP 申请了两条专线，带宽都是 2 Mbps，要求两条链路互为备份，且对公司的流量进行负载分担，如何配置静态路由，满足上述条件，请写出具体的配置步骤。

具体配置如下所述。

步骤一：路由器基本配置

完全同场景八的配置。

步骤二：配置等价静态路由

在路由器 R1 上配置静态等价路由。

R1(config)#**ip route 0.0.0.0 0.0.0.0 s0/0/0**
R1(config)#**ip route 0.0.0.0 0.0.0.0 s0/0/1**

在 ISP 路由器上配置静态等价路由。

ISP(config)#**ip route 216.8.19.0 255.255.255.0 s0/0/0**
ISP(config)#**ip route 216.8.19.0 255.255.255.0 s0/0/1**

步骤三：查看路由表

查看路由器 R1 的路由表。

R1#**show ip route**
Codes: C - connected, S - static, I - IGRP, R - RIP, M - mobile, B - BGP

Gateway of last resort is 0.0.0.0 to network 0.0.0.0

```
            198.7.13.0/30 is subnetted, 1 subnets
C           198.7.13.0 is directly connected, Serial0/0/0
            200.5.19.0/30 is subnetted, 1 subnets
C           200.5.19.0 is directly connected, Serial0/0/1
C           216.8.19.0/24 is directly connected, FastEthernet0/0
S*          0.0.0.0/0 is directly connected, Serial0/0/0
            is directly connected, Serial0/0/1
```

从 R1 的路由表输出结果可见，两条路径均显示在路由表中，只不过是对网络地址进行了合并，但依然是两条静态路由，说明两者是等价的，无优劣之分。接下来查看路由器 ISP 的路由表，结果和 R1 的一致。

```
ISP#show ip route
Codes: C - connected, S - static, I - IGRP, R - RIP, M - mobile, B - BGP

Gateway of last resort is not set

            198.7.13.0/30 is subnetted, 1 subnets
C           198.7.13.0 is directly connected, Serial0/0/0
            200.5.19.0/30 is subnetted, 1 subnets
C           200.5.19.0 is directly connected, Serial0/0/1
C           200.200.200.0/24 is directly connected, FastEthernet0/0
S           216.8.19.0/24 is directly connected, Serial0/0/0
            is directly connected, Serial0/0/1
```

由以上路由表的输出可知，两条去往同一目的的路径是等价的，即路径的管理距离是一样的，这就是我们所谓的等价路径的负载均衡。

5.3.7　场景十：配置递归路由

🔍**场景十**：本堂课 CFL 老师将带领大家一起学习最后一个特殊静态路由"递归路由"。学习递归路由之前，CFL 老师当着全班同学的面又一次表扬了 XL 同学。老师说这堂课是专门来解决一周前 XL 电子邮件中提出的问题。老师将问题投在大屏幕上，内容如下：

老师，我看书上写了一句话：在配置静态路由时，一般使用实际下一跳路由器的 IP 地址，但实质上可以是任意 IP 地址，只要它可以在路由表中解析。这句话是什么意思？难道下一跳还可以随便指定吗？

看到这个问题，全班同学都将目光投向了 XL，大家都很佩服 XL 总能提出很经典的问题，还有的同学说老师又该给 XL 加分了。老师笑着问同学们是否都打过黑司机的车，那些司机为了赚钱，载着乘客绕弯子而不走正常路线，最终还是把乘客送到了目的地。大多数同学都回应有类似的经历。CFL 老师说递归路由和黑司机载客绕弯子是一回事，只要能把数据包送往目的地就可以了，但这是不可取的，对路由器来讲，势必会降低路由器的性能。本节课就是让大家开阔一下眼界，知道静态路由是可以非常灵活去控制的。

下面就请大家跟 CFL 老师依据图 5-22 所示，在路由器 R1 上通过配置递归路由实现到 R2 网段数据的转发，R2 正常配置静态路由。

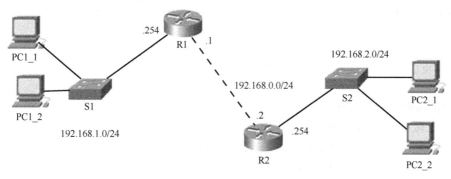

图 5-22 配置递归路由

具体配置如下。

步骤一：路由器基本配置

确保路由器均有直连路由。

R1#show ip interface brief | include manual			
FastEthernet0/0	192.168.1.254	YES manual up	up
FastEthernet0/1	192.168.0.1	YES manual up	up

R2#show ip interface brief | include manual			
FastEthernet0/0	192.168.2.254	YES manual up	up
FastEthernet0/1	192.168.0.2	YES manual up	up

步骤二：配置静态路由

在路由器 R2 上配置标准静态路由，如下所示。

R2(config)#**ip route 192.168.1.0 255.255.255.0 192.168.0.1**

在路由器 R1 上配置递归路由。

R1(config)#**ip route 192.168.2.0 255.255.255.0 10.0.0.2**　　// 10.0.0.2 是任意下一跳 IP 地址
R1(config)#**ip route 10.0.0.0 255.0.0.0 172.16.1.2**　　//去 10.0.0.0/8 采用任意下一跳 IP 地址
R1(config)#**ip route 172.16.0.0 255.255.0.0 217.3.2.4**　　//去 172.16.0.0/16 采用任意下一跳 IP 地址
R1(config)#**ip route 217.3.2.0 255.255.255.0 192.168.0.2**　　//去 217.3.2.0/24 采用了真正的下一跳，可解析

注意：我们可以写 n 条静态路由，因为前 n-1 条都是无法正确解析的，因为在路由表中找不到对应的送出接口。所以，在第 n 条命令输入之前，前 n-1 条路由不会被写进路由表，仅当输入第 n 条命令且可以在路由表中正确解析，即能找到对应的送出接口时，n 条路由才会被一起写进路由表。

步骤三：检查路由器的路由表

```
R1#show ip route
Codes: C - connected, S - static, I - IGRP, R - RIP, M - mobile, B - BGP

Gateway of last resort is not set

S    10.0.0.0/8 [1/0] via 172.16.1.2
S    172.16.0.0/16 [1/0] via 217.3.2.4
C    192.168.0.0/24 is directly connected, FastEthernet0/1
C    192.168.1.0/24 is directly connected, FastEthernet0/0
S    192.168.2.0/24 [1/0] via 10.0.0.2
S    217.3.2.0/24 [1/0] via 192.168.0.2
```

由 R1 路由表的输出可知，新增了 4 条静态路由，这样，若要转发去 192.168.2.0/24 网段的数据包，则需递归查询 5 次。

```
R2#show ip route
Codes: C - connected, S - static, I - IGRP, R - RIP, M - mobile, B - BGP

Gateway of last resort is not set

C    192.168.0.0/24 is directly connected, FastEthernet0/1
S    192.168.1.0/24 [1/0] via 192.168.0.1
C    192.168.2.0/24 is directly connected, FastEthernet0/0
```

R1 的路由表是递归路由，R2 的路由表正常。

步骤四：测试全网连通性

全网均已实现互通，测试结果如图 5-23 所示。

Fire	Last Status	Source	Destination	Type	Color	Time(sec)	Periodic	Num	Edit	Delete
●	Successful	PC1_1	PC1_2	ICMP		0.000	N	0	(edit)	
●	Successful	PC2_1	PC2_2	ICMP		0.000	N	1	(edit)	
●	Successful	PC1_1	PC2_1	ICMP		0.000	N	2	(edit)	
●	Successful	PC1_1	PC2_2	ICMP		0.000	N	3	(edit)	
●	Successful	PC1_2	PC2_1	ICMP		0.000	N	4	(edit)	
●	Successful	PC1_2	PC2_2	ICMP		0.000	N	5	(edit)	

图 5-23　全网连通性测试结果

5.3.8 特殊静态路由小结

本节的特殊静态路由共介绍了 7 种，其配置都是基于如下命令：
Router(config)#ip route {destination-address} {mask} {next-hop-address | exit- interface} [distance]
特殊静态路由的特点及应用场合总结如表 5-6 所示。

表 5-6 特殊静态路由特点及应用

编　号	特殊路由	特殊路由的特点	特殊路由的应用场合
1	默认路由	掩码为/0，网络地址全 0	末节路由器和边缘路由器
2	主机路由	掩码为/32 的路由	对特定主机访问的场合
3	汇总路由	有相同的网络地址位，出接口或下一跳一致	需要优化路由表的应用场合
4	黑洞路由	采用 Null0 接口丢弃数据包	未达到精确汇总的应用场合
5	浮动路由	作为主路由的备份路径，隐形显示	冗余的高可靠性应用场合
6	等价路由	相同目的地址，AD 相同，同时出现在路由表中	冗余且要求负载分担的场合
7	递归路由	下一跳不是真正的下一跳	不建议使用

5.4 企业网综合配置案例

5.4.1 任务背景

YTBH 公司是一家中规模的 IT 公司，目前公司已在三座城市设有分支机构。公司新部署的企业互连网拓扑如图 5-24 所示。三个分支路由器分别为 BR1、BR2 和 BR3，核心路由器为 RegionA，企业边界路由器为 HQ。公司向 ISP 申请了两条 10 Mbps 专线将企业网连接到了 Internet，既可以互为备份，又可以均衡负载。分支路由器 BR2 到核心路由器 RegionA 之间采用双线冗余链路，主链路是光纤链路，备用链路为 WAN 链路。假设你工作在 ISP，并兼任 YTBH 公司企业网规划的项目经理，请你用静态路由实现全网互通。为确保你在 ISP 工作期间也能访问到公司内部网络，请在 ISP 路由器上临时添加一条静态路由以便远程施工。

5.4.2 网络拓扑

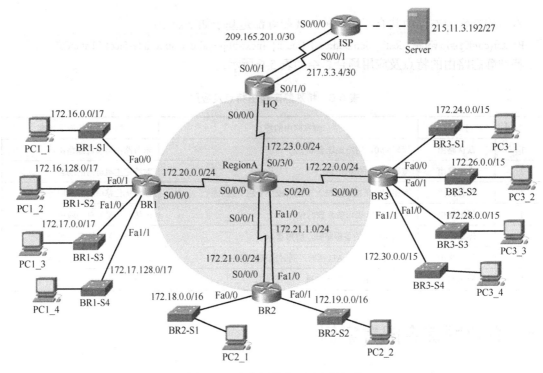

图 5-24 企业网综合配置案例拓扑

5.4.3 配置任务

请打开 Packet Tracer 模拟器，按照图 5-24 所示搭建网络拓扑。

任务一：完成路由器的基本配置

- 按照图中所示为所有路由器命名；
- 分支路由器 BR1、BR2、BR3 的以太口均采用本网段最大可用 IP 地址；
- RegionA 的所有接口均采用对应网段第一个可用 IP 地址；
- 分支路由器 BR1、BR2、BR3 以及边界路由器 HQ 的串行接口均采用本网段第二个可用 IP 地址；
- 路由器 ISP 的所有接口均采用本网段最小可用 IP 地址

任务二：配置静态路由实现全网连通

- 边界路由器采用静态默认路由；
- 区域间路由器采用静态汇总路由；
- 请采用最优方案进行配置；

任务三：终端 PC 的配置

- 请为终端 PC 配置本网段内合法 IP 地址、掩码、网关。

任务四：网络连通性测试

- PC 间可以相互 ping 通；
- PC 与路由器间可以相互 ping 通。

5.4.4　任务实施

任务实施的具体步骤如下。

步骤一：路由器基本配置

确保路由器均有直连路由。

```
BR1>show ip interface brief | include manual
FastEthernet0/0      172.16.127.254    YES manual up                    up
FastEthernet0/1      172.16.255.254    YES manual up                    up
Serial0/0/0          172.20.0.2        YES manual up                    up
FastEthernet1/0      172.17.127.254    YES manual up                    up
FastEthernet1/1      172.17.255.254    YES manual up                    up
```

```
BR2>show ip interface brief | include manual
FastEthernet0/0      172.18.255.254    YES manual up                    up
FastEthernet0/1      172.19.255.254    YES manual up                    up
Serial0/0/0          172.21.0.2        YES manual up                    up
FastEthernet1/0      172.21.1.2        YES manual up                    up
```

```
BR3>show ip interface brief | include manual
FastEthernet0/0      172.25.255.254    YES manual up                    up
FastEthernet0/1      172.27.255.254    YES manual up                    up
Serial0/0/0          172.22.0.2        YES manual up                    up
FastEthernet1/0      172.29.255.254    YES manual up                    up
FastEthernet1/1      172.31.255.254    YES manual up                    up
```

```
RegionA>show ip interface brief | include manual
Serial0/0/0            172.20.0.1        YES manual up                      up
Serial0/0/1            172.21.0.1        YES manual up                      up
Serial0/2/0            172.22.0.1        YES manual up                      up
Serial0/3/0            172.23.0.1        YES manual up                      up
FastEthernet1/0        172.21.1.1        YES manual up                      up

HQ>show ip interface brief | include manual
Serial0/0/0            172.23.0.2        YES manual up                      up
Serial0/0/1            209.165.201.2     YES manual up                      up
Serial0/1/0            217.3.3.6         YES manual up                      up

ISP>show ip interface brief | include manual
FastEthernet0/0        215.11.3.193      YES manual up                      up
Serial0/0/0            209.165.201.1     YES manual up                      up
Serial0/0/1            217.3.3.5         YES manual up                      up
```

步骤二：配置分支路由器的静态路由

```
BR1(config)#ip route 0.0.0.0 0.0.0.0 s0/0/0

BR1>show ip route | begin Gateway
Gateway of last resort is 0.0.0.0 to network 0.0.0.0

     172.16.0.0/17 is subnetted, 2 subnets
C       172.16.0.0 is directly connected, FastEthernet0/0
C       172.16.128.0 is directly connected, FastEthernet0/1
     172.17.0.0/17 is subnetted, 2 subnets
C       172.17.0.0 is directly connected, FastEthernet1/0
C       172.17.128.0 is directly connected, FastEthernet1/1
     172.20.0.0/24 is subnetted, 1 subnets
C       172.20.0.0 is directly connected, Serial0/0/0
S*   0.0.0.0/0 is directly connected, Serial0/0/0

BR2(config)#ip route 0.0.0.0 0.0.0.0 s0/0/0 220
BR2(config)#ip route 0.0.0.0 0.0.0.0 172.21.1.1

BR2>show ip route | begin Gateway
Gateway of last resort is 172.21.1.1 to network 0.0.0.0
```

| C | 172.18.0.0/16 is directly connected, FastEthernet0/0 |
| C | 172.19.0.0/16 is directly connected, FastEthernet0/1 |

172.21.0.0/24 is subnetted, 2 subnets

C	172.21.0.0 is directly connected, Serial0/0/0
C	172.21.1.0 is directly connected, FastEthernet1/0
S*	0.0.0.0/0 [1/0] via 172.21.1.1

BR3(config)#**ip route 0.0.0.0 0.0.0.0 s0/0/0**

BR3>**show ip route | begin Gateway**
Gateway of last resort is 0.0.0.0 to network 0.0.0.0

172.22.0.0/24 is subnetted, 1 subnets

C	172.22.0.0 is directly connected, Serial0/0/0
C	172.24.0.0/15 is directly connected, FastEthernet0/0
C	172.26.0.0/15 is directly connected, FastEthernet0/1
C	172.28.0.0/15 is directly connected, FastEthernet1/0
C	172.30.0.0/15 is directly connected, FastEthernet1/1
S*	0.0.0.0/0 is directly connected, Serial0/0/0

步骤三：配置核心路由器的静态路由

RegionA(config)#**ip route 172.16.0.0 255.254.0.0 s0/0/0**
RegionA(config)#**ip route 172.18.0.0 255.254.0.0 172.21.1.2**
RegionA(config)#**ip route 172.18.0.0 255.254.0.0 s0/0/1 220**
RegionA(config)#**ip route 172.24.0.0 255.248.0.0 s0/2/0**
RegionA(config)#**ip route 0.0.0.0 0.0.0.0 s0/3/0**

RegionA>**show ip route | begin Gateway**
Gateway of last resort is 0.0.0.0 to network 0.0.0.0

| S | 172.16.0.0/15 is directly connected, Serial0/0/0 |
| S | 172.18.0.0/15 [1/0] via 172.21.1.2 |

172.20.0.0/24 is subnetted, 1 subnets

| C | 172.20.0.0 is directly connected, Serial0/0/0 |

172.21.0.0/24 is subnetted, 2 subnets

| C | 172.21.0.0 is directly connected, Serial0/0/1 |
| C | 172.21.1.0 is directly connected, FastEthernet1/0 |

172.22.0.0/24 is subnetted, 1 subnets

| C | 172.22.0.0 is directly connected, Serial0/2/0 |

```
           172.23.0.0/24 is subnetted, 1 subnets
C          172.23.0.0 is directly connected, Serial0/3/0
S          172.24.0.0/13 is directly connected, Serial0/2/0
S*         0.0.0.0/0 is directly connected, Serial0/3/0
```

步骤四：配置边界路由器的静态路由

HQ(config)#**ip route 0.0.0.0 0.0.0.0 s0/0/1**
HQ(config)#**ip route 0.0.0.0 0.0.0.0 s0/1/0**
HQ(config)#**ip route 172.16.0.0 255.240.0.0 s0/0/0**

HQ>**show ip route | begin Gateway**
Gateway of last resort is 0.0.0.0 to network 0.0.0.0

```
S      172.16.0.0/12 is directly connected, Serial0/0/0
       172.23.0.0/24 is subnetted, 1 subnets
C          172.23.0.0 is directly connected, Serial0/0/0
       209.165.201.0/30 is subnetted, 1 subnets
C          209.165.201.0 is directly connected, Serial0/0/1
       217.3.3.0/30 is subnetted, 1 subnets
C          217.3.3.4 is directly connected, Serial0/1/0
S*     0.0.0.0/0 is directly connected, Serial0/0/1
                  is directly connected, Serial0/1/0
```

步骤五：配置 ISP 静态路由

ISP(config)#**ip route 172.16.0.0 255.240.0.0 s0/0/0**
ISP(config)#**ip route 172.16.0.0 255.240.0.0 s0/0/1**

ISP>**show ip route | begin Gateway**
Gateway of last resort is not set

```
S      172.16.0.0/12 is directly connected, Serial0/0/0
                     is directly connected, Serial0/0/1
       209.165.201.0/30 is subnetted, 1 subnets
C          209.165.201.0 is directly connected, Serial0/0/0
       215.11.3.0/27 is subnetted, 1 subnets
C          215.11.3.192 is directly connected, FastEthernet0/0
       217.3.3.0/30 is subnetted, 1 subnets
C          217.3.3.4 is directly connected, Serial0/0/1
```

5.4.5 连通测试

同分支内的 PC 间实现了互通,如图 5-25 所示。

Fire	Last Status	Source	Destination	Type	Color	Time(sec)	Periodic	Num	Edit	Delete
●	Successful	PC1_1	PC1_2	ICMP		0.000	N	0	(edit)	
●	Successful	PC1_2	PC1_3	ICMP		0.000	N	1	(edit)	
●	Successful	PC1_3	PC1_4	ICMP		0.000	N	2	(edit)	
●	Successful	PC1_4	PC1_1	ICMP		0.000	N	3	(edit)	
●	Successful	PC2_1	PC2_2	ICMP		0.000	N	4	(edit)	
●	Successful	PC3_1	PC3_2	ICMP		0.000	N	5	(edit)	
●	Successful	PC3_2	PC3_3	ICMP		0.000	N	6	(edit)	
●	Successful	PC3_3	PC3_4	ICMP		0.000	N	7	(edit)	
●	Successful	PC3_4	PC3_1	ICMP		0.000	N	8	(edit)	

图 5-25 分支内 PC 连通测试

各分支 PC 间实现了互通,如图 5-26 所示。

Fire	Last Status	Source	Destination	Type	Color	Time(sec)	Periodic	Num	Edit	Delete
●	Successful	PC1_1	PC2_1	ICMP		0.000	N	0	(edit)	
●	Successful	PC1_2	PC2_2	ICMP		0.000	N	1	(edit)	
●	Successful	PC1_3	PC3_1	ICMP		0.000	N	2	(edit)	
●	Successful	PC1_4	PC3_2	ICMP		0.000	N	3	(edit)	
●	Successful	PC2_1	PC3_3	ICMP		0.000	N	4	(edit)	
●	Successful	PC2_2	PC3_4	ICMP		0.000	N	5	(edit)	

图 5-26 分支间 PC 连通测试

分支 PC 均可以成功访问外网的服务器,实现企业网对外网的访问,如图 5-27 所示。

Fire	Last Status	Source	Destination	Type	Color	Time(sec)	Periodic	Num	Edit	Delete
●	Successful	PC1_1	Server	ICMP		0.000	N	0	(edit)	
●	Successful	PC1_2	Server	ICMP		0.000	N	1	(edit)	
●	Successful	PC1_3	Server	ICMP		0.000	N	2	(edit)	
●	Successful	PC1_4	Server	ICMP		0.000	N	3	(edit)	
●	Successful	PC2_1	Server	ICMP		0.000	N	4	(edit)	
●	Successful	PC2_2	Server	ICMP		0.000	N	5	(edit)	
●	Successful	PC3_1	Server	ICMP		0.000	N	6	(edit)	
●	Successful	PC3_2	Server	ICMP		0.000	N	7	(edit)	
●	Successful	PC3_3	Server	ICMP		0.000	N	8	(edit)	
●	Successful	PC3_4	Server	ICMP		0.000	N	9	(edit)	

图 5-27 内网到外网连通测试

通过外网服务器也可以成功访问内网 PC,实现了在 ISP 上远程访问企业网,如图 5-28 所示。

Fire	Last Status	Source	Destination	Type	Color	Time(sec)	Periodic	Num	Edit	Delete
	Successful	Server	PC1_1	ICMP		0.000	N	0	(edit)	
	Successful	Server	PC1_2	ICMP		0.000	N	1	(edit)	
	Successful	Server	PC1_3	ICMP		0.000	N	2	(edit)	
	Successful	Server	PC1_4	ICMP		0.000	N	3	(edit)	
	Successful	Server	PC2_1	ICMP		0.000	N	4	(edit)	
	Successful	Server	PC2_2	ICMP		0.000	N	5	(edit)	
	Successful	Server	PC3_1	ICMP		0.000	N	6	(edit)	
	Successful	Server	PC3_2	ICMP		0.000	N	7	(edit)	
	Successful	Server	PC3_3	ICMP		0.000	N	8	(edit)	
	Successful	Server	PC3_4	ICMP		0.000	N	9	(edit)	

图 5-28　外网到内网连通测试

PC 与路由器接口间也顺利通过连通性测试，至此，全网实现了互联互通。

5.4.6　案例小结

本企业网案例，使重要的几种静态路由类型得到了充分应用，现将具体的应用总结如下：

表 5-7　静态路由类型在企业网中应用

编号	设备名称	静态路由的应用类型
1	BR1	静态默认路由
2	BR2	静态默认路由、静态浮动路由
3	BR3	静态默认路由
4	RegionA	静态汇总路由、静态默认路由、静态浮动路由
5	HQ	静态汇总路由、静态默认路由、静态等价路由
6	ISP	静态汇总路由、静态等价路由

5.5　拓展思维案例

5.5.1　案例一：网络配置挑战

> 🔍挑战背景：YTXH 是一家拥有三个子公司的水产食品有限公司。公司企业网拓扑如图 5-29 所示。三个子公司的路由器分别为 BR1、BR2 和 BR3，核心路由器为 RegionA。企业网通过边界路由器 HQ 连接到了 Internet。子公司 BR1 和 BR2 通过专线实现互连，各子公司都与核心路由器直接互连。各子公司都有双线冗余链路，所有 WAN 出口的链路带宽一致。请打开 Packet Tracer 模拟器，按照图 5-29 所示搭建网络拓扑并实施。

要求采用静态路由实现全网互通，并采用最优配置方案；要处处考虑冗余。请你赶快动手接受挑战吧。

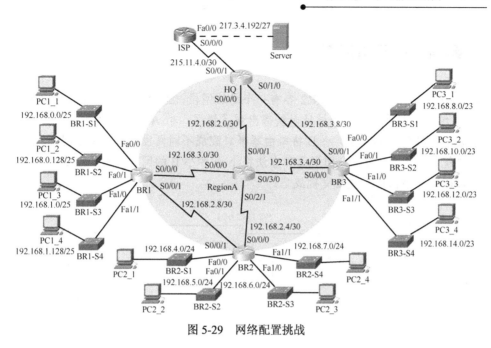

图 5-29 网络配置挑战

5.5.2 案例二：潜精研思挑战

挑战背景：请根据图 5-30 所示拓扑用 Packet Tracer 搭建网络，5 台路由器实现了两两互连，链路出口带宽完全一致。用本章所学知识配置并调通网络，确保终端 PC 间可以直接通信。请小伙伴们挑战一下谁最快完成，谁的配置方案最优。

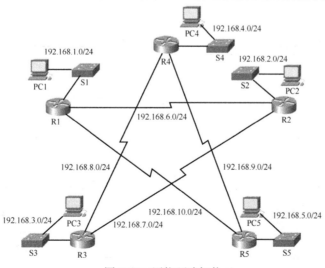

图 5-30 网络环路拓扑

5.5.3 案例三：开拓创新挑战

> 🔍 **挑战背景**：路由器 R1 和 R2 各有一个局域网段，通过串行接口实现了互连。小 A 和小 B 同学在配置路由器串口地址时候都出现了错误。小 A 将路由器两个串口配成了同一个地址（如图 5-31 所示），小 B 将路由器串口配成不同网段的地址（如图 5-32 所示）。请你将错就错，开动脑筋，分别用两种方法将小 A 和小 B 的网络调通，只要终端 PC 可以实现互访即可。小伙伴赶快接受挑战吧！

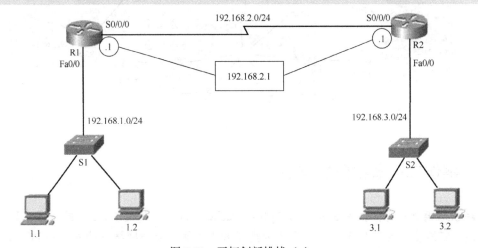

图 5-31 开拓创新挑战（a）

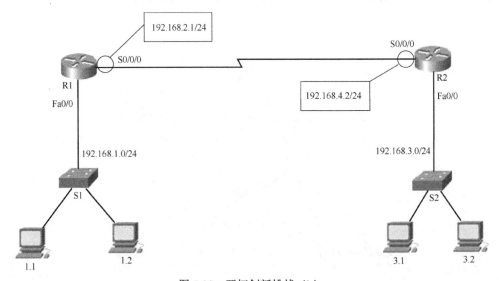

图 5-32 开拓创新挑战（b）

5.5.4 案例四：面试闯关挑战

> **挑战背景**：X 公司是一家小型网络公司，公司刚接手一个大项目，急需招聘多名有工作经验的 CCNA 认证工程师。你是这次来公司的应聘者之一，下面是你抽取到的技术面试题。

如下显示的是 Cisco 某一型号路由器的路由表，请根据路由表进行答题。

```
R1#show ip route
Codes:   C - connected, S - static, I - IGRP, R - RIP, M - mobile, B - BGP
         D - EIGRP, EX - EIGRP external, O - OSPF, IA - OSPF inter area
         N1 - OSPF NSSA external type 1, N2 - OSPF NSSA external type 2
         E1 - OSPF external type 1, E2 - OSPF external type 2, E - EGP
         i - IS-IS, L1 - IS-IS level-1, L2 - IS-IS level-2, ia - IS-IS inter area
         * - candidate default, U - per-user static route, o - ODR
         P - periodic downloaded static route

Gateway of last resort is not set

S    10.0.0.0/8 [1/0] via 172.16.40.2
     64.0.0.0/16 is subnetted, 1 subnets
C       64.100.0.0 is directly connected, Serial0/1/0
C    128.107.0.0/16 is directly connected, Loopback2
     172.16.0.0/24 is subnetted, 1 subnets
S       172.16.40.0 is directly connected, Serial0/2/1
C    192.168.1.0/24 is directly connected, FastEthernet0/0
C    192.168.2.0/24 [1/0] via 172.16.40.2
C    198.133.219.0/24 is directly connected, Loopback0
```

① 请找出路由表中的错误并进行改正。
② 请打开 Packet Tracer 模拟器，依据你改正后的路由表搭建网络拓扑并进行配置，要求输出与其一致的路由表。

祝你闯关成功！

"学习静态路由"一章即将结束。本章通过设置的 10 个应用场景、1 个综合配置案例和 4 个拓展思维案例，将第 5 章所学知识进行了有机融合。其中 10 个场景承载了两种标准静态路由、7 种特殊静态路由以及 ARP 代理等知识及其典型应用，是课堂的缩影。场景多以问答的形式引出了技术的重点和难点，并对知识进行了梳理，有助于读者理解和记忆。本章采用的企业网配置案例是对所学技术的一个综合应用，并将课堂所学知识与实际工程进行了紧密结合。最后的 4 个拓展案例，旨在提高读者对本章所学知识的活学活用能力、综合分析能力以及创新能力，并进一步开阔视野，拓展思维。

第6章

学习 RIP 路由协议

本章要点

- 认识 RIP 路由协议
- 配置多个场景的网络
- RIPv2 解决 RIPv1 存在的问题
- 验证 RIP 协议特性
- 挑战闯关训练

第 5 章，"学习静态路由"，是用户或管理员通过手工方式来添加远程路由，从而实现远程连接。本章，"学习 RIP 路由协议"，是通过在路由器上配置路由协议，让路由器动态学习远程路由，从而方便网络后期的维护与管理，即可以动态适应网络拓扑的改变。RIP 是最早的因特网标准协议，是距离矢量路由协议。深入学习 RIP 协议可为后期学好其他高级路由协议奠定坚实的基础。为此，本章设计了 4 个应用场景，2 个问题解决方案，5 个实验和 1 个挑战闯关训练。其中 4 个应用场景将 RIPv1 的缺陷暴露无遗，如何解决这些问题，需要升级协议至 RIPv2，由后续的 2 个解决方案来实现；6 个实验进一步验证了 RIP 的相关特性，如度量、选路原则、跳步数限制、RIP 更新等；最后通过 1 个挑战闯关训练进一步检验对本章知识的理解与应用。本章共设计 12 个教学案例可让你轻松掌握知识，深入理解所学内容，避免枯燥的强化记忆。

6.1 认识 RIP 路由协议

6.1.1 认识动态路由协议

1. 动态路由协议的定义

动态路由协议是路由器之间交换路由信息的协议。通过路由协议，路由器可以动态学习有关远程网络的信息，并自动将信息添加到各自的路由表中。路由协议可以确定到达各个目的网络的最佳路径。

2. 动态路由协议的好处

动态路由协议的一个主要好处是，只要网络拓扑结构发生了变化，路由器之间就会交换路由信息，通过信息交换，路由器不仅可以自动获知新增加的网络，也可以在当前网络失败时自动找出备用路径。

3. 动态与静态路由的比较

动态路由与静态路由的比较如表 6-1 所示。

表 6-1　动态路由与静态路由的比较

特 性 比 较	动 态 路 由	静 态 路 由
资源占用情况	占用 CPU、内存和链路带宽资源多	占 CPU 和内存资源少，不占链路带宽
拓扑结构改变	动态适应网络拓扑的改变	需要管理员参与，手动调整配置
配置的复杂性	通常不受网络规模的限制	伴随网络规模扩大，配置趋向复杂
对网管员要求	较高，需掌握高级的知识与技能	较低，不需要掌握高深的专业知识
网络的扩展性	扩展性好，简单和复杂网络均可	扩展性差，适合简单网络拓扑
网络的安全性	不够安全，因需交换路由信息	更安全，因无须交换路由信息

尽管动态路由协议有诸多好处，但静态路由仍有其用武之地。有的网络适合使用静态路由，有的网络更适合动态路由，也有的网络需要动态和静态路由相结合。

4. 动态路由协议的发展

动态路由协议自 20 世纪 80 年代初期开始应用于网络。RIP（路由信息协议）是最早的路由协议之一，目前已经演变到 RIPv2。但 RIPv2 仍旧不具有扩展性，无法用于较大型的网络。为满足大型网络的需求，两种高级路由协议 OSPF（开放式最短路径优先）和 IS-IS（中间系统到中间系统）应运而生。思科也推出了面向大型网络的 IGRP（内部网关路由协议）和 EIGRP（增强型 IGRP）协议。此外不同网际网之间也提出了网间路由的需求，在各 ISP 间以及 ISP 与大型专有客户之间采用 BGP（边界网关协议）来交换路由信息。伴随着 IPv6 的出现，新的 IPv6 路由协议的诞生用于支持 IPv6 通信，如表 6-2 所示。

表 6-2 路由协议的发展及分类

内部网关协议（IGP）				外部网关协议（EGP）	
距离矢量路由协议（D-V）		链路状态路由协议（L-S）		路径矢量路由协议（P-V）	
RIP	IGRP			EGP	有类
RIPv2	EIGRP	OSPFv2	IS-IS	BGPv4	无类
RIPng	EIGRP for IPv6	OSPFv3	IS-IS for IPv6	BGPv4 for IPv6	IPv6

6.1.2 RIP 路由协议的特点

RIP（Routing Information Protocol）协议，是由美国 Xerox 公司在 20 世纪 70 年代开发的，最初在 RFC1058 中定义，是最早的距离矢量（Distance Vector，D-V）路由协议，是因特网的标准协议。距离矢量以距离和方向构成的矢量来通告路由信息。RIP 协议的"距离"按"跳数"度量，方向则是下一跳的路由器或送出接口。RIP 协议会定期向直连的邻居路由器发送完整的路由表。在大型网络中，路由更新的数据量愈趋庞大，因而会在链路上产生大规模的通信流量。

6.1.3 RIP 路由协议的版本

尽管 RIP 缺少许多更为高级路由协议所具备的复杂功能，但其最大优点是简单。RIP 前后发布的两个版本 v1 和 v2 都用于 IPv4 通信，通常所说的 RIP 协议指的是版本 1，版本 2 通常表示为 RIPv2，RIP 只适用于小型网络。两个版本的消息格式如图 6-1 和图 6-2 所示。两个版本的 RIP 消息都包含字段命令、版本号和路由条目（最多 25 条）。每个路由条目都包含地址类型标识符、路由可达的 IP 地址及其度量值。RIP 消息的头部占 4 个八位组字节，每个路由条目占

20个8位组字节（5个32位字长）。

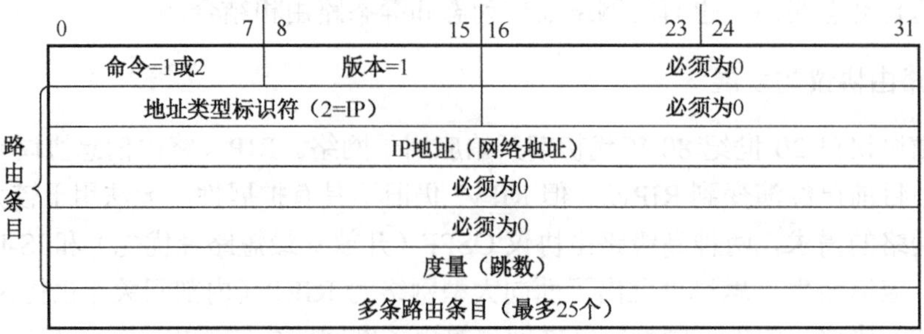

图 6-1 RIPv1 的消息格式

图 6-2 RIPv2 的消息格式

- **命令字段**：取值为 1 或 2。若为 1，表示该消息是请求消息；若为 2，表示该消息是响应消息。起初，RIP 从每个启用 RIP 协议的接口广播或组播带有请求消息的数据包。接着 RIP 进程进入一个循环状态，不断地侦听来自其他路由器的 RIP 请求或响应消息，而接收请求的路由器则回送包含其路由表的响应消息。
- **IP 地址字段**：路由的目的地址。可以是主网地址、子网地址等。
- **度量字段**：RIP 的跳数，取值范围为 1~16。接收到度量为 16 的路由条目会忽略该条消息。
- **子网掩码字段**：RIPv1 没有，RIPv2 有。RIPv1 是有类路由协议，RIPv2 是无类路由协议，其本质区别在于 RIPv2 的消息格式有子网掩码字段，打破对有类网络（A 类/8，B 类/16，C 类/24）采用默认子网掩码的限制，故 RIPv2 支持 VLSM 和 CIDR，而 RIPv1 不支持，其根本原因在于没有子网掩码字段。

6.1.4　RIP 路由协议的原理

默认情况下，每个开启 RIP 协议的路由器会每隔 30 秒利用 UDP 的 520 端口向其直连的邻居路由器广播或组播路由更新信息，初始情况下路由器不知道网络的全局情况。若路由更新信息在网络上传播较慢，路由表达到一致的过程就慢，从而会导致路由环路。RIP 采用水平分割、毒化反转、定义最大跳数、触发更新和抑制计时器等机制来避免环路的产生。RIP 要求网络中每一台路由器都要维护从其自身到每个目的网络的距离记录（这一组记录，即"距离矢量"）。RIP 将"距离"定义为从本地路由器到直连网络的距离为 0，到非直接网络的距离为每经过一个路由器值加 1，这里的"距离"通常被称为"跳数"。

当路由器冷启动或通电开机时，它完全不了解网络拓扑结构，甚至不知道在其链路另一端是否存在其他设备，它唯一能了解的信息来自其 NVRAM 中存储的配置文件。路由器成功启动之后，将会加载配置文件。若路由器正确配置了接口 IP 地址，则它会首先发现其接口的直连路由，并将其添加到路由表中，这是路由器的初始信息，如图 6-3 所示。

图 6-3　发现直连路由

有了直连路由，路由器彼此间才能交换路由信息，才可以接收用户静态配置信息或通过动态路由协议学习路由。

配置路由协议后，路由器就会开始交换路由更新信息。一开始，这些更新仅仅包含有关其直连网络的信息。收到更新后，路由器会检查更新，从中找出新信息，任何当前路由表中没有的且可达的路由都会被添加到路由表中。如图 6-4 所示的就是路由器 R1、R2 和 R3 初次交换路由信息的过程。三台路由器都发送仅仅含有直连网络的路由表给其直连邻居，每台路由器都会根据从邻居接收到的路由表来更新自己的路由表。

图 6-4　初次路由信息交换

经过第一轮更新后，每台路由器都学习到其直连邻居的直连网络。你是否发现路由器 R1 还不知道如何到达 10.4.0.0 网络，R3 也不知道如何到达 10.1.0.0 网络，因此还需要经过一次路由信息交换，网络中路由器的路由表才能够达到完全一致，我们把这个过程叫作收敛。所谓的收敛，就是指网络中所有路由器的路由表达到一致的过程，当所有路由器都获取到有关全网的完整而准确的网络信息时，网络即完成了收敛。

在图 6-4 中，路由器在获知其直连网络基础上，又学习到其直连邻居的直连网络，如何获知直连邻居的邻居的直连网络呢？例如，R1 如何获知其直连邻居 R2 的直连邻居 R3 的直连网络呢？这就要依赖路由器的下一轮更新之后，网络才可以收敛，如图 6-5 所示的网络在第二轮更新之后完成了收敛。

图 6-5 第二轮路由更新后网络收敛

3 台直连路由器经两轮更新后完成网络收敛，那 4 台直连路由器需要几轮更新才可以完成收敛呢？如图 6-6 所示，主要描述了 4 台路由器的网络是如何完成收敛的。

图 6-6 所示的网络，经过了三轮更新后，达到了收敛。在 t_0 时刻，路由器彼此学习到了自己的直连路由，直连路由最先加入路由表；在 t_1 时刻，路由器与直连邻居开始交换信息，彼此学习到对方的直连路由，完成第一轮的更新；在 t_2 时刻，直连邻居之间互相发送自己第一轮更新后的路由表，此时路由器 RouterB 和 RouterC 在第二轮更新之后路由表完整；在 t_3 时刻，路由器直连邻居之间再一次交换第二轮更新后的路由表，在第三轮更新中，路由器 RouterA 和 RouterD 学习到了完整的路由条目，网络达到收敛。

那么，请大家来思考一个问题，若有 n 台（$2 \leqslant n \leqslant 16$）路由器串联在一起，路由器运行 RIP 协议，经几轮更新后，网络才可以收敛？ 答案应该是 n-1 轮更新。

```
                10.1.1.0            10.1.2.0            10.1.3.0            10.1.4.0            10.1.5.0
                        .1       .1         .1       .2         .2       .1         .2       .1
                        Router A              Router B              Router C              Router D
```

	网络	下一跳	跳数	网络	下一跳	跳数	网络	下一跳	跳数	网络	下一跳	跳数
t_0	10.1.1.0	---	0	10.1.2.0	---	0	10.1.3.0	---	0	10.1.4.0	---	0
	10.1.2.0	---	0	10.1.3.0	---	0	10.1.4.0	---	0	10.1.5.0	---	0

	网络	下一跳	跳数	网络	下一跳	跳数	网络	下一跳	跳数	网络	下一跳	跳数
t_1	10.1.1.0	---	0	10.1.2.0	---	0	10.1.3.0	---	0	10.1.4.0	---	0
	10.1.2.0	---	0	10.1.3.0	---	0	10.1.4.0	---	0	10.1.5.0	---	0
	10.1.3.0	10.1.2.2	1	10.1.1.0	10.1.2.1	1	10.1.2.0	10.1.3.1	1	10.1.3.0	10.1.4.1	1
				10.1.4.0	10.1.3.2	1	10.1.5.0	10.1.4.2	1			

	网络	下一跳	跳数	网络	下一跳	跳数	网络	下一跳	跳数	网络	下一跳	跳数
t_2	10.1.1.0	---	0	10.1.2.0	---	0	10.1.3.0	---	0	10.1.4.0	---	0
	10.1.2.0	---	0	10.1.3.0	---	0	10.1.4.0	---	0	10.1.5.0	---	0
	10.1.3.0	10.1.2.2	1	10.1.1.0	10.1.2.1	1	10.1.2.0	10.1.3.1	1	10.1.3.0	10.1.4.1	1
	10.1.4.0	10.1.2.2	2	10.1.4.0	10.1.3.2	1	10.1.5.0	10.1.4.2	1	10.1.2.0	10.1.4.1	2
				10.1.5.0	10.1.3.2	2	10.1.1.0	10.1.3.1	2			

	网络	下一跳	跳数	网络	下一跳	跳数	网络	下一跳	跳数	网络	下一跳	跳数
t_3	10.1.1.0	---	0	10.1.2.0	---	0	10.1.3.0	---	0	10.1.4.0	---	0
	10.1.2.0	---	0	10.1.3.0	---	0	10.1.4.0	---	0	10.1.5.0	---	0
	10.1.3.0	10.1.2.2	1	10.1.1.0	10.1.2.1	1	10.1.2.0	10.1.3.1	1	10.1.3.0	10.1.4.1	1
	10.1.4.0	10.1.2.2	2	10.1.4.0	10.1.3.2	1	10.1.5.0	10.1.4.2	1	10.1.2.0	10.1.4.1	2
	10.1.5.0	10.1.2.2	3	10.1.5.0	10.1.3.2	2	10.1.1.0	10.1.3.1	2	10.1.1.0	10.1.4.1	3

图 6-6 网络收敛过程

6.1.5 RIP 路由协议的配置

RIP 协议的配置语法如下所述。

- Router(config)#**router rip**：启用 RIP 协议（无进程号选项，即每台路由器只支持一个 RIP 进程）；
- Router(config-router)#**network** *network-address*：在指定接口上启用 RIP 进程并通告直连路由；
- 配置 RIP 协议只需要两步：启动 RIP 进程，指定需要运行 RIP 协议的主网；
- no router rip：禁用 RIP 协议（关闭 RIP 进程）；
- network-address：有类网络地址即主网号（family address），若 network-address 被指定为接口 IP 地址或子网地址，则路由器能自动将其转换为相应的主网地址，如 network 10.1.2.1 或 network 10.1.2.0，路由器会自动转换为 network 10.0.0.0。

协议模式的部分配置命令如下：

```
Router(config-router)#?
auto-summary        Enter Address Family command mode        //开启自动汇总功能
```

default-information	Control distribution of default information	//设置默认路由传播
distance	Define an administrative distance	//修改协议管理距离
exit	from routing protocol configuration mode	//退出协议模式
network	Enable routing on an IP network	
no	Negate a command or set its defaults	
passive-interface	Suppress routing updates on an interface	//设置被动接口
redistribute	Redistribute information from another routing protocol	//路由重分发
timers	Adjust routing timers	//设置 RIP 计时器
version	Set routing protocol version	//设置路由协议版本

6.2 配置多个场景的网络

6.2.1 场景一：配置有类网络

🔑**场景一**：如图 6-7 所示网络，3 台路由器 R1、R2 和 R3 各自连接一个局域网，路由器之间通过 WAN 接口实现互连，图中网络采用 5 个 C 类的主网地址进行规划，要求采用 RIPv1 实现全网互通。

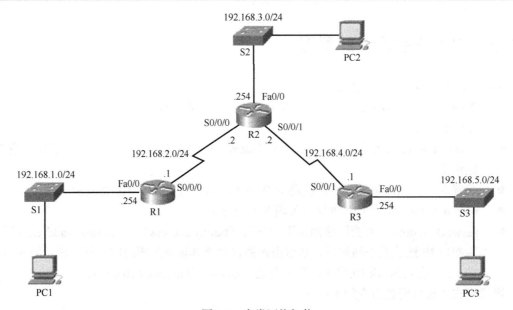

图 6-7　有类网络拓扑

具体配置步骤如下。

步骤一：路由器基本配置

确保路由器的直连路由出现在路由表中，否则需要继续排错，方可进行下一步。

步骤二：配置 RIP 协议

在路由器 R1 上配置 RIP 协议。

R1(config)#**router rip**
R1(config-router)#**network 192.168.1.0**
R1(config-router)#**network 192.168.2.0**

在路由器 R2 上配置 RIP 协议。

R2(config)#**router rip**
R2(config-router)#**network 192.168.2.0**
R2(config-router)#**network 192.168.3.0**
R2(config-router)#**network 192.168.4.0**

在路由器 R3 上配置 RIP 协议。

R3(config)#**router rip**
R3(config-router)#**network 192.168.4.0**
R3(config-router)#**network 192.168.5.0**

步骤三：查看路由表

查看路由器 R1 的路由表。

R1#**show ip route**
Codes: C - connected, S - static, I - IGRP, R - RIP, M - mobile, B - BGP
Gateway of last resort is not set

C   192.168.1.0/24 is directly connected, FastEthernet0/0
C   192.168.2.0/24 is directly connected, Serial0/0/0
R   192.168.3.0/24 [120/1] via 192.168.2.2, 00:00:16, Serial0/0/0
R   192.168.4.0/24 [120/1] via 192.168.2.2, 00:00:16, Serial0/0/0
R   192.168.5.0/24 [120/2] via 192.168.2.2, 00:00:16, Serial0/0/0

路由器 R1 的路由表完整，经接口 S0/0/0 通过邻居路由器 192.168.2.2 学习到了 3 条路由。
查看路由器 R2 的路由表。

R2#**show ip route**
Codes: C - connected, S - static, I - IGRP, R - RIP, M - mobile, B - BGP
Gateway of last resort is not set

```
R       192.168.1.0/24 [120/1] via 192.168.2.1, 00:00:19, Serial0/0/0
C       192.168.2.0/24 is directly connected, Serial0/0/0
C       192.168.3.0/24 is directly connected, FastEthernet0/0
C       192.168.4.0/24 is directly connected, Serial0/0/1
R       192.168.5.0/24 [120/1] via 192.168.4.1, 00:00:26, Serial0/0/1
```

路由器 R2 的路由表完整，通过本地接口 S0/0/0 经邻居路由器 192.168.2.1 学习到了路由 192.168.1.0/24，并通过接口 S0/0/1 经邻居路由器 192.168.4.1 学习到了路由 192.168.5.0/24。再查看路由器 R3 的路由表。

```
R3#show ip route
Codes: C - connected, S - static, I - IGRP, R - RIP, M - mobile, B - BGP
Gateway of last resort is not set

R       192.168.1.0/24 [120/2] via 192.168.4.2, 00:00:07, Serial0/0/1
R       192.168.2.0/24 [120/1] via 192.168.4.2, 00:00:07, Serial0/0/1
R       192.168.3.0/24 [120/1] via 192.168.4.2, 00:00:07, Serial0/0/1
C       192.168.4.0/24 is directly connected, Serial0/0/1
C       192.168.5.0/24 is directly connected, FastEthernet0/0
```

路由器 R3 的路由表也完全正确，经接口 S0/0/1 学习到下一跳路由器（192.168.4.2）的 3 条路由 192.168.1.0/24、192.168.2.0/24 和 192.168.3.0/24，全网收敛。

步骤四：网络连通性测试

测试顺利通过，PC 间可互访，全网实现了互联互通。

6.2.2 场景二：配置不连续网络

场景二： 如图 6-8 所示，3 台路由器 R1、R2 和 R3 各自连接一个局域网，路由器之间通过 WAN 接口实现了互连，其中路由器 R1 和 R3 连接的局域网是同一个主网 192.168.1.0/24 下的两个子网，它们被主网 192.168.2.0/24 和 192.168.4.0/24 分割，形成了不连续网络（Discontiguous），请采用 RIPv1 配置网络。

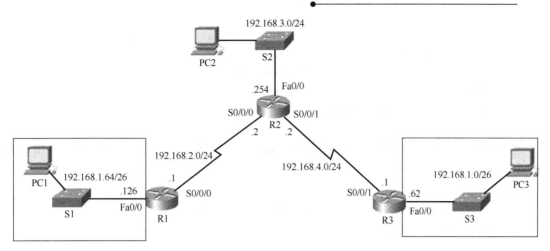

图 6-8 不连续网络拓扑

具体配置步骤如下。

步骤一：路由器基本配置

确保路由器的直连路由出现在路由表中。

步骤二：配置 RIP 协议

在路由器 R1 上配置 RIP 协议。

R1(config)#**router rip**
R1(config-router)#**network 192.168.1.0**
R1(config-router)#**network 192.168.2.0**

在路由器 R2 上配置 RIP 协议。

R2(config)#**router rip**
R2(config-router)#**network 192.168.2.0**
R2(config-router)#**network 192.168.3.0**
R2(config-router)#**network 192.168.4.0**

在路由器 R3 上配置 RIP 协议。

R3(config)#**router rip**
R3(config-router)#**network 192.168.1.0**
R3(config-router)#**network 192.168.4.0**

步骤三：查看路由表

R1#**show ip route**
Codes: C - connected, S - static, I - IGRP, R - RIP, M - mobile, B - BGP
Gateway of last resort is not set

```
              192.168.1.0/26 is subnetted, 1 subnets
C                192.168.1.64 is directly connected, FastEthernet0/0
C         192.168.2.0/24 is directly connected, Serial0/0/0
R         192.168.3.0/24 [120/1] via 192.168.2.2, 00:00:11, Serial0/0/0
R         192.168.4.0/24 [120/1] via 192.168.2.2, 00:00:11, Serial0/0/0
```

通过查看 R1 路由表，我们发现缺少 192.168.1.0/26 的路由，接下来我们去查看路由器 R2 的路由表。

```
R2#show ip route
Codes: C - connected, S - static, I - IGRP, R - RIP, M - mobile, B - BGP
Gateway of last resort is not set

R         192.168.1.0/24 [120/1] via 192.168.2.1, 00:00:21, Serial0/0/0
                         [120/1] via 192.168.4.1, 00:00:12, Serial0/0/1
C         192.168.2.0/24 is directly connected, Serial0/0/0
C         192.168.3.0/24 is directly connected, FastEthernet0/0
C         192.168.4.0/24 is directly connected, Serial0/0/1
```

结果我们发现 R2 的路由表缺少子网 192.168.1.0/26 和 192.168.1.64/26 的路由。但是 R2 的路由表却显示了主网 192.168.1.0/24 的路由，且显示了两条等价路径均可到达主网 192.168.1.0/24。R3 的路由表是怎样的，我们继续查看。

```
R3#show ip route
Codes: C - connected, S - static, I - IGRP, R - RIP, M - mobile, B - BGP

Gateway of last resort is not set


              192.168.1.0/26 is subnetted, 1 subnets
C                192.168.1.0 is directly connected, FastEthernet0/0
R         192.168.2.0/24 [120/1] via 192.168.4.2, 00:00:20, Serial0/0/1
R         192.168.3.0/24 [120/1] via 192.168.4.2, 00:00:20, Serial0/0/1
C         192.168.4.0/24 is directly connected, Serial0/0/1
```

通过查看发现路由器 R3 的路由表和 R1 的非常相似，都缺少对方的子网路由。

为什么路由器会显示不完整的路由条目呢？这主要源于拓扑中出现了不连续网络，而 RIPv1 协议对不连续网络是不支持的。所谓的不连续网络就是一个主网的多个子网被至少一个主网分割，导致网络不连续。在该场景中，主网 192.168.1.0/24 的两个子网 192.168.1.0/26 和 192.168.1.64/26 被主网 192.168.2.0/24 和 192.168.4.0/24 分割，产生了不连续网络。

步骤四：网络连通性测试

在 PC1 上 ping 主机 PC2，结果如下：

C:\>**ping 192.168.3.1**

Pinging 192.168.3.1 with 32 bytes of data:
Reply from 192.168.3.1: bytes=32 time=17ms TTL=126
Request timed out.
Reply from 192.168.3.1: bytes=32 time=11ms TTL=126
Request timed out.

Ping statistics for 192.168.3.1:
  Packets: Sent = 4, Received = 2, Lost = 2 (50% loss),
Approximate round trip times in milli-seconds:
  Minimum = 11ms, Maximum = 17ms, Average = 14ms

PC1 和 PC2 之间出现间歇性通信，成功率 50%。在主机 PC2 上 ping 主机 PC3，结果如下：

C:\>**ping 192.168.1.1**

Pinging 192.168.1.1 with 32 bytes of data:
Reply from 192.168.2.1: Destination host unreachable.
Reply from 192.168.1.1: bytes=32 time=12ms TTL=126
Reply from 192.168.2.1: Destination host unreachable.
Reply from 192.168.1.1: bytes=32 time=12ms TTL=126

Ping statistics for 192.168.1.1:
  Packets: Sent = 4, Received = 2, Lost = 2 (50% loss),
Approximate round trip times in milli-seconds:
  Minimum = 2ms, Maximum = 15ms, Average = 20ms

PC2 和 PC3 之间也是时断时续，成功率是 50%。在主机 PC3 上 ping 主机 PC1，结果如下：

C:\>**ping 192.168.1.65**

Pinging 192.168.1.65 with 32 bytes of data:
Reply from 192.168.1.62: Destination host unreachable.
Reply from 192.168.1.62: Destination host unreachable.
Request timed out.
Reply from 192.168.1.62: Destination host unreachable.

Ping statistics for 192.168.1.65:
  Packets: Sent = 4, Received = 0, Lost = 4 (100% loss),

PC1 不能与 PC3 通信，丢包率 100%。

显然在场景二中，通过 RIPv1 并没有实现全网络互通，那如何解决网络不通问题？我们将在 6.3.1 节中介绍。

6.2.3 场景三：配置等长掩码网络

> **场景三**：如图 6-9 所示，3 台路由器 R1、R2 和 R3 各自连接一个局域网，路由器通过 WAN 接口实现了互连，整个网络地址的规划采用了相同主网 172.30.0.0/16 的子网，且是等长子网掩码，均为/24，网间互连互通 IPv1 实现。

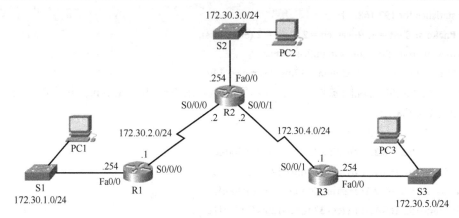

图 6-9 等长子网掩码网络拓扑

具体配置步骤如下。

步骤一：路由器基本配置

确保路由器的直连路由出现在路由表中。

步骤二：配置 RIP 协议

在路由器 R1 上配置 RIP 协议。

R1(config)#**router rip**
R1(config-router)#**network 172.30.0.0**

在路由器 R2 上配置 RIP 协议。

R2(config)#**router rip**
R2(config-router)#**network 172.30.0.0**

在路由器 R3 上配置 RIP 协议。

R3(config)#**router rip**
R3(config-router)#**network 172.30.0.0**

步骤三：查看路由表

```
R1#show ip route
Codes: C - connected, S - static, I - IGRP, R - RIP, M - mobile, B - BGP
Gateway of last resort is not set

     172.30.0.0/24 is subnetted, 5 subnets
C       172.30.1.0 is directly connected, FastEthernet0/0
C       172.30.2.0 is directly connected, Serial0/0/0
R       172.30.3.0 [120/1] via 172.30.2.2, 00:00:21, Serial0/0/0
R       172.30.4.0 [120/1] via 172.30.2.2, 00:00:21, Serial0/0/0
R       172.30.5.0 [120/2] via 172.30.2.2, 00:00:21, Serial0/0/0
```

路由器 R1 的路由表是完整的，通过同一接口 S0/0/0 经邻居路由器 172.30.2.2（下一跳）学习到 3 条远程路由 172.30.3.0/24、172.30.4.0/24 和 172.30.5.0/24。从 R1 的路由表中，我们不难推断出 R1 只有一个直连的三层邻居 172.30.2.2。

```
R2#show ip route
Codes: C - connected, S - static, I - IGRP, R - RIP, M - mobile, B - BGP
Gateway of last resort is not set

     172.30.0.0/24 is subnetted, 5 subnets
R       172.30.1.0 [120/1] via 172.30.2.1, 00:00:18, Serial0/0/0
C       172.30.2.0 is directly connected, Serial0/0/0
C       172.30.3.0 is directly connected, FastEthernet0/0
C       172.30.4.0 is directly connected, Serial0/0/1
R       172.30.5.0 [120/1] via 172.30.4.1, 00:00:26, Serial0/0/1
```

路由器 R2 的路由表也是完整的，通过接口 S0/0/0 经邻居路由器 172.30.2.1 学习到远程网络 172.30.1.0/24 的路由；通过接口 S0/0/1 经邻居路由器 172.30.4.1 学习到远程网络 172.30.5.0/24 的路由。我们也可以推断出 R2 有两个直连的三层邻居 172.30.2.1 和 172.30.4.1。

```
R3#show ip route
Codes: C - connected, S - static, I - IGRP, R - RIP, M - mobile, B - BGP
Gateway of last resort is not set

     172.30.0.0/24 is subnetted, 5 subnets
R       172.30.1.0 [120/2] via 172.30.4.2, 00:00:22, Serial0/0/1
R       172.30.2.0 [120/1] via 172.30.4.2, 00:00:22, Serial0/0/1
R       172.30.3.0 [120/1] via 172.30.4.2, 00:00:22, Serial0/0/1
C       172.30.4.0 is directly connected, Serial0/0/1
C       172.30.5.0 is directly connected, FastEthernet0/0
```

最后，我们查看路由器 R3 的路由表，发现路由表也准确无误。经接口 S0/0/1 从邻居路由器 172.30.4.2 学习到去往 172.30.1.0/24（2 跳可达）、172.30.2.0/24（1 跳可达）和 172.30.3.0/24（1 跳可达）3 条路由，全网达到收敛状态。3 张路由表的共同特征是有 1 条一级父路由，5 条二级子路由。

步骤四：网络连通性测试

测试顺利通过，PC 间可互访，全网实现了互连互通。

6.2.4 场景四：配置 VLSM 与 CIDR 网络

> **场景四**：如图 6-10 所示，4 个局域网 172.30.1.0/24、172.30.3.0/24、192.168.0.0/23 和 172.30.5.0/27 通过 3 台路由器 R1、R2 和 R3 的 WAN 接口实现了互连，整个拓扑规划采用了相同主网 172.30.0.0/16 的 5 个子网（172.30.1.0/24、172.30.2.0/24、172.30.3.0/24、172.30.4.0/25 和 172.30.5.0/27）及 192.168.0.0/23 的超网。172.30.0.0/16 子网有/24、/25 和/27 三个子网掩码，属于 VLSM（Varible Length Subnet Mask），即不变化的子网的掩码网间互通采用 RIPv1 实现。

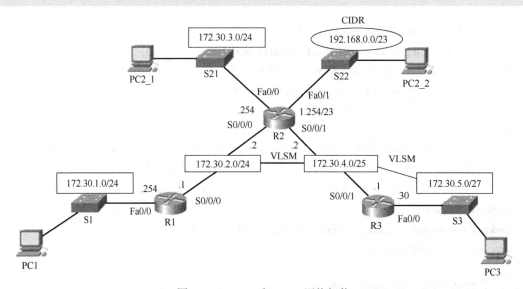

图 6-10 VLSM 和 CIDR 网络拓扑

具体配置步骤如下。

步骤一：路由器基本配置

确保路由器的直连路由出现在路由表中。

步骤二：配置 RIP 协议

在路由器 R1 上配置 RIP 协议。

R1(config)#**router rip**
R1(config-router)#**network 172.30.0.0**

在路由器 R2 上配置 RIP 协议。

R2(config)#**router rip**
R2(config-router)#**network 172.30.0.0**
R2(config-router)#**network 192.168.0.0**

在路由器 R3 上配置 RIP 协议。

R3(config)#**router rip**
R3(config-router)#**network 172.30.0.0**

步骤三：查看路由表

R1#**show ip route**

Codes: C - connected, S - static, I - IGRP, R - RIP, M - mobile, B - BGP

Gateway of last resort is not set

      172.30.0.0/24 is subnetted, 3 subnets
C   172.30.1.0 is directly connected, FastEthernet0/0
C   172.30.2.0 is directly connected, Serial0/0/0
R   **172.30.3.0 [120/1] via 172.30.2.2, 00:00:24, Serial0/0/0**

R1 的路由表只学习到 172.30.3.0/24 的路由条目，缺失远程路由 192.168.0.0/23、172.30.4.0/25 和 172.30.5.0/27。

R2#**show ip route**

Codes: C - connected, S - static, I - IGRP, R - RIP, M - mobile, B - BGP

Gateway of last resort is not set

      172.30.0.0/16 is variably subnetted, 4 subnets, 2 masks
R   **172.30.1.0/24 [120/1] via 172.30.2.1, 00:00:11, Serial0/0/0**
C   172.30.2.0/24 is directly connected, Serial0/0/0
C   172.30.3.0/24 is directly connected, FastEthernet0/0
C   172.30.4.0/25 is directly connected, Serial0/0/1
C   192.168.0.0/23 is directly connected, FastEthernet0/1

R2 的路由表缺失路由 172.30.5.0/27。

R3#**show ip route**

Codes: C - connected, S - static, I - IGRP, R - RIP, M - mobile, B - BGP

```
Gateway of last resort is not set

        172.30.0.0/16 is variably subnetted, 2 subnets, 2 masks
C       172.30.4.0/25 is directly connected, Serial0/0/1
C       172.30.5.0/27 is directly connected, FastEthernet0/0
```

R3 的路由表除了自身的直连路由外没有学习到任何路由条目。

步骤四：网络连通性测试

只有 PC1 和 PC2_1 可以通信，其他任意主机之间都无法通信。

在本场景中，由于路由器学习不到完整的路由表而影响全网互通，因为 RIPv1 不支持 VLSM 和 CIDR，因此导致路由表路由的缺失。如何实现本场景中 VLSM 和 CIDR 网络连通呢？我们将在 6.3.2 节解决这个问题。

6.2.5 总结一：收发更新原则

在前面的 4 个场景中，场景一和场景三经 RIPv1 配置完后，路由表准确无误，但是场景二和场景四的路由表出现了问题，路由表不完整，不能实现全网互通。通过下面的表 6-3 我们把 4 个场景的地址规划特点及产生结果做了比对，对场景二和场景四中出现的问题给出了相应的解决办法。

表 6-3　场景一至场景四地址规划特点及产生结果比较

场景	网络地址规划的特点	路由条目	路由表	说明
场景一	5 个主类网络	R1(5)R2(5) R3 (5)	完整	RIPv1 是有类网络协议
场景二	3 个主类网络，2 个不连续子网	R1(4)R2(5) R3 (4)	不完整	RIPv1 不支持不连续网络，子网发生了自动汇总 解决办法：更换 RIPv2 来配置网络且必须关闭自动汇总
场景三	同一主网的 5 个子网，且子网掩码等长	R1(5)R2(5) R3 (5)	完整	发送接口所在主网与发送路由条目的主网相同，且掩码相同，该路由条目就被发送
场景四	同一主网的 5 个子网，且掩码不等长（VLSM），1 个超网（CIDR）	R1(3)R2(5) R3 (2)	不完整	RIPv1 不支持 VLSM 和 CIDR 解决办法：采用 RIPv2

因为 RIPv1 是一种有类路由协议，它在发送路由更新时不带子网掩码，但是为什么路由表中的路由条目却都有子网掩码呢？运行 RIPv1 的路由器如何判断添加路由的子网掩码呢？下面我们一起来了解 RIPv1 收发更新的原则。

1. RIPv1 接收更新原则

RIPv1 接收更新信息要依赖接收接口。

① 若某条路由更新与其接收接口属于相同的主网，则对该路由条目应用该接口的子网掩码。

② 若某条路由更新后与其接收接口属于不同的主网，则对该路由条目应用其主网的有类掩码。

2. RIPv1 发送更新原则

① 若要发送的某条路由更新信息是主网地址（有类网络地址），则直接发送该主网路由信息。

② 若要发送的某条路由更新信息是超网地址，则不会发送该超网路由信息。

③ 若要发送的某条路由更新信息是子网地址，则要看发送接口的地址：

- 若属于相同主网，掩码等长，则发送子网路由信息；若掩码不等长，则不发送。
- 若属于不同主网，则发送其有类网络地址的主网路由信息，即自动汇总后再发送。

RIPv1 路由协议发送的更新信息会包含网络地址及相关度量，但不会在更新信息中发送子网掩码信息。因此路由器将使用本地接收接口的子网掩码或根据地址类别应用默认有类子网掩码。

6.3 RIPv2 解决 RIPv1 存在的问题

6.3.1 问题一：解决不连续网络问题

1. 问题描述

在 6.2.2 节的场景二中，我们采用了 RIPv1 配置不连续（Discontiguous）网络，结果导致路由器的路由表不完整。主要原因是同一主网 192.168.1.0/24 的两个子网 192.168.1.64/26 和 192.168.1.0/26 被主网 192.168.2.0/24 和 192.168.4.0/24 分割，导致网络不连续，如图 6-11 所示。RIPv1 对该场景的网络是不支持的，如何解决这个不连续网络导致的路由表不完整问题呢？请给出相应的解决方案。

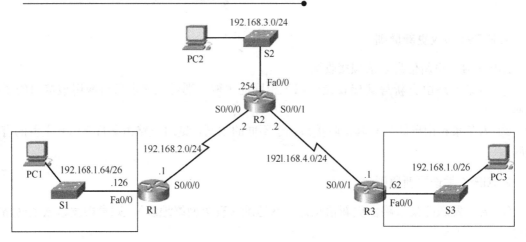

图 6-11 不连续（Discontiguous）网络

2. 解决方案

在所有路由器上升级路由协议为 RIPv2，并在路由器上关闭自动汇总功能。

步骤一：升级路由协议为 RIPv2

R1(config)#**router rip**
R1(config-router)#**version 2**
R1(config-router)#**no auto-summary**

R2(config)#**router rip**
R2(config-router)#**version 2**
R2(config-router)#**no auto-summary**

R3(config)#**router rip**
R3(config-router)#**version 2**
R3(config-router)#**no auto-summary**

步骤二：查看路由表

R1#**show ip route**
Codes: C - connected, S - static, I - IGRP, R - RIP, M - mobile, B - BGP
Gateway of last resort is not set

     192.168.1.0/26 is subnetted, 2 subnets
R   192.168.1.0 [120/2] via 192.168.2.2, 00:00:12, Serial0/0/0
C   192.168.1.64 is directly connected, FastEthernet0/0
C   192.168.2.0/24 is directly connected, Serial0/0/0

| R | 192.168.3.0/24 [120/1] via 192.168.2.2, 00:00:12, Serial0/0/0 |
| R | 192.168.4.0/24 [120/1] via 192.168.2.2, 00:00:12, Serial0/0/0 |

通过查看 R1 的路由表，我们发现另一子网 192.168.1.0/26 被正确安装到了 R1 的路由表中，路由表完整。

R2#**show ip route**
Codes: C - connected, S - static, I - IGRP, R - RIP, M - mobile, B - BGP
Gateway of last resort is not set

    192.168.1.0/26 is subnetted, 2 subnets
R   192.168.1.0 [120/1] via 192.168.4.1, 00:00:12, Serial0/0/1
R   192.168.1.64 [120/1] via 192.168.2.1, 00:00:14, Serial0/0/0
C   192.168.2.0/24 is directly connected, Serial0/0/0
C   192.168.3.0/24 is directly connected, FastEthernet0/0
C   192.168.4.0/24 is directly connected, Serial0/0/1

通过查看 R2 的路由表，我们发现两个子网路由均被正确安装到 R2 的路由表中。

R3#**show ip route**
Codes: C - connected, S - static, I - IGRP, R - RIP, M - mobile, B - BGP
Gateway of last resort is not set

    192.168.1.0/26 is subnetted, 2 subnets
C   192.168.1.0 is directly connected, FastEthernet0/0
R   192.168.1.64 [120/2] via 192.168.4.2, 00:00:05, Serial0/0/1
R   192.168.2.0/24 [120/1] via 192.168.4.2, 00:00:05, Serial0/0/1
R   192.168.3.0/24 [120/1] via 192.168.4.2, 00:00:05, Serial0/0/1
C   192.168.4.0/24 is directly connected, Serial0/0/1

路由器 R3 也学习到了路由器 R1 的子网 192.168.1.64/26 的路由，路由表完整，网络收敛，不连续网络问题得到了彻底解决。

6.3.2　问题二：解决 VLSM&CIDR 问题

1. 问题描述

在 6.2.4 节的场景四中，我们采用 RIPv1 来配置规划 VLSM 和 CIDR 的网络，结果导致路由器的路由表不完整。路由器 R1 仅学习到 R2 的 172.30.3.0/24 的路由条目，没有学到 R2 的路由 192.168.0.0/23 和 172.30.4.0/25，也没有学到 R3 的路由 172.30.5.0/27。路由器 R2 的路由表中缺少 R3 的路由 172.30.5.0/27。

R3 的路由表除了自身的直连路由，竟然没有学习到任何路由条目。这是因为 RIPv1 对 VLSM 和 CIDR 的网络不支持，如何解决这 3 台路由器路由表不完整的问题？请给出相应解决方案。

VLSM 和 CIDR 网络如图 6-12 所示。

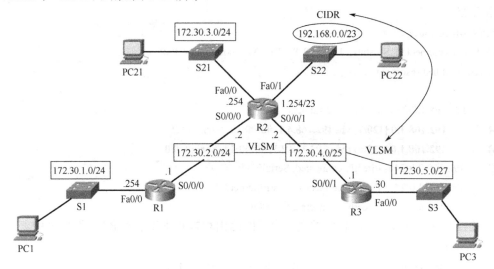

图 6-12　VLSM 和 CIDR 网络

2. 解决方案

在 3 台路由器 R1、R2 和 R3 上升级 RIPv1 协议为 RIPv2。

步骤一：升级协议为 RIPv2

R1(config)#**router rip**
R1(config-router)#**version 2**

R2(config)#**router rip**
R2(config-router)#**version 2**

R3(config)#**router rip**
R3(config-router)#**version 2**

注意：若升级至 RIPv2 后查看路由表，我们会发现路由器 R1 和 R3 都学习到了 172.30.0.0/24 的其他子网路由，解决了 VLSM 的问题。但是，我们看到路由器并没有学习到 192.168.0.0/23 的超网路由，要解决这个问题，需要在路由器 R2 上添加命令 **network192.168.1.0** 来完成。

步骤二：查看路由表

R1#**show ip route**
Codes: C - connected, S - static, I - IGRP, R - RIP, M - mobile, B - BGP
Gateway of last resort is not set

```
            172.30.0.0/16 is variably subnetted, 5 subnets, 3 masks
C              172.30.1.0/24 is directly connected, FastEthernet0/0
C              172.30.2.0/24 is directly connected, Serial0/0/0
R              172.30.3.0/24 [120/1] via 172.30.2.2, 00:00:19, Serial0/0/0
R              172.30.4.0/25 [120/1] via 172.30.2.2, 00:00:19, Serial0/0/0
R              172.30.5.0/27 [120/2] via 172.30.2.2, 00:00:19, Serial0/0/0
R              192.168.0.0/23 [120/1] via 172.30.2.2, 00:00:19, Serial0/0/0
```

通过查看 R1 的路由表，我们发现原来缺失的 3 条路由 172.30.4.0/25、172.30.5.0/27 和 192.168.0.0/23 被正确安装到了 R1 的路由表中，R1 的路由表完整。

```
R2#show ip route
Codes: C - connected, S - static, I - IGRP, R - RIP, M - mobile, B - BGP
Gateway of last resort is not set

            172.30.0.0/16 is variably subnetted, 5 subnets, 3 masks
R              172.30.1.0/24 [120/1] via 172.30.2.1, 00:00:00, Serial0/0/0
C              172.30.2.0/24 is directly connected, Serial0/0/0
C              172.30.3.0/24 is directly connected, FastEthernet0/0
C              172.30.4.0/25 is directly connected, Serial0/0/1
R              172.30.5.0/27 [120/1] via 172.30.4.1, 00:00:24, Serial0/0/1
C           192.168.0.0/23 is directly connected, FastEthernet0/1
```

通过查看 R2 的路由表，我们发现 R2 原来缺失的路由 172.30.5.0/27 出现在路由表中，R2 的路由表完整。

```
R3#show ip route
Codes: C - connected, S - static, I - IGRP, R - RIP, M - mobile, B - BGP
Gateway of last resort is not set

            172.30.0.0/16 is variably subnetted, 5 subnets, 3 masks
R              172.30.1.0/24 [120/2] via 172.30.4.2, 00:00:18, Serial0/0/1
R              172.30.2.0/24 [120/1] via 172.30.4.2, 00:00:18, Serial0/0/1
R              172.30.3.0/24 [120/1] via 172.30.4.2, 00:00:18, Serial0/0/1
C              172.30.4.0/25 is directly connected, Serial0/0/1
C              172.30.5.0/27 is directly connected, FastEthernet0/0
R           192.168.0.0/23 [120/1] via 172.30.4.2, 00:00:18, Serial0/0/1
```

通过查看 R3 路由表的输出，我们发现有 4 条路由安装进 R3 的路由表中，R3 的路由表完整，网络完成了收敛。在规划 VLSM 和 CIDR 的网络中，路由表不完整问题被 RIPv2 彻底解决。

6.3.3 总结二：两个版本的区别与联系

RIPv1 和 RIPv2 的区别如表 6-4 所示。

表 6-4 RIPv1 和 RIPv2 的区别

	RIPv1	RIPv2
协议分类	有类（Classful）路由协议	无类（Classless）路由协议
处理子网掩码	发送更新时，不发送子网掩码	发送更新时，发送子网掩码
不连续网络	不支持	支持
关闭自动汇总	不支持	支持
安全认证	不支持	支持
VLSM&CIDR	不支持	支持
更新方式	广播更新（255.255.255.255）	组播更新（224.0.0.9）

RIPv1 和 RIPv2 的共性如下所述。

- 都是距离矢量（D-V）路由协议，采用"跳数"作为唯一度量；
- 都将 15 跳作为最大度量，16 跳则不可达；
- 默认路由更新周期都为 30 s，默认管理距离（AD）为 120；
- 都支持水平分割、毒化反转、抑制计时器、触发更新等机制避免路由环路；
- 都支持等价路径的负载均衡，默认为 4；
- 都是应用层协议，源端口和目的端口为 UDP 520 端口。
- 一个更新数据包最多可以包含 25 个路由条目，数据包最大为 512 B（UDP 报头 8 B，RIP 报头 4 B，路由条目 25×20=500 B）。

6.4 验证 RIP 协议特性

6.4.1 实验一：验证用跳数度量最佳路径

1. RIP 协议度量实验 1

实验描述：3 台路由器 R1、R2 和 R3 之间，两两互连，其中 R1 与 R3 间链路的带宽是 56 kbps，R1 与 R2、R2 与 R3 间链路带宽均为 1.544 Mbps。

实验目的：验证 RIP 协议衡量最佳路径的唯一度量标准为跳数。根据 RIP 协议的度量标准，实验拓扑中 PC1 访问 PC2 应选择 R1→R3 路径，不会因高带宽而选择 R1→R2→R3 路径。

实验要求：搭建如图 6-13 所示拓扑，配置路由器和终端设备的地址并配置 RIP 协议使全网互通，修改链路出口带宽为图中所示。请采用相关的命令验证 PC1 访问 PC2 的路径。

实验拓扑：如图 6-13 所示。

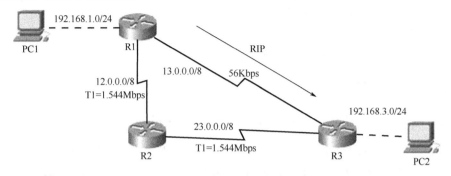

图 6-13　RIP 协议的度量（A）

如图 6-13 所示的 3 台路由器的接口地址已经配置好，终端 PC 地址也已经配置好。

配置 RIP 协议，修改 R1 至 R3 链路带宽。其他两条链路带宽为 1.544 Mbps，是默认带宽。

```
R1(config)#router rip
R1(config-router)#network 12.0.0.0
R1(config-router)#network 13.0.0.0
R1(config-router)#network 192.168.1.0
R1(config-router)#interface Serial0/0/1
R1(config-if)#bandwidth 56
```

```
R2(config)#router rip
R2(config-router)#network 12.0.0.0
R2(config-router)#network 23.0.0.0
```

```
R3(config)#router rip
R3(config-router)#network 13.0.0.0
R3(config-router)#network 23.0.0.0
R3(config-router)#network 192.168.3.0
R3(config-router)#interface Serial0/0/1
R3(config-if)#bandwidth 56
```

测试网络已经实现了互通，下面我们用两种方法验证 PC1 访问 PC2 数据包所经路径。

方法一：通过查看路由器的路由表确定下一跳

```
R1#show ip route
Codes: C - connected, S - static, I - IGRP, R - RIP, M - mobile, B - BGP
Gateway of last resort is not set
```

```
C    12.0.0.0/8 is directly connected, Serial0/0/0
C    13.0.0.0/8 is directly connected, Serial0/0/1
R    23.0.0.0/8 [120/1] via 13.0.0.3, 00:00:11, Serial0/0/1
C    192.168.1.0/24 is directly connected, FastEthernet0/0
R    192.168.3.0/24 [120/1] via 13.0.0.3, 00:00:11, Serial0/0/1
```

通过查看路由器 R1 的路由表可知，路由 192.168.3.0/24 的下一跳为 13.0.0.3 即 R3，所以 PC1 的数据包选择经 R1→R3 路径到达 PC2。

方法二：通过对源主机到目的主机的进行跟踪确定路径

跟踪至从源主机 192.168.1.2 到目的主机 192.168.3.2 的数据包结果如图 6-14 所示。

```
PC>ipconfig

FastEthernet0 Connection:(default port)

   Link-local IPv6 Address.........: ::
   IP Address......................: 192.168.1.2
   Subnet Mask.....................: 255.255.255.0
   Default Gateway.................: 192.168.1.1

PC>tracert 192.168.3.2

Tracing route to 192.168.3.2 over a maximum of 30 hops:

  1   0 ms       0 ms       0 ms       192.168.1.1
  2   1 ms       3 ms       1 ms       13.0.0.3
  3   0 ms       0 ms       1 ms       192.168.3.2

Trace complete.
```

图 6-14 跟踪结果

跟踪结果表明，数据包先后经网关 192.168.1.1（R1），再经 13.0.0.3（R3）到达目的主机 192.168.3.2（PC2），所以路径为 R1→R3 间的链路。

2. RIP 协议度量实验 2

实验描述： 如图 6-15 所示网络拓扑，两台 PC 间的互访要经过广域网，广域网链路出口带宽均一致，采用默认带宽。

实验目的： 验证 RIP 协议衡量最佳路径的唯一度量标准为跳数。根据 RIP 协议的度量标准，实验拓扑中 PC1 访问 PC2 有 4 条路径可以选择，理论上应选择路径 R1→R2，而不会选择其他三条路径。

实验要求： 搭建如图 6-15 所示拓扑，配置路由器和终端设备的地址，配置 RIP 协议使全网互通，然后采用相关命令验证 PC1 访问 PC2 的路径，尝试破坏当前的最佳路径，观察路由器如何选路。

实验拓扑：如图 6-15 所示。

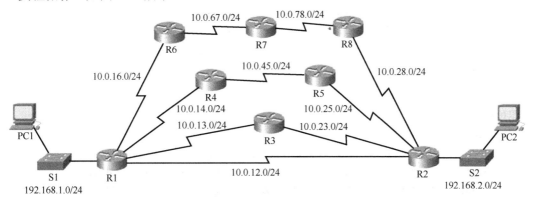

图 6-15　RIP 协议的度量（B）

本实验验证了从路由器 R1 到达目的网络 192.168.2.0/24 的最佳路径，通过查看路由表我们来分析路由器是如何选路的。从图 6-15 所示拓扑我们可以看到，R1 通往目的网络 192.168.2.0/24 共有 4 条路径，分别是 R1→R2、R1→R3→R2、R1→R4→R5→R2、R1→R6→R7→R8→R2。因为通过这 4 条路径到达目的网络的度量分别为 1 跳、2 跳、3 跳、4 跳，正常情况下路由器会将度量为 1 跳的路径安装到路由表中。我们通过查看 R1 的路由表来验证。

完成路由器的基本配置、协议配置后，再经过一段时间网络达到收敛状态，我们来查看当前路由器 R1 的路由表。

```
R1#show ip route
Codes: C - connected, S - static, I - IGRP, R - RIP, M - mobile, B - BGP
Gateway of last resort is not set

     10.0.0.0/24 is subnetted, 10 subnets
C       10.0.12.0 is directly connected, Serial0/1/1
C       10.0.13.0 is directly connected, Serial0/1/0
C       10.0.14.0 is directly connected, Serial0/0/1
C       10.0.16.0 is directly connected, Serial0/0/0
R       10.0.23.0 [120/1] via 10.0.13.3, 00:00:20, Serial0/1/0
                  [120/1] via 10.0.12.2, 00:00:17, Serial0/1/1
R       10.0.25.0 [120/1] via 10.0.12.2, 00:00:17, Serial0/1/1
R       10.0.28.0 [120/1] via 10.0.12.2, 00:00:17, Serial0/1/1
R       10.0.45.0 [120/1] via 10.0.14.4, 00:00:15, Serial0/0/1
R       10.0.67.0 [120/1] via 10.0.16.6, 00:00:15, Serial0/0/0
R       10.0.78.0 [120/2] via 10.0.12.2, 00:00:17, Serial0/1/1
```

                    [120/2] via 10.0.16.6, 00:00:15, Serial0/0/0
C   192.168.1.0/24 is directly connected, FastEthernet0/0
R   192.168.2.0/24 [120/1] via 10.0.12.2, 00:00:17, Serial0/1/1

从 R1 的路由表可以看出，通往 192.168.2.0/24 的最佳路径是度量为 1 跳的路径，下一跳为 10.0.12.2（R2）。若让度量为 2 跳的路由写进路由表中，除非它是最佳路径，我们可以人为使度量为 1 跳的链路出现故障，可以通过如下命令完成。

R1#**configure terminal**
R1(config)#**int s0/1/1**
R1(config-if)#**shutdown**
%LINK-5-CHANGED: Interface Serial0/1/1, changed state to administratively down
%LINEPROTO-5-UPDOWN: Line protocol on Interface Serial0/1/1, changed state to down

此时，S0/1/1 接口为非活动接口，路由器必须重新选路来适应网络拓扑的变化，可以查看此时的 R1 路由表发生了怎样的变化。

R1(config-if)#**do show ip route**
Codes: C - connected, S - static, I - IGRP, R - RIP, M - mobile, B - BGP
Gateway of last resort is not set

         10.0.0.0/24 is subnetted, 9 subnets
C   10.0.13.0 is directly connected, Serial0/1/0
C   10.0.14.0 is directly connected, Serial0/0/1
C   10.0.16.0 is directly connected, Serial0/0/0
R   10.0.23.0 [120/1] via 10.0.13.3, 00:00:16, Serial0/1/0
R   10.0.25.0 [120/2] via 10.0.14.4, 00:00:12, Serial0/0/1
                    [120/2] via 10.0.13.3, 00:00:16, Serial0/1/0
R   10.0.28.0 [120/2] via 10.0.13.3, 00:00:16, Serial0/1/0
R   10.0.45.0 [120/1] via 10.0.14.4, 00:00:12, Serial0/0/1
R   10.0.67.0 [120/1] via 10.0.16.6, 00:00:10, Serial0/0/0
R   10.0.78.0 [120/2] via 10.0.16.6, 00:00:10, Serial0/0/0
C   192.168.1.0/24 is directly connected, FastEthernet0/0
R   192.168.2.0/24 [120/2] via 10.0.13.3, 00:00:16, Serial0/1/0

通过查看，我们发现通往目的网络地址 192.168.2.0/24 的下一跳改为了 10.0.13.3（R3），度量为 2 跳。接着，我们进一步制造故障，让路由器 R1 重新选择并安装最佳路径，此时我们又 down 掉接口 S0/1/0。

R1(config-if)# **int s0/1/0**
R1(config-if)#**shutdown**
R1(config-if)#
%LINK-5-CHANGED: Interface Serial0/1/0, changed state to administratively down

%LINEPROTO-5-UPDOWN: Line protocol on Interface Serial0/1/0, changed state to down

S0/1/0 口二层协议 down 掉后，其 10.0.13.0/24 的直连路由将会从路由表中消失，与此同时与该接口关联的 192.168.2.0/24 的 2 跳路由会随之消失。接下来，路由器会从其他邻居中获取消息来更新其路由表。我们进一步查看当前路由器的路由表。

R1(config-if)#**do show ip route**
Codes: C - connected, S - static, I - IGRP, R - RIP, M - mobile, B - BGP
Gateway of last resort is not set

      10.0.0.0/24 is subnetted, 8 subnets
C   10.0.14.0 is directly connected, Serial0/0/1
C   10.0.16.0 is directly connected, Serial0/0/0
R   10.0.23.0 [120/3] via 10.0.14.4, 00:00:13, Serial0/0/1
R   10.0.25.0 [120/2] via 10.0.14.4, 00:00:13, Serial0/0/1
R   10.0.28.0 [120/3] via 10.0.14.4, 00:00:13, Serial0/0/1
            [120/3] via 10.0.16.6, 00:00:09, Serial0/0/0
R   10.0.45.0 [120/1] via 10.0.14.4, 00:00:13, Serial0/0/1
R   10.0.67.0 [120/1] via 10.0.16.6, 00:00:09, Serial0/0/0
R   10.0.78.0 [120/2] via 10.0.16.6, 00:00:09, Serial0/0/0
C   192.168.1.0/24 is directly connected, FastEthernet0/0
R   192.168.2.0/24 [120/3] via 10.0.14.4, 00:00:13, Serial0/0/1

我们看到了此时 R1 选择了通过度量为 3 跳的邻居 10.0.14.4（R4）到达目的网络地址 192.168.2.0/24。目前通往目的网络地址还有一条备用路径，即经 R6 到达的路径，若我们破坏了经 R4 到达的路径，路由器势必会选择经 R6 到达，因为它是唯一一条通往 192.168.2.0/24 的路径，再 down 掉 S0/0/1 接口，查看路由表来观察度量为 4 跳的路由能否写进路由表。

R1(config-if)# **int s0/0/1**
R1(config-if)#**shutdown**
%LINK-5-CHANGED: Interface Serial0/0/1, changed state to administratively down
%LINEPROTO-5-UPDOWN: Line protocol on Interface Serial0/0/1, changed state to down

R1(config-if)#**do show ip route**
Codes: C - connected, S - static, I - IGRP, R - RIP, M - mobile, B - BGP
Gateway of last resort is not set

      10.0.0.0/24 is subnetted, 7 subnets
C   10.0.16.0 is directly connected, Serial0/0/0
R   10.0.23.0 [120/4] via 10.0.16.6, 00:00:01, Serial0/0/0
R   10.0.25.0 [120/4] via 10.0.16.6, 00:00:01, Serial0/0/0

R   10.0.28.0 [120/3] via 10.0.16.6, 00:00:01, Serial0/0/0
R   10.0.45.0 [120/5] via 10.0.16.6, 00:00:01, Serial0/0/0
R   10.0.67.0 [120/1] via 10.0.16.6, 00:00:01, Serial0/0/0
R   10.0.78.0 [120/2] via 10.0.16.6, 00:00:01, Serial0/0/0
C   192.168.1.0/24 is directly connected, FastEthernet0/0
R   192.168.2.0/24 [120/4] via 10.0.16.6, 00:00:01, Serial0/0/0

当前在 3 条路径都出现故障的情况下，度量为 4 跳的路径显然是最佳路径，因为别无选择。以上实验，我们验证了路由器选择最佳路径使用的度量是"跳数"。

6.4.2 实验二：验证最大负载均衡路径数

实验描述：互联网络中共有 11 台路由器，从路由器 R1 访问局域网 192.168.2.0/24，共有 5 条路径。其中 3 条路径经过广域网通过 R2、R3 和 R4 到达，2 条路径经过局域网通过 R5、R6 到达。

实验目的：验证 RIP 协议默认最多选择 4 条等价路径，实现负载均衡。在本实验拓扑中 R1 访问 192.168.2.0/24 有 5 条等价路径，理论上会选择 4 条，最佳路径选择的度量仅仅与"跳数"有关，而与带宽无关。

实验要求：搭建如图 6-16 所示拓扑，配置路由器接口地址以及 RIP 协议，在网络收敛的情况下，采用相关命令验证 R1 去往目标网络 192.168.2.0/24 最多支持 4 条等价路径，即能在 4 条等价路径上实现负载均衡。

实验拓扑：如图 6-16 所示。

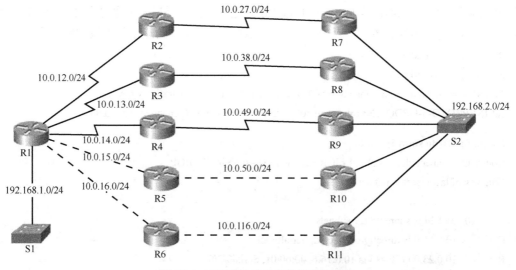

图 6-16　RIP 默认最大负载均衡路径数

如图 6-16 所示的 11 台路由器的接口地址以及 RIP 协议已配好，我们来查看路由器 R1 的路由表。

```
R1#show ip route
Codes: C - connected, S - static, I - IGRP, R - RIP, M - mobile, B - BGP
Gateway of last resort is not set

     10.0.0.0/24 is subnetted, 10 subnets
C       10.0.12.0 is directly connected, Serial0/0/0
C       10.0.13.0 is directly connected, Serial0/0/1
C       10.0.14.0 is directly connected, Serial0/1/0
C       10.0.15.0 is directly connected, FastEthernet0/0
C       10.0.16.0 is directly connected, FastEthernet0/1
R       10.0.27.0 [120/1] via 10.0.12.2, 00:00:07, Serial0/0/0
R       10.0.38.0 [120/1] via 10.0.13.3, 00:00:07, Serial0/0/1
R       10.0.49.0 [120/1] via 10.0.14.4, 00:00:08, Serial0/1/0
R       10.0.50.0 [120/1] via 10.0.15.5, 00:00:07, FastEthernet0/0
R       10.0.116.0 [120/1] via 10.0.16.6, 00:00:09, FastEthernet0/1
C    192.168.1.0/24 is directly connected, Ethernet0/3/0
R    192.168.2.0/24 [120/2] via 10.0.13.3, 00:00:08, Serial0/0/1
                    [120/2] via 10.0.14.4, 00:00:08, Serial0/1/0
                    [120/2] via 10.0.12.2, 00:00:10, Serial0/0/0
                    [120/2] via 10.0.16.6, 00:00:07, FastEthernet0/1
```

通过查看路由器 R1 的输出结果，我们发现去 192.168.2.0/24 网段只有 4 条等价路径，分别经 R3、R4、R2 和 R6 到达目的网络。down 掉 R2 的链路，如下所示：

```
R1(config-if)#int s0/0/0
R1(config-if)#shutdown
%LINK-5-CHANGED: Interface Serial0/0/0, changed state to administratively down
%LINEPROTO-5-UPDOWN: Line protocol on Interface Serial0/0/0, changed state to down
```

我们再继续观察 R1 的路由表

```
R1#show ip route
Codes: C - connected, S - static, I - IGRP, R - RIP, M - mobile, B - BGP
Gateway of last resort is not set

     10.0.0.0/24 is subnetted, 9 subnets
C       10.0.13.0 is directly connected, Serial0/0/1
C       10.0.14.0 is directly connected, Serial0/1/0
```

```
C      10.0.15.0 is directly connected, FastEthernet0/0
C      10.0.16.0 is directly connected, FastEthernet0/1
R      10.0.27.0 [120/3] via 10.0.13.3, 00:00:06, Serial0/0/1
                 [120/3] via 10.0.15.5, 00:00:04, FastEthernet0/0
                 [120/3] via 10.0.16.6, 00:00:03, FastEthernet0/1
R      10.0.38.0 [120/1] via 10.0.13.3, 00:00:06, Serial0/0/1
R      10.0.49.0 [120/1] via 10.0.14.4, 00:00:09, Serial0/1/0
R      10.0.50.0 [120/1] via 10.0.15.5, 00:00:04, FastEthernet0/0
R      10.0.116.0 [120/1] via 10.0.16.6, 00:00:03, FastEthernet0/1
C      192.168.1.0/24 is directly connected, Ethernet0/3/0
R      192.168.2.0/24 [120/2] via 10.0.13.3, 00:00:07, Serial0/0/1
                      [120/2] via 10.0.14.4, 00:00:09, Serial0/1/0
                      [120/2] via 10.0.16.6, 00:00:03, FastEthernet0/1
                      [120/2] via 10.0.15.5, 00:00:04, FastEthernet0/0
```

我们发现此时的路由表中增加了一条经 10.0.15.5（R5）到达目的网络的路由，路由器安装了 4 条等价路径，即可证明路由器默认支持 4 条路径的等价负载均衡。

6.4.3　实验三：验证度量的最大跳数限制

实验描述：如图 6-17 所示网络拓扑，共有 17 台路由器，路由器间通过串行接口实现互连，其中每台路由器都有一个环回接口。

实验目的：验证 RIP 协议的最大度量值为 15，16 跳不可达。从路由器 R1 到达目标网络 R17 的 17.0.0.0/8 共 16 跳，理论上讲 R1 的路由表中不会出现 17.0.0.0/8 的路由条目；同理路由器 R17 到达目标网络 R1 的 1.0.0.0/8 也是 16 跳，理论上 R17 的路由表中也不会出现 1.0.0.0/8 的路由条目。除了这两台边界路由器之外，其他任何一台路由器的数据包均能路由到图中所有网络。

实验要求：搭建如图 6-17 所示的拓扑，配置路由器接口地址并采用 RIP 协议实现全网互通。然后检验相应路由器的路由表来验证 RIP 协议最大度量值为 15。请打开 debug 调试命令，验证更新为 16 跳的消息能被路由器收到，但因度量为 16 跳的路由不可达，所以不被添加至路由表。

实验拓扑：如图 6-17 所示。

第 6 章 学习 RIP 路由协议

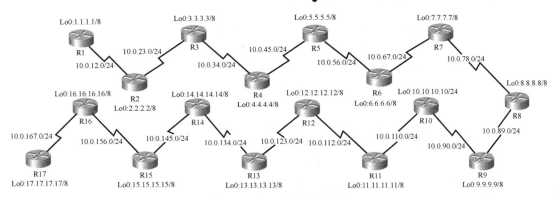

图 6-17 RIP 最大跳数限制

如图 6-17 所示，17 台路由器的接口地址已经配置好，RIP 进程也已经开启，我们来查看路由器 R1 的路由表。

```
R1#show ip route
Codes: C - connected, S - static, I - IGRP, R - RIP, M - mobile, B - BGP
Gateway of last resort is not set

C    1.0.0.0/8 is directly connected, Loopback0
R    2.0.0.0/8 [120/1] via 10.0.12.2, 00:00:16, Serial0/0/0
R    3.0.0.0/8 [120/2] via 10.0.12.2, 00:00:16, Serial0/0/0
R    4.0.0.0/8 [120/3] via 10.0.12.2, 00:00:16, Serial0/0/0
R    5.0.0.0/8 [120/4] via 10.0.12.2, 00:00:16, Serial0/0/0
R    6.0.0.0/8 [120/5] via 10.0.12.2, 00:00:16, Serial0/0/0
R    7.0.0.0/8 [120/6] via 10.0.12.2, 00:00:16, Serial0/0/0
R    8.0.0.0/8 [120/7] via 10.0.12.2, 00:00:16, Serial0/0/0
R    9.0.0.0/8 [120/8] via 10.0.12.2, 00:00:16, Serial0/0/0
     10.0.0.0/24 is subnetted, 17 subnets
C       10.0.12.0 is directly connected, Serial0/0/0
R       10.0.23.0 [120/1] via 10.0.12.2, 00:00:16, Serial0/0/0
R       10.0.34.0 [120/2] via 10.0.12.2, 00:00:16, Serial0/0/0
R       10.0.45.0 [120/3] via 10.0.12.2, 00:00:16, Serial0/0/0
R       10.0.56.0 [120/4] via 10.0.12.2, 00:00:16, Serial0/0/0
R       10.0.67.0 [120/5] via 10.0.12.2, 00:00:16, Serial0/0/0
R       10.0.78.0 [120/6] via 10.0.12.2, 00:00:16, Serial0/0/0
R       10.0.89.0 [120/7] via 10.0.12.2, 00:00:16, Serial0/0/0
R       10.0.90.0 [120/8] via 10.0.12.2, 00:00:16, Serial0/0/0
R       10.0.110.0 [120/9] via 10.0.12.2, 00:00:16, Serial0/0/0
R       10.0.112.0 [120/10] via 10.0.12.2, 00:00:16, Serial0/0/0
```

R   10.0.123.0 [120/11] via 10.0.12.2, 00:00:16, Serial0/0/0
R   10.0.134.0 [120/12] via 10.0.12.2, 00:00:16, Serial0/0/0
R   10.0.145.0 [120/13] via 10.0.12.2, 00:00:16, Serial0/0/0
R   10.0.156.0 [120/14] via 10.0.12.2, 00:00:16, Serial0/0/0
R   10.0.167.0 [120/15] via 10.0.12.2, 00:00:16, Serial0/0/0
R   10.10.10.0 [120/9] via 10.0.12.2, 00:00:16, Serial0/0/0
R   11.0.0.0/8 [120/10] via 10.0.12.2, 00:00:16, Serial0/0/0
R   12.0.0.0/8 [120/11] via 10.0.12.2, 00:00:16, Serial0/0/0
R   13.0.0.0/8 [120/12] via 10.0.12.2, 00:00:16, Serial0/0/0
R   14.0.0.0/8 [120/13] via 10.0.12.2, 00:00:16, Serial0/0/0
R   15.0.0.0/8 [120/14] via 10.0.12.2, 00:00:16, Serial0/0/0
R   16.0.0.0/8 [120/15] via 10.0.12.2, 00:00:16, Serial0/0/0

通过查看 R1 的路由表，我们发现缺少 17.0.0.0/8 网段路由。因为 RIP 协议度量值最大为 15 跳，16 跳不可达，那么 17.0.0.0/8 网段的路由是否传递到了当前路由器呢？接下来我们通过 debug ip rip 命令来查看。

R1#**debug ip rip**
RIP protocol debugging is on
RIP: received v2 update from 10.0.12.2 on Serial0/0/0
        2.0.0.0/8 via 0.0.0.0 in 1 hops
        3.0.0.0/8 via 0.0.0.0 in 2 hops
        4.0.0.0/8 via 0.0.0.0 in 3 hops
        5.0.0.0/8 via 0.0.0.0 in 4 hops
        6.0.0.0/8 via 0.0.0.0 in 5 hops
        7.0.0.0/8 via 0.0.0.0 in 6 hops
        8.0.0.0/8 via 0.0.0.0 in 7 hops
        9.0.0.0/8 via 0.0.0.0 in 8 hops
        10.0.23.0/24 via 0.0.0.0 in 1 hops
        10.0.34.0/24 via 0.0.0.0 in 2 hops
        10.0.45.0/24 via 0.0.0.0 in 3 hops
        10.0.56.0/24 via 0.0.0.0 in 4 hops
        10.0.67.0/24 via 0.0.0.0 in 5 hops
        10.0.78.0/24 via 0.0.0.0 in 6 hops
        10.0.89.0/24 via 0.0.0.0 in 7 hops
        10.0.90.0/24 via 0.0.0.0 in 8 hops
        10.0.110.0/24 via 0.0.0.0 in 9 hops
        10.0.112.0/24 via 0.0.0.0 in 10 hops
        10.0.123.0/24 via 0.0.0.0 in 11 hops

```
            10.0.134.0/24 via 0.0.0.0 in 12 hops
            10.0.145.0/24 via 0.0.0.0 in 13 hops
            10.0.156.0/24 via 0.0.0.0 in 14 hops
            10.0.167.0/24 via 0.0.0.0 in 15 hops
            10.10.10.0/24 via 0.0.0.0 in 9 hops
            11.0.0.0/8 via 0.0.0.0 in 10 hops
RIP: received v2 update from 10.0.12.2 on Serial0/0/0
            12.0.0.0/8 via 0.0.0.0 in 11 hops
            13.0.0.0/8 via 0.0.0.0 in 12 hops
            14.0.0.0/8 via 0.0.0.0 in 13 hops
            15.0.0.0/8 via 0.0.0.0 in 14 hops
            16.0.0.0/8 via 0.0.0.0 in 15 hop
            17.0.0.0/8 via 0.0.0.0 in 16 hops    //到达 17.0.0.0/8 为 16 跳，不可达
```

在路由器 R1 上通过 debug，我们确实看到了来自邻居 10.0.12.2 的更新信息中包含 17.0.0.0/16 的路由条目，因为路由器判断 16 跳是不可达的，所以就没有写进路由表。

采用同样的办法我们查看最远端的路由器 R17 的路由表，发现它也缺少 1.0.0.0/8 网段的路由，2.0.0.0/15 的路由已经是最大跳数。

```
R17#show ip route
Codes: C - connected, S - static, I - IGRP, R - RIP, M - mobile, B - BGP
Gateway of last resort is not set

R    2.0.0.0/8 [120/15] via 10.0.167.16, 00:00:25, Serial0/0/1   //路由达到了最大跳数
R    3.0.0.0/8 [120/14] via 10.0.167.16, 00:00:25, Serial0/0/1
R    4.0.0.0/8 [120/13] via 10.0.167.16, 00:00:25, Serial0/0/1
R    5.0.0.0/8 [120/12] via 10.0.167.16, 00:00:25, Serial0/0/1
R    6.0.0.0/8 [120/11] via 10.0.167.16, 00:00:25, Serial0/0/1
R    7.0.0.0/8 [120/10] via 10.0.167.16, 00:00:25, Serial0/0/1
R    8.0.0.0/8 [120/9] via 10.0.167.16, 00:00:25, Serial0/0/1
R    9.0.0.0/8 [120/8] via 10.0.167.16, 00:00:25, Serial0/0/1
     10.0.0.0/24 is subnetted, 17 subnets
R       10.0.12.0 [120/15] via 10.0.167.16, 00:00:25, Serial0/0/1
R       10.0.23.0 [120/14] via 10.0.167.16, 00:00:25, Serial0/0/1
R       10.0.34.0 [120/13] via 10.0.167.16, 00:00:25, Serial0/0/1
R       10.0.45.0 [120/12] via 10.0.167.16, 00:00:25, Serial0/0/1
R       10.0.56.0 [120/11] via 10.0.167.16, 00:00:25, Serial0/0/1
R       10.0.67.0 [120/10] via 10.0.167.16, 00:00:25, Serial0/0/1
R       10.0.78.0 [120/9] via 10.0.167.16, 00:00:25, Serial0/0/1
```

```
R        10.0.89.0 [120/8] via 10.0.167.16, 00:00:25, Serial0/0/1
R        10.0.90.0 [120/7] via 10.0.167.16, 00:00:25, Serial0/0/1
R        10.0.110.0 [120/6] via 10.0.167.16, 00:00:25, Serial0/0/1
R        10.0.112.0 [120/5] via 10.0.167.16, 00:00:25, Serial0/0/1
R        10.0.123.0 [120/4] via 10.0.167.16, 00:00:25, Serial0/0/1
R        10.0.134.0 [120/3] via 10.0.167.16, 00:00:25, Serial0/0/1
R        10.0.145.0 [120/2] via 10.0.167.16, 00:00:25, Serial0/0/1
R        10.0.156.0 [120/1] via 10.0.167.16, 00:00:25, Serial0/0/1
C        10.0.167.0 is directly connected, Serial0/0/1
R        10.10.10.0 [120/7] via 10.0.167.16, 00:00:25, Serial0/0/1
R     11.0.0.0/8 [120/6] via 10.0.167.16, 00:00:25, Serial0/0/1
R     12.0.0.0/8 [120/5] via 10.0.167.16, 00:00:25, Serial0/0/1
R     13.0.0.0/8 [120/4] via 10.0.167.16, 00:00:25, Serial0/0/1
R     14.0.0.0/8 [120/3] via 10.0.167.16, 00:00:25, Serial0/0/1
R     15.0.0.0/8 [120/2] via 10.0.167.16, 00:00:25, Serial0/0/1
R     16.0.0.0/8 [120/1] via 10.0.167.16, 00:00:25, Serial0/0/1
C     17.0.0.0/8 is directly connected, Loopback0
```

接下来，我们需要进一步确认 1.0.0.0/8 的路由更新信息是否被 R17 收到，我们需要在 R17 上打开 debug 进行查看。

```
R17#debug ip rip
RIP protocol debugging is on
RIP: received v2 update from 10.0.167.16 on Serial0/0/1
        1.0.0.0/8 via 0.0.0.0 in 16 hops    //到达 1.0.0.0/8 为 16 跳，不可达
        2.0.0.0/8 via 0.0.0.0 in 15 hops
        3.0.0.0/8 via 0.0.0.0 in 14 hops
        4.0.0.0/8 via 0.0.0.0 in 13 hops
        5.0.0.0/8 via 0.0.0.0 in 12 hops
        6.0.0.0/8 via 0.0.0.0 in 11 hops
        7.0.0.0/8 via 0.0.0.0 in 10 hops
        8.0.0.0/8 via 0.0.0.0 in 9 hops
        9.0.0.0/8 via 0.0.0.0 in 8 hops
        10.0.12.0/24 via 0.0.0.0 in 15 hops
        10.0.23.0/24 via 0.0.0.0 in 14 hops
        10.0.34.0/24 via 0.0.0.0 in 13 hops
        10.0.45.0/24 via 0.0.0.0 in 12 hops
        10.0.56.0/24 via 0.0.0.0 in 11 hops
        10.0.67.0/24 via 0.0.0.0 in 10 hops
```

```
            10.0.78.0/24 via 0.0.0.0 in 9 hops
            10.0.89.0/24 via 0.0.0.0 in 8 hops
            10.0.90.0/24 via 0.0.0.0 in 7 hops
            10.0.110.0/24 via 0.0.0.0 in 6 hops
            10.0.112.0/24 via 0.0.0.0 in 5 hops
            10.0.123.0/24 via 0.0.0.0 in 4 hops
            10.0.134.0/24 via 0.0.0.0 in 3 hops
            10.0.145.0/24 via 0.0.0.0 in 2 hops
            10.0.156.0/24 via 0.0.0.0 in 1 hops
            10.10.10.0/24 via 0.0.0.0 in 7 hops
RIP: received v2 update from 10.0.167.16 on Serial0/0/1
            11.0.0.0/8 via 0.0.0.0 in 6 hops
            12.0.0.0/8 via 0.0.0.0 in 5 hops
            13.0.0.0/8 via 0.0.0.0 in 4 hops
            14.0.0.0/8 via 0.0.0.0 in 3 hops
            15.0.0.0/8 via 0.0.0.0 in 2 hops
            16.0.0.0/8 via 0.0.0.0 in 1 hops
```

在边缘路由器 R1 和 R17 上彼此都学习不到最远端度量为 16 跳的路由，16 跳的路由是不可以写进路由表的。除了这两台路由器之外的任何一台路由器的路由表都应该是完整的，我们用 show ip route 命令在路由器 R2 上进一步确认。

```
R2#show ip route
Codes: C - connected, S - static, I - IGRP, R - RIP, M - mobile, B - BGP
Gateway of last resort is not set

R       1.0.0.0/8 [120/1] via 10.0.12.1, 00:00:04, Serial0/0/0
C       2.0.0.0/8 is directly connected, Loopback0
R       3.0.0.0/8 [120/1] via 10.0.23.3, 00:00:25, Serial0/0/1
R       4.0.0.0/8 [120/2] via 10.0.23.3, 00:00:25, Serial0/0/1
R       5.0.0.0/8 [120/3] via 10.0.23.3, 00:00:25, Serial0/0/1
R       6.0.0.0/8 [120/4] via 10.0.23.3, 00:00:25, Serial0/0/1
R       7.0.0.0/8 [120/5] via 10.0.23.3, 00:00:25, Serial0/0/1
R       8.0.0.0/8 [120/6] via 10.0.23.3, 00:00:25, Serial0/0/1
R       9.0.0.0/8 [120/7] via 10.0.23.3, 00:00:25, Serial0/0/1
        10.0.0.0/24 is subnetted, 17 subnets
C       10.0.12.0 is directly connected, Serial0/0/0
C       10.0.23.0 is directly connected, Serial0/0/1
R       10.0.34.0 [120/1] via 10.0.23.3, 00:00:25, Serial0/0/1
```

```
R    10.0.45.0 [120/2] via 10.0.23.3, 00:00:25, Serial0/0/1
R    10.0.56.0 [120/3] via 10.0.23.3, 00:00:25, Serial0/0/1
R    10.0.67.0 [120/4] via 10.0.23.3, 00:00:25, Serial0/0/1
R    10.0.78.0 [120/5] via 10.0.23.3, 00:00:25, Serial0/0/1
R    10.0.89.0 [120/6] via 10.0.23.3, 00:00:25, Serial0/0/1
R    10.0.90.0 [120/7] via 10.0.23.3, 00:00:25, Serial0/0/1
R    10.0.110.0 [120/8] via 10.0.23.3, 00:00:25, Serial0/0/1
R    10.0.112.0 [120/9] via 10.0.23.3, 00:00:25, Serial0/0/1
R    10.0.123.0 [120/10] via 10.0.23.3, 00:00:25, Serial0/0/1
R    10.0.134.0 [120/11] via 10.0.23.3, 00:00:25, Serial0/0/1
R    10.0.145.0 [120/12] via 10.0.23.3, 00:00:25, Serial0/0/1
R    10.0.156.0 [120/13] via 10.0.23.3, 00:00:25, Serial0/0/1
R    10.0.167.0 [120/14] via 10.0.23.3, 00:00:25, Serial0/0/1
R    10.10.10.0 [120/8] via 10.0.23.3, 00:00:25, Serial0/0/1
R    11.0.0.0/8 [120/9] via 10.0.23.3, 00:00:25, Serial0/0/1
R    12.0.0.0/8 [120/10] via 10.0.23.3, 00:00:25, Serial0/0/1
R    13.0.0.0/8 [120/11] via 10.0.23.3, 00:00:25, Serial0/0/1
R    14.0.0.0/8 [120/12] via 10.0.23.3, 00:00:25, Serial0/0/1
R    15.0.0.0/8 [120/13] via 10.0.23.3, 00:00:25, Serial0/0/1
R    16.0.0.0/8 [120/14] via 10.0.23.3, 00:00:25, Serial0/0/1
R    17.0.0.0/8 [120/15] via 10.0.23.3, 00:00:25, Serial0/0/1
```

经查看发现当前路由器 R2 可以到达任何一台路由器，路由表是完整的。

RIP 协议将最大度量值设置为 15 以防止度量无限大，这样可以有效地控制收敛时间。将度量值设置为 15 的另外一个原因是 RIP 协议是为中小型网络设计的，它适用的网络规模在 15 跳以内，也就是最远的两个网段之间不超过 16 台路由器。

6.4.4　实验四：验证更新的最大路由条数

实验描述：路由器 Router1 和 Router2 之间通过局域网接口实现网络互连，两路由器间需要开启 RIP 协议，在路由器 Router1 上共创建 61 个环回接口，在路由表中产生 61 条直连路由。

实验目的：验证 RIP 协议发送更新信息的最多路由条数为 25。在本实验拓扑中，Router1 在发送更新信息给 Router2 时，理论上，其路由表中的 61 条记录会发送 3 次更新包。

实验要求：配置路由器接口地址，配置路由协议 RIP，网络完成收敛。请采用相关命令验证 Router1 发送更新信息给 Router2 的最多路由条数为 25，Router2 接收 Router1 更新的最多路由条数为 25。

实验拓扑：如图 6-18 所示。

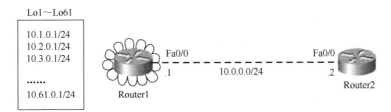

图 6-18 RIP 更新最大路由条数验证拓扑

图 6-18 所示的路由器 Router1 和 Router2 的接口 IP 地址以及 RIP 协议已经配置好，我们先来查看一下路由器 Router1 的路由表，看看路由条目是否显示完整。

```
Router1#show ip route
Codes: C - connected, S - static, I - IGRP, R - RIP, M - mobile, B - BGP
Gateway of last resort is not set


     10.0.0.0/24 is subnetted, 62 subnets
C       10.0.0.0 is directly connected, FastEthernet0/0
C       10.1.0.0 is directly connected, Loopback1
C       10.2.0.0 is directly connected, Loopback2
C       10.3.0.0 is directly connected, Loopback3
C       10.4.0.0 is directly connected, Loopback4
C       10.5.0.0 is directly connected, Loopback5
C       10.6.0.0 is directly connected, Loopback6
C       10.7.0.0 is directly connected, Loopback7
C       10.8.0.0 is directly connected, Loopback8
C       10.9.0.0 is directly connected, Loopback9
C       10.10.0.0 is directly connected, Loopback10
C       10.11.0.0 is directly connected, Loopback11
C       10.12.0.0 is directly connected, Loopback12
C       10.13.0.0 is directly connected, Loopback13
C       10.14.0.0 is directly connected, Loopback14
C       10.15.0.0 is directly connected, Loopback15
C       10.16.0.0 is directly connected, Loopback16
C       10.17.0.0 is directly connected, Loopback17
C       10.18.0.0 is directly connected, Loopback18
C       10.19.0.0 is directly connected, Loopback19
C       10.20.0.0 is directly connected, Loopback20
C       10.21.0.0 is directly connected, Loopback21
C       10.22.0.0 is directly connected, Loopback22
```

C	10.23.0.0 is directly connected, Loopback23	
C	10.24.0.0 is directly connected, Loopback24	
C	10.25.0.0 is directly connected, Loopback25	
C	10.26.0.0 is directly connected, Loopback26	
C	10.27.0.0 is directly connected, Loopback27	
C	10.28.0.0 is directly connected, Loopback28	
C	10.29.0.0 is directly connected, Loopback29	
C	10.30.0.0 is directly connected, Loopback30	
C	10.31.0.0 is directly connected, Loopback31	
C	10.32.0.0 is directly connected, Loopback32	
C	10.33.0.0 is directly connected, Loopback33	
C	10.34.0.0 is directly connected, Loopback34	
C	10.35.0.0 is directly connected, Loopback35	
C	10.36.0.0 is directly connected, Loopback36	
C	10.37.0.0 is directly connected, Loopback37	
C	10.38.0.0 is directly connected, Loopback38	
C	10.39.0.0 is directly connected, Loopback39	
C	10.40.0.0 is directly connected, Loopback40	
C	10.41.0.0 is directly connected, Loopback41	
C	10.42.0.0 is directly connected, Loopback42	
C	10.43.0.0 is directly connected, Loopback43	
C	10.44.0.0 is directly connected, Loopback44	
C	10.45.0.0 is directly connected, Loopback45	
C	10.46.0.0 is directly connected, Loopback46	
C	10.47.0.0 is directly connected, Loopback47	
C	10.48.0.0 is directly connected, Loopback48	
C	10.49.0.0 is directly connected, Loopback49	
C	10.50.0.0 is directly connected, Loopback50	
C	10.51.0.0 is directly connected, Loopback51	
C	10.52.0.0 is directly connected, Loopback52	
C	10.53.0.0 is directly connected, Loopback53	
C	10.54.0.0 is directly connected, Loopback54	
C	10.55.0.0 is directly connected, Loopback55	
C	10.56.0.0 is directly connected, Loopback56	
C	10.57.0.0 is directly connected, Loopback57	
C	10.58.0.0 is directly connected, Loopback58	
C	10.59.0.0 is directly connected, Loopback59	

C		10.60.0.0 is directly connected, Loopback60
C		10.61.0.0 is directly connected, Loopback61

经查看发现 Router1 的路由表显示了所有直连的路由条目，该路由器没有学到任何路由，因为 Router2 是 Router1 唯一的直连邻居。

```
Router2#show ip route
Codes: C - connected, S - static, I - IGRP, R - RIP, M - mobile, B - BGP
Gateway of last resort is not set

     10.0.0.0/24 is subnetted, 62 subnets
C       10.0.0.0 is directly connected, FastEthernet0/0
R       10.1.0.0 [120/1] via 10.0.0.1, 00:00:28, FastEthernet0/0
R       10.2.0.0 [120/1] via 10.0.0.1, 00:00:28, FastEthernet0/0
R       10.3.0.0 [120/1] via 10.0.0.1, 00:00:28, FastEthernet0/0
R       10.4.0.0 [120/1] via 10.0.0.1, 00:00:28, FastEthernet0/0
R       10.5.0.0 [120/1] via 10.0.0.1, 00:00:28, FastEthernet0/0
R       10.6.0.0 [120/1] via 10.0.0.1, 00:00:28, FastEthernet0/0
R       10.7.0.0 [120/1] via 10.0.0.1, 00:00:28, FastEthernet0/0
R       10.8.0.0 [120/1] via 10.0.0.1, 00:00:28, FastEthernet0/0
R       10.9.0.0 [120/1] via 10.0.0.1, 00:00:28, FastEthernet0/0
R       10.10.0.0 [120/1] via 10.0.0.1, 00:00:28, FastEthernet0/0
R       10.11.0.0 [120/1] via 10.0.0.1, 00:00:28, FastEthernet0/0
R       10.12.0.0 [120/1] via 10.0.0.1, 00:00:28, FastEthernet0/0
R       10.13.0.0 [120/1] via 10.0.0.1, 00:00:28, FastEthernet0/0
R       10.14.0.0 [120/1] via 10.0.0.1, 00:00:28, FastEthernet0/0
R       10.15.0.0 [120/1] via 10.0.0.1, 00:00:28, FastEthernet0/0
R       10.16.0.0 [120/1] via 10.0.0.1, 00:00:28, FastEthernet0/0
R       10.17.0.0 [120/1] via 10.0.0.1, 00:00:28, FastEthernet0/0
R       10.18.0.0 [120/1] via 10.0.0.1, 00:00:28, FastEthernet0/0
R       10.19.0.0 [120/1] via 10.0.0.1, 00:00:28, FastEthernet0/0
R       10.20.0.0 [120/1] via 10.0.0.1, 00:00:28, FastEthernet0/0
R       10.21.0.0 [120/1] via 10.0.0.1, 00:00:28, FastEthernet0/0
R       10.22.0.0 [120/1] via 10.0.0.1, 00:00:28, FastEthernet0/0
R       10.23.0.0 [120/1] via 10.0.0.1, 00:00:28, FastEthernet0/0
R       10.24.0.0 [120/1] via 10.0.0.1, 00:00:28, FastEthernet0/0
R       10.25.0.0 [120/1] via 10.0.0.1, 00:00:28, FastEthernet0/0
R       10.26.0.0 [120/1] via 10.0.0.1, 00:00:27, FastEthernet0/0
R       10.27.0.0 [120/1] via 10.0.0.1, 00:00:27, FastEthernet0/0
```

R   10.28.0.0 [120/1] via 10.0.0.1, 00:00:27, FastEthernet0/0
R   10.29.0.0 [120/1] via 10.0.0.1, 00:00:27, FastEthernet0/0
R   10.30.0.0 [120/1] via 10.0.0.1, 00:00:27, FastEthernet0/0
R   10.31.0.0 [120/1] via 10.0.0.1, 00:00:27, FastEthernet0/0
R   10.32.0.0 [120/1] via 10.0.0.1, 00:00:27, FastEthernet0/0
R   10.33.0.0 [120/1] via 10.0.0.1, 00:00:27, FastEthernet0/0
R   10.34.0.0 [120/1] via 10.0.0.1, 00:00:27, FastEthernet0/0
R   10.35.0.0 [120/1] via 10.0.0.1, 00:00:27, FastEthernet0/0
R   10.36.0.0 [120/1] via 10.0.0.1, 00:00:27, FastEthernet0/0
R   10.37.0.0 [120/1] via 10.0.0.1, 00:00:27, FastEthernet0/0
R   10.38.0.0 [120/1] via 10.0.0.1, 00:00:27, FastEthernet0/0
R   10.39.0.0 [120/1] via 10.0.0.1, 00:00:27, FastEthernet0/0
R   10.40.0.0 [120/1] via 10.0.0.1, 00:00:27, FastEthernet0/0
R   10.41.0.0 [120/1] via 10.0.0.1, 00:00:27, FastEthernet0/0
R   10.42.0.0 [120/1] via 10.0.0.1, 00:00:27, FastEthernet0/0
R   10.43.0.0 [120/1] via 10.0.0.1, 00:00:27, FastEthernet0/0
R   10.44.0.0 [120/1] via 10.0.0.1, 00:00:27, FastEthernet0/0
R   10.45.0.0 [120/1] via 10.0.0.1, 00:00:27, FastEthernet0/0
R   10.46.0.0 [120/1] via 10.0.0.1, 00:00:27, FastEthernet0/0
R   10.47.0.0 [120/1] via 10.0.0.1, 00:00:27, FastEthernet0/0
R   10.48.0.0 [120/1] via 10.0.0.1, 00:00:27, FastEthernet0/0
R   10.49.0.0 [120/1] via 10.0.0.1, 00:00:27, FastEthernet0/0
R   10.50.0.0 [120/1] via 10.0.0.1, 00:00:27, FastEthernet0/0
R   10.51.0.0 [120/1] via 10.0.0.1, 00:00:27, FastEthernet0/0
R   10.52.0.0 [120/1] via 10.0.0.1, 00:00:27, FastEthernet0/0
R   10.53.0.0 [120/1] via 10.0.0.1, 00:00:27, FastEthernet0/0
R   10.54.0.0 [120/1] via 10.0.0.1, 00:00:27, FastEthernet0/0
R   10.55.0.0 [120/1] via 10.0.0.1, 00:00:27, FastEthernet0/0
R   10.56.0.0 [120/1] via 10.0.0.1, 00:00:27, FastEthernet0/0
R   10.57.0.0 [120/1] via 10.0.0.1, 00:00:27, FastEthernet0/0
R   10.58.0.0 [120/1] via 10.0.0.1, 00:00:27, FastEthernet0/0
R   10.59.0.0 [120/1] via 10.0.0.1, 00:00:27, FastEthernet0/0
R   10.60.0.0 [120/1] via 10.0.0.1, 00:00:27, FastEthernet0/0
R   10.61.0.0 [120/1] via 10.0.0.1, 00:00:27, FastEthernet0/0

通过查看路由器 Router2 的路由表，我们发现该路由器学习到 Router1 的 61 个网段，即 Router1 的 61 个环回接口全部被 Router2 学习到。

在路由器 Router1 上，通过 debug 命令查看发送的更新信息最多包含多少条记录。

第 6 章 学习 RIP 路由协议

```
Router1#debug ip rip
RIP protocol debugging is on
```

RIP: sending v2 update to 224.0.0.9 via FastEthernet0/0 (10.0.0.1)
RIP: build update entries
      10.1.0.0/24 via 0.0.0.0, metric 1, tag 0
      10.2.0.0/24 via 0.0.0.0, metric 1, tag 0
      10.3.0.0/24 via 0.0.0.0, metric 1, tag 0
      10.4.0.0/24 via 0.0.0.0, metric 1, tag 0
      10.5.0.0/24 via 0.0.0.0, metric 1, tag 0
      10.6.0.0/24 via 0.0.0.0, metric 1, tag 0
      10.7.0.0/24 via 0.0.0.0, metric 1, tag 0
      10.8.0.0/24 via 0.0.0.0, metric 1, tag 0
      10.9.0.0/24 via 0.0.0.0, metric 1, tag 0
      10.10.0.0/24 via 0.0.0.0, metric 1, tag 0
      10.11.0.0/24 via 0.0.0.0, metric 1, tag 0
      10.12.0.0/24 via 0.0.0.0, metric 1, tag 0
      10.13.0.0/24 via 0.0.0.0, metric 1, tag 0
      10.14.0.0/24 via 0.0.0.0, metric 1, tag 0
      10.15.0.0/24 via 0.0.0.0, metric 1, tag 0
      10.16.0.0/24 via 0.0.0.0, metric 1, tag 0
      10.17.0.0/24 via 0.0.0.0, metric 1, tag 0
      10.18.0.0/24 via 0.0.0.0, metric 1, tag 0
      10.19.0.0/24 via 0.0.0.0, metric 1, tag 0
      10.20.0.0/24 via 0.0.0.0, metric 1, tag 0
      10.21.0.0/24 via 0.0.0.0, metric 1, tag 0
      10.22.0.0/24 via 0.0.0.0, metric 1, tag 0
      10.23.0.0/24 via 0.0.0.0, metric 1, tag 0
      10.24.0.0/24 via 0.0.0.0, metric 1, tag 0
      10.25.0.0/24 via 0.0.0.0, metric 1, tag 0   //发送的第 1 个更新信息包含 25 条记录

RIP: sending v2 update to 224.0.0.9 via FastEthernet0/0 (10.0.0.1)
RIP: build update entries
      10.26.0.0/24 via 0.0.0.0, metric 1, tag 0
      10.27.0.0/24 via 0.0.0.0, metric 1, tag 0
      10.28.0.0/24 via 0.0.0.0, metric 1, tag 0
      10.29.0.0/24 via 0.0.0.0, metric 1, tag 0
      10.30.0.0/24 via 0.0.0.0, metric 1, tag 0
      10.31.0.0/24 via 0.0.0.0, metric 1, tag 0

        10.32.0.0/24 via 0.0.0.0, metric 1, tag 0
        10.33.0.0/24 via 0.0.0.0, metric 1, tag 0
        10.34.0.0/24 via 0.0.0.0, metric 1, tag 0
        10.35.0.0/24 via 0.0.0.0, metric 1, tag 0
        10.36.0.0/24 via 0.0.0.0, metric 1, tag 0
        10.37.0.0/24 via 0.0.0.0, metric 1, tag 0
        10.38.0.0/24 via 0.0.0.0, metric 1, tag 0
        10.39.0.0/24 via 0.0.0.0, metric 1, tag 0
        10.40.0.0/24 via 0.0.0.0, metric 1, tag 0
        10.41.0.0/24 via 0.0.0.0, metric 1, tag 0
        10.42.0.0/24 via 0.0.0.0, metric 1, tag 0
        10.43.0.0/24 via 0.0.0.0, metric 1, tag 0
        10.44.0.0/24 via 0.0.0.0, metric 1, tag 0
        10.45.0.0/24 via 0.0.0.0, metric 1, tag 0
        10.46.0.0/24 via 0.0.0.0, metric 1, tag 0
        10.47.0.0/24 via 0.0.0.0, metric 1, tag 0
        10.48.0.0/24 via 0.0.0.0, metric 1, tag 0
        10.49.0.0/24 via 0.0.0.0, metric 1, tag 0
        10.50.0.0/24 via 0.0.0.0, metric 1, tag 0   //发送的第 2 个更新信息包含 25 条记录
RIP: sending v2 update to 224.0.0.9 via FastEthernet0/0 (10.0.0.1)
RIP: build update entries
        10.51.0.0/24 via 0.0.0.0, metric 1, tag 0
        10.52.0.0/24 via 0.0.0.0, metric 1, tag 0
        10.53.0.0/24 via 0.0.0.0, metric 1, tag 0
        10.54.0.0/24 via 0.0.0.0, metric 1, tag 0
        10.55.0.0/24 via 0.0.0.0, metric 1, tag 0
        10.56.0.0/24 via 0.0.0.0, metric 1, tag 0
        10.57.0.0/24 via 0.0.0.0, metric 1, tag 0
        10.58.0.0/24 via 0.0.0.0, metric 1, tag 0
        10.59.0.0/24 via 0.0.0.0, metric 1, tag 0
        10.60.0.0/24 via 0.0.0.0, metric 1, tag 0
        10.61.0.0/24 via 0.0.0.0, metric 1, tag 0   //发送的第 3 个更新消息包含剩下的 11 条记录

  本实验证明，RIP 协议发送的更新信息所包含的最多路由数是 25 条，若超出 25 条，会分成多个更新信息包发送。本实验的 61 条记录被分成了 3 个信息包发送，发送的条数分别为 25、25 和 11。在路由器 Router2 上，通过 debug 命令查看接收的更新信息条目与发送的更新信息条目是一致的。

```
RIP: received v2 update from 10.0.0.1 on FastEthernet0/0
      10.1.0.0/24 via 0.0.0.0 in 1 hops
      10.2.0.0/24 via 0.0.0.0 in 1hops
      10.3.0.0/24 via 0.0.0.0 in 1 hops
      10.4.0.0/24 via 0.0.0.0 in 1 hops
      10.5.0.0/24 via 0.0.0.0 in 1 hops
      10.6.0.0/24 via 0.0.0.0 in 1 hops
      10.7.0.0/24 via 0.0.0.0 in 1 hops
      10.8.0.0/24 via 0.0.0.0 in 1 hops
      10.9.0.0/24 via 0.0.0.0 in 1 hops
      10.10.0.0/24 via 0.0.0.0 in 1 hops
      10.11.0.0/24 via 0.0.0.0 in 1 hops
      10.12.0.0/24 via 0.0.0.0 in 1 hops
      10.13.0.0/24 via 0.0.0.0 in 1 hops
      10.14.0.0/24 via 0.0.0.0 in 1 hops
      10.15.0.0/24 via 0.0.0.0 in 1 hops
      10.16.0.0/24 via 0.0.0.0 in 1 hops
      10.17.0.0/24 via 0.0.0.0 in 1 hops
      10.18.0.0/24 via 0.0.0.0 in 1 hops
      10.19.0.0/24 via 0.0.0.0 in 1 hops
      10.20.0.0/24 via 0.0.0.0 in 1 hops
      10.21.0.0/24 via 0.0.0.0 in 1 hops
      10.22.0.0/24 via 0.0.0.0 in 1 hops
      10.23.0.0/24 via 0.0.0.0 in 1 hops
      10.24.0.0/24 via 0.0.0.0 in 1 hops
      10.25.0.0/24 via 0.0.0.0 in 1 hops      //接收的第 1 个更新信息包含 25 条记录
RIP: received v2 update from 10.0.0.1 on FastEthernet0/0
      10.26.0.0/24 via 0.0.0.0 in 1 hops
      10.27.0.0/24 via 0.0.0.0 in 1 hops
      10.28.0.0/24 via 0.0.0.0 in 1 hops
      10.29.0.0/24 via 0.0.0.0 in 1 hops
      10.30.0.0/24 via 0.0.0.0 in 1 hops
      10.31.0.0/24 via 0.0.0.0 in 1 hops
      10.32.0.0/24 via 0.0.0.0 in 1 hops
      10.33.0.0/24 via 0.0.0.0 in 1 hops
      10.34.0.0/24 via 0.0.0.0 in 1 hops
      10.35.0.0/24 via 0.0.0.0 in 1 hops
```

```
            10.36.0.0/24 via 0.0.0.0 in 1 hops
            10.37.0.0/24 via 0.0.0.0 in 1 hops
            10.38.0.0/24 via 0.0.0.0 in 1 hops
            10.39.0.0/24 via 0.0.0.0 in 1 hops
            10.40.0.0/24 via 0.0.0.0 in 1 hops
            10.41.0.0/24 via 0.0.0.0 in 1 hops
            10.42.0.0/24 via 0.0.0.0 in 1 hops
            10.43.0.0/24 via 0.0.0.0 in 1 hops
            10.44.0.0/24 via 0.0.0.0 in 1 hops
            10.45.0.0/24 via 0.0.0.0 in 1 hops
            10.46.0.0/24 via 0.0.0.0 in 1 hops
            10.47.0.0/24 via 0.0.0.0 in 1 hops
            10.48.0.0/24 via 0.0.0.0 in 1 hops
            10.49.0.0/24 via 0.0.0.0 in 1 hops
            10.50.0.0/24 via 0.0.0.0 in 1 hops    //接收的第 2 个更新信息包含 25 条记录
RIP: received v2 update from 10.0.0.1 on FastEthernet0/0
            10.51.0.0/24 via 0.0.0.0 in 1 hops
            10.52.0.0/24 via 0.0.0.0 in 1 hops
            10.53.0.0/24 via 0.0.0.0 in 1 hops
            10.54.0.0/24 via 0.0.0.0 in 1 hops
            10.55.0.0/24 via 0.0.0.0 in 1 hops
            10.56.0.0/24 via 0.0.0.0 in 1 hops
            10.57.0.0/24 via 0.0.0.0 in 1 hops
            10.58.0.0/24 via 0.0.0.0 in 1 hops
            10.59.0.0/24 via 0.0.0.0 in 1 hops
            10.60.0.0/24 via 0.0.0.0 in 1 hops
            10.61.0.0/24 via 0.0.0.0 in 1 hops    //接收的第 3 个更新信息包含 11 条记录
```

6.4.5　实验五：验证两个版本更新的不同

实验描述：如图 6-19 所示网络拓扑，路由器 R1、R2 及 R3 之间通过广域网实现了互连，3 台路由器间先后运行 RIP 协议的版本 1 和版本 2。

实验目的：验证 RIPv1 和 RIPv2 发送更新方式的不同。RIPv1 采用广播方式更新，更新信息中不包含子网掩码；RIPv2 采用组播方式更新，更新信息中包含子网掩码。

实验要求：搭建如图 6-19 所示拓扑，配置路由器接口地址和 RIP 协议，实现全网互通。请采用相关命令验证路由器 R1 采用的协议版本，以及发送更新的特点；再改成 RIPv2，验证协议的版本并进一步验证更新信息的特点，将前后两个实验结果做一下比对，证明 RIPv1 和

RIPv2 更新信息不同。

实验拓扑：如图 6-19 所示：

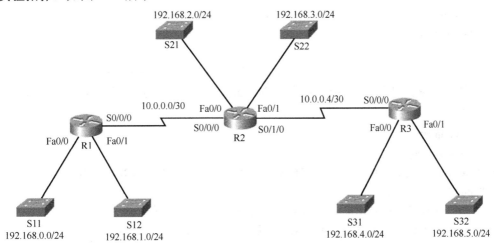

图 6-19　RIPv1&RIPv2 更新的比较

路由器已经配置好，通过 RIP 协议实现了互通，通过 show ip protocols 命令可以确认当前路由器使用的 RIP 协议的版本号。

```
R1#show ip protocols
Routing Protocol is "rip"
Sending updates every 30 seconds, next due in 25 seconds
Invalid after 180 seconds, hold down 180, flushed after 240
Outgoing update filter list for all interfaces is not set
Incoming update filter list for all interfaces is not set
Redistributing: rip
Default version control: send version 1, receive any version
  Interface             Send  Recv  Triggered RIP  Key-chain
  FastEthernet0/0       1     2 1
  FastEthernet0/1       1     2 1
  Serial0/0/0           1     2 1
Automatic network summarization is in effect
Maximum path: 4
Routing for Networks:
    10.0.0.0
    192.168.0.0
    192.168.1.0
Passive Interface(s):
```

Routing Information Sources:
   Gateway   Distance   Last Update
   10.0.0.2   120   00:00:04
Distance: (default is 120)

通过显示可知，当前路由器发送 v1 版本的更新信息时，同时接收 v1 和 v2 的更新信息（本实验采用的是 RIP 默认版本）。接下来我们通过 debug ip rip 命令进一步查看当前路由器发送更新信息的形式。

R1#**debug ip rip**
RIP protocol debugging is on
RIP: sending v1 update to 255.255.255.255 via FastEthernet0/0 (192.168.0.254)   //RIPv1 采用广播更新
RIP: build update entries
        network 10.0.0.0 metric 1   //发送更新信息时，路由度量值加 1 后再发
        network 192.168.1.0 metric 1   //RIPv1 发送的更新信息中不携带子网掩码
        network 192.168.2.0 metric 2
        network 192.168.3.0 metric 2
        network 192.168.4.0 metric 3
        network 192.168.5.0 metric 3
RIP: sending v1 update to 255.255.255.255 via FastEthernet0/1 (192.168.1.254)
RIP: build update entries
        network 10.0.0.0 metric 1
        network 192.168.0.0 metric 1
        network 192.168.2.0 metric 2
        network 192.168.3.0 metric 2
        network 192.168.4.0 metric 3
        network 192.168.5.0 metric 3
RIP: **sending v1 update to 255.255.255.255 via Serial0/0/0 (10.0.0.1)**
RIP: build update entries
        network 192.168.0.0 metric 1
        network 192.168.1.0 metric 1
RIP: received v1 update from 10.0.0.2 on Serial0/0/0
        10.0.0.4 in 1 hops   //RIPv1 接收的更新信息中，包含子网信息，不携带子网掩码
        192.168.2.0 in 1 hops
        192.168.3.0 in 1 hops
        192.168.4.0 in 2 hops
        192.168.5.0 in 2 hops

下面我们在 3 台路由器上开启 RIPv2 协议，并进一步验证 RIPv2 发送更新信息的特点。

R1(config)#**router rip**
R1(config-router)#**version 2**

R2(config)#**router rip**
R2(config-router)#**version 2**

R3(config)#**router rip**
R3(config-router)#**version 2**

通过查看路由器上运行的协议，我们可以确认当前路由器已经开始运行 RIPv2。

R1#**show ip protocols**
Routing Protocol is "rip"
Sending updates every 30 seconds, next due in 14 seconds
Invalid after 180 seconds, hold down 180, flushed after 240
Outgoing update filter list for all interfaces is not set
Incoming update filter list for all interfaces is not set
Redistributing: rip
Default version control: send version 2, receive 2

Interface	Send	Recv	Triggered RIP	Key-chain
FastEthernet0/0	2	2		
FastEthernet0/1	2	2		
Serial0/0/0	2	2		

Automatic network summarization is in effect
Maximum path: 4
Routing for Networks:
    10.0.0.0
    192.168.0.0
    192.168.1.0
Passive Interface(s):
Routing Information Sources:

Gateway	Distance	Last Update
10.0.0.2	120	00:00:16

Distance: (default is 120)

通过 debug ip rip 命令，我们可以看到 RIPv2 使用地址 224.0.0.9 来发送组播更新信息，更新的路由条目中包含了子网掩码，显然与 RIPv1 不同。

R1#**debug ip rip**
RIP protocol debugging is on
RIP: sending v2 update to 224.0.0.9 via FastEthernet0/0 (192.168.0.254)　　//RIPv2 采用组播更新
RIP: build update entries

```
        10.0.0.0/8 via 0.0.0.0, metric 1, tag 0           //边界路由器，子网汇总成主类网络 10.0.0.0/8
        192.168.1.0/24 via 0.0.0.0, metric 1, tag 0       //RIPv2 发送的更新信息中，携带子网掩码
        192.168.2.0/24 via 0.0.0.0, metric 2, tag 0
        192.168.3.0/24 via 0.0.0.0, metric 2, tag 0
        192.168.4.0/24 via 0.0.0.0, metric 3, tag 0
        192.168.5.0/24 via 0.0.0.0, metric 3, tag 0
RIP: sending v2 update to 224.0.0.9 via FastEthernet0/1 (192.168.1.254)
RIP: build update entries
        10.0.0.0/8 via 0.0.0.0, metric 1, tag 0
        192.168.0.0/24 via 0.0.0.0, metric 1, tag 0
        192.168.2.0/24 via 0.0.0.0, metric 2, tag 0
        192.168.3.0/24 via 0.0.0.0, metric 2, tag 0
        192.168.4.0/24 via 0.0.0.0, metric 3, tag 0
        192.168.5.0/24 via 0.0.0.0, metric 3, tag 0
RIP: sending v2 update to 224.0.0.9 via Serial0/0/0 (10.0.0.1)
RIP: build update entries
        192.168.0.0/24 via 0.0.0.0, metric 1, tag 0
        192.168.1.0/24 via 0.0.0.0, metric 1, tag 0
RIP: received v2 update from 10.0.0.2 on Serial0/0/0
        10.0.0.4/30 via 0.0.0.0 in 1 hops              //接收的更新信息中，携带子网掩码
        192.168.2.0/24 via 0.0.0.0 in 1 hops
        192.168.3.0/24 via 0.0.0.0 in 1 hops
        192.168.4.0/24 via 0.0.0.0 in 2 hops
        192.168.5.0/24 via 0.0.0.0 in 2 hops
```

6.5 挑战闯关训练

6.5.1 挑战任务

请用本章所学协议配置网络拓扑 A 和 B，如图 6-20 和图 6-21 所示。请在配置网络之前，先对拓扑 A 和拓扑 B 产生的路由条数进行分析和预判，并按照以下格式记录结果：

拓扑 A	拓扑 B
R1: *n*=	R1: *n*=
R2: *n*=	R2: *n*=
R3: *n*=	R3: *n*=

接下来，开始实验配置，要求实现全网互通，并验证如上结论。

请记录：
① 实验结果和以上记录的结果一致吗？
② 在实验过程中，你遇到了什么问题？如何解决的？
③ 拓扑 A 和拓扑 B 的设计有何区别？

若你实现全网互通，实验的结果和你事先分析的结果完全一致，那么恭喜你顺利闯关！若遇到问题，并能自行解决，完全理解实验结果和记录结果的偏差，你已经很棒了！

6.5.2 挑战闯关练习拓扑

挑战闯关练习拓扑如图 6-20 和图 6-21 所示。

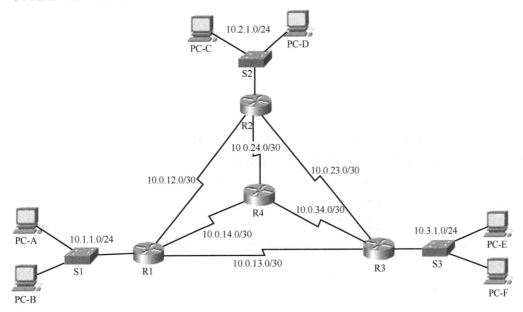

图 6-20 挑战闯关练习拓扑 A

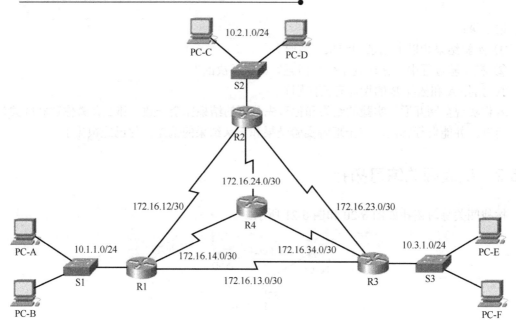

图 6-21 挑战闯关练习拓扑 B

"学习 RIP 路由协议"一章临近尾声。本章通过设置的 4 个应用场景、2 个问题解决方案、5 个实验和 1 个挑战闯关训练，贯穿本章的教学内容。其中 4 个场景暴露了 RIPv1 的缺陷，不支持不连续网络，不支持 VLSM 和 CIDR，正因为 RIPv1 的消息格式没有子网掩码字段，所以它是有类路由协议。如何解决这些问题，只有升级协议，采用 RIPv2，因为它的消息格式采用了子网掩码字段，所有问题迎刃而解。本章设计的 5 个实验进一步验证了 RIP 选路原则——采用唯一的跳数作为度量值、默认最多支持 4 条等价路径、16 跳不可达、更新信息最多包含 25 条记录等。本章的 12 个教学案例旨在向读者传达一种学习理念，多用实验去验证一些枯燥的理论，在实验中找不足，学习就成为了一种乐趣。

第 7 章

RIP 网络实战

本章要点

- RIP 网络故障排错案例
- 企业网综合配置案例
- 园区网络规划案例
- 课外拓展训练

第 6 章 "学习 RIP 路由协议"，通过设计的应用场景揭示了 RIPv1 的缺陷以及 RIPv2 的优势，通过大量的实验进一步验证 RIP 协议的相关特性，便于读者理解和记忆。第 7 章 "RIP 网络实战"重点在于对协议的综合应用，第 6 章重"点"，第 7 章重"面"。通过实验室小组在实验过程中碰到的问题来展开故障排错学习，体现了在"做中学"的教学思路。通过企业网的综合配置案例，让我们对网络的整体实施有一个全面的认识，把课堂所学知识与实际工程进行了紧密结合。本章深入剖析园区网规划的 4 个方案，让我们对网络整体架构、规划设计、路由优化有一个全方位的认识和思考。本章最后设计的 4 个课外拓展训练能进一步拓宽我们的视野，对路由表有更深入的认识。

7.1 RIP 网络故障排错案例

7.1.1 故障原因分析

1. 故障定位分析

总体来说，网络的故障分析应该基于 OSI RM。当故障定位在路由协议层面时，问题可能会涉及协议运行过程中的每一环节。RIP 协议的故障分析方法就是按照协议的工作过程，逐项检查每一环节的工作是否正常。这些环节包括：是否启用 RIP 协议，是否发送路由更新信息，能否收到更新信息并正确安装到路由表中。按照 RIP 协议的执行顺序，启用 RIP 协议后路由器就应该周期性地向邻居路由器发送路由更新信息，这是路由学习的基础。如果路由器不能正确地发送路由更新信息，则会导致其邻居路由器无法学习到它的路由信息；当然，如果路由器能正常发送路由更新信息，而其邻居路由器却不能正常接收，这也会导致邻居路由器无法安装路由条目。所以针对路由协议层面的故障排错，故障可能会定位在发送端路由器上，也可能会定位在接收端路由器上。

2. 常用 show 排错工具

在检查和排除路由协议故障时，请首先使用路由表查询命令 show ip route 来验证路由表是否安装了预期的路由条目，若有缺失路由，则进一步使用 show ip protocols 命令来检验路由协议的配置是否正确。show ip protocols 命令可以检验几项重要的配置，包括 RIP 协议是否被启用、RIP 协议的版本、RIP 自动汇总功能是否开启、哪些接口被设置成了被动接口以及 network 语句是否包含应该宣告的网络。请记住，在配置任何路由（无论静态或动态）时，都应该使用 show ip interface brief 命令确保所有直连的接口均处于 up 状态，也就是在确保直连路由都出现在路由表的前提条件下，再去排查相应缺失的路由条目。在某些情况下，当通过查看路由表及协议的配置信息无法对故障进行定位时，show running-config 这个命令就会非常有效。它是用于检查当前配置的几乎所有命令，并能确定配置中是否有明显遗漏或是否有误，对故障排查非常有帮助。

3. debug 实时诊断工具

若想进一步了解有关协议发送或接收路由更新的动态信息,可以使用 debug ip rip 命令来查看。debug 命令是一个用来诊断和发现网络问题的实用工具,它能提供实时、可持续的有用信息。因为在 CPU 中,debug 命令的输出被分配有很高的优先级,所以 debug 命令可能导致系统瘫痪。基于上述原因,只有在对某个特定问题排错时才建议使用 debug 命令。

下面,我们将对一些故障案例使用相应的排错工具来分析故障原因,排除故障。

7.1.2 场景一:故障排错案例

> 🔍**场景一**:在 YTVC 思科网络实验室里,被分成 6 个小组的 15NET1 班同学正在紧张地进行 RIPv2 的实验。实验室的电子屏幕上显示了本次实验的网络拓扑,如图 7-1 所示。路由器 R1 和 R2 各自连接了两个局域网,两台路由器之间通过 WAN 接口实现互连,要求通过配置 RIPv2 协议实现全网互连互通。实验过程中,有 4 个小组进行得比较顺利,但是一组和五组的同学讨论十分激烈,他们的实验碰到了不同的问题,纷纷寻求老师前去指导。

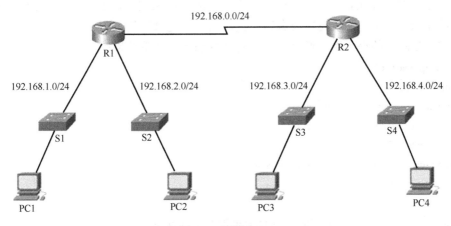

图 7-1 配置 RIPv2

一组的组长向老师反馈的问题是:PC1 和 PC2 都不能访问 PC3,而其他 PC 间的通信没有问题。

接下来,老师要求一组的同学检查两台路由器各自的路由表是否存在问题。下面是路由器 R1 的路由表输出结果。

```
R1#show ip route
Codes: C - connected, S - static, I - IGRP, R - RIP, M - mobile, B - BGP
Gateway of last resort is not set
```

```
C    192.168.0.0/24 is directly connected, Serial0/0/0
C    192.168.1.0/24 is directly connected, FastEthernet0/0
C    192.168.2.0/24 is directly connected, FastEthernet0/1
R    192.168.4.0/24 [120/1] via 192.168.0.2, 00:00:25, Serial0/0/0
```

从 R1 的路由表输出结果来看，我们发现 R1 只学习到了 R2 的一个网段 192.168.4.0/24，缺少另一个网段 192.168.3.0/24。

R2#show ip route
Codes: C - connected, S - static, I - IGRP, R - RIP, M - mobile, B - BGP
Gateway of last resort is not set

```
C    192.168.0.0/24 is directly connected, Serial0/0/0
R    192.168.1.0/24 [120/1] via 192.168.0.1, 00:00:24, Serial0/0/0
R    192.168.2.0/24 [120/1] via 192.168.0.1, 00:00:24, Serial0/0/0
C    192.168.3.0/24 is directly connected, FastEthernet0/0
C    192.168.4.0/24 is directly connected, FastEthernet0/1
```

一组组长告诉老师 R2 路由表的输出是完全正确的。老师提示一组的学生动脑筋思考故障可能发生在哪台路由器上，XL 同学说："既然 R2 的路由表完整，说明 R1 肯定将更新信息正确地发送给了 R2，R1 的配置肯定没有错，我认为可能是 R2 的问题"。

组长在 XL 同学的提示下，通过协议查看命令输出了如下结果：

R2#show ip protocols
Routing Protocol is "rip"
Sending updates every 30 seconds, next due in 19 seconds
Invalid after 180 seconds, hold down 180, flushed after 240
Outgoing update filter list for all interfaces is not set
Incoming update filter list for all interfaces is not set
Redistributing: rip
Default version control: send version 2, receive 2

Interface	Send	Recv	Triggered RIP	Key-chain
FastEthernet0/1	2	2		
Serial0/0/0	2	2		

Automatic network summarization is in effect
Maximum path: 4
Routing for Networks:
    192.168.0.0
    192.168.4.0
Passive Interface(s):
Routing Information Sources:

Gateway	Distance	Last Update
192.168.0.1	120	00:00:11

Distance: (default is 120)

果然，在命令的输出结果中我们发现路由器 R2 确实漏宣告 192.168.3.0/24，导致邻居路由器 R1 无法学习到该路由。

一组问题解决了，接下来老师开始询问五组同学遇到了什么困难。五组组长向老师反馈他们组实验很不顺利，只有连接到同一台路由器的两台 PC 之间才可以互访。

老师通过五组的实验现象分析两台路由器 R1 和 R2 连接的局域网直连路由是没有问题的，会不会是两台路由器互连的 WAN 链路出现了问题，组长告诉老师 WAN 链路层协议都是 up 状态。老师要求先查看两台路由器的路由表。

```
R1#show ip route
Codes: C - connected, S - static, I - IGRP, R - RIP, M - mobile, B - BGP
       D - EIGRP, EX - EIGRP external, O - OSPF, IA - OSPF inter area
       N1 - OSPF NSSA external type 1, N2 - OSPF NSSA external type 2
       E1 - OSPF external type 1, E2 - OSPF external type 2, E - EGP

Gateway of last resort is not set

C    192.168.0.0/24 is directly connected, Serial0/0/0
C    192.168.1.0/24 is directly connected, FastEthernet0/0
C    192.168.2.0/24 is directly connected, FastEthernet0/1
R    192.168.3.0/24 [120/1] via 192.168.0.2, 00:00:07, Serial0/0/0
R    192.168.4.0/24 [120/1] via 192.168.0.2, 00:00:07, Serial0/0/0
```

从以上 R1 路由表的输出结果来看，没有任何问题，路由条目是完整的，R1 学习到了全部的路由。接下来老师要求查看 R2 的路由表。

```
R2#show ip route
Codes: C - connected, S - static, I - IGRP, R - RIP, M - mobile, B - BGP
       D - EIGRP, EX - EIGRP external, O - OSPF, IA - OSPF inter area
       N1 - OSPF NSSA external type 1, N2 - OSPF NSSA external type 2
       E1 - OSPF external type 1, E2 - OSPF external type 2, E - EGP

Gateway of last resort is not set

C    192.168.0.0/24 is directly connected, Serial0/0/0
C    192.168.3.0/24 is directly connected, FastEthernet0/0
C    192.168.4.0/24 is directly connected, FastEthernet0/1
```

路由器 R2 的路由表输出让五组的同学惊呆了，竟然只有直连路由，R2 没有学习到 R1 的两个局域网段。老师问组长 R1 的两个局域网段是不是没有宣告，组长回答肯定宣告了。他马上通过 show ip protocols 命令将结果输出给老师看，正如组长所言直连网络全部正确宣告。

```
R1#show ip protocols
Routing Protocol is "rip"
Sending updates every 30 seconds, next due in 16 seconds
Invalid after 180 seconds, hold down 180, flushed after 240
Outgoing update filter list for all interfaces is not set
Incoming update filter list for all interfaces is not set
Redistributing: rip
Default version control: send version 2, receive 2
    Interface              Send    Recv    Triggered RIP    Key-chain
    FastEthernet0/0         2       2
    FastEthernet0/1         2       2
Automatic network summarization is in effect
Maximum path: 4
Routing for Networks:
    192.168.0.0
    192.168.1.0
    192.168.2.0
Passive Interface(s):
    Serial0/0/0
Routing Information Sources:
    Gateway            Distance        Last Update
    192.168.0.2        120             00:00:21
Distance: (default is 120)
```

此时五组的 HY 同学突然间很不好意思地向大家道歉，他忘记自己曾敲了一条不该敲的命令，想借实验验证一下命令的效果。同学们很好奇 HY 到底敲了一条什么命令。一旁的老师非常欣赏 HY 同学在实验中敢于去大胆尝试，老师让五组同学观察 show ip protocols 命令的输出，在宣告的直连网络 192.168.2.0 的最下方有个 Passive-interface(s):Serial0/0/0，这就是 HY 同学敲下那条命令的效果。答案就在这里了。同学们恍然大悟，HY 同学原来把 WAN 接口设置成了被动接口，这样路由器 R1 只能被动地接收邻居路由器 R2 发来的更新消息，却不能主动发更新信息给路由器 R2。这堂实验课两个故障让同学们收获了很多。

7.1.3　故障排错总结

路由器直连路由正常工作是确保 RIP 协议正常运行的前提。首先要保证启动 RIP 协议后路由器能够向外正常发送路由更新信息，这是路由器之间相互学习路由的基础。若路由器不能正常发送路由更新信息，究其原因，主要可能有以下几点：
- 缺少或错误配置了 network 命令；
- passive-interface <接口>命令阻止发送路由更新信息；
- RIP 协议的水平分割机制抑制了某些路由条目更新；
- 在规划 VLSM 和 CIDR 的网络中使用了 RIPv1，阻止了某些路由条目向外发送更新信息；
- 自动汇总功能开启，导致更新信息中不包含某些子网路由。

当发送方将路由更新信息正确发送出去之后，就需要接收方检查能否收到这些更新信息，以及能否计算出最佳路由并安装至路由表中。在确保直连路由正常工作且不存在链路连通性问题以及接收方已经收到路由更新信息的前提下，却没有安装相应的路由至路由表，可能的原因如下：
- 缺少或错误配置了 network 命令；
- RIP 协议的版本不兼容；
- 在不连续网络中使用了 RIPv1；
- 度量值达到了 RIP 的度量限制。

常用 RIP 排错工具汇总如下：
- show ip route；
- show ip protocols；
- show running-config；
- debug ip rip。

7.2 企业网综合配置案例

1. 项目背景

BoSea 公司是 ChangDa 公司的子公司，目前公司在不断发展，业务量也在不断扩大，同时对计算机网络应用的依赖程度与日俱增。为适应互联网时代的发展，目前公司正在面临转型，急需成立 IT 部门。你作为几年前入职的网络工程师被任命为 IT 部门的技术经理，并担任本次网络规划的项目经理。你接到的任务是将 Bosea 子公司与 Changda 公司实现互连，并通过总公司接入 ISP。因为近几年对公司的计算机网络技术没有做到及时更新，所以公司要求你采用尽可能简单的网络协议规划公司网络，尽快让网络运行起来。子公司和总公司网间互连采用静态路由。

2. 项目拓扑

企业网综合配置项目拓扑如图 7-2 所示，IP 规划表如表 7-1 所示。

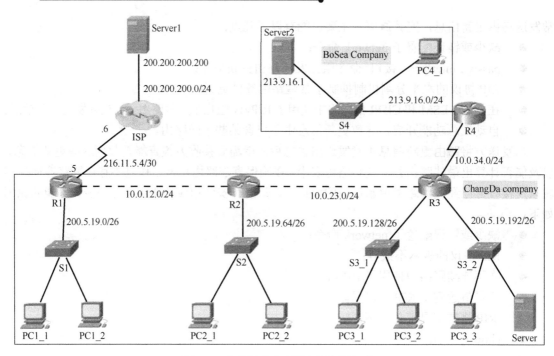

图 7-2　企业网综合配置案例项目拓扑

表 7-1　IP 规划表

设备名称	接口	IP 地址	子网掩码	描述
R1	G0/0	200.5.19.62	/26	Link to S1 G0/1
	G0/1	10.0.12.1	/24	Link to R2 G0/0
	S0/0/0	216.11.5.5	/30	Link to ISP S0/0/0
R2	G0/0	10.0.12.2	/24	Link to R1 G0/1
	G0/2	10.0.23.2	/24	Link to R3 G0/0
	G0/1	200.5.19.126	/26	Link to S2 G0/1
R3	G0/0	10.0.23.3	/24	Link to R2 G0/2
	G0/1	200.5.19.190	/26	Link to S3_1 G0/1
	G0/2	200.5.19.254	/26	Link to S3_2 G0/1
	S0/0/0	10.0.34.3	/24	Link to R4 S0/0/0
R4	S0/0/0	10.0.34.4	/24	Link to R3 S0/0/0
	G0/1	213.9.16.254	/24	Link to S4 G0/1
ISP	S0/0/0	216.11.5.6	/30	Link to R1 S0/0/0
	G0/0	200.200.200.254	/24	Link to Server1 G0
Server1	NIC	200.200.200.200	/24	Link to ISP G0/0
Server	NIC	200.5.19.194	/26	Link to S3_2 F0/2
Server2	NIC	213.9.16.1	/24	Link to S4 F0/2

3. 项目需求

公司项目经理已经按照上述要求对网络设备进行了相应地址规划。要求先对网络设备进行配置使之可以实现互连互通，具体要求如下：

- 按照图 7-2 和表 7-1 配置路由器的接口 IP 地址；
- 对 Changda 总公司的 3 台路由器配置 RIPv2 协议，宣告相应的网络，但不宣告连接 ISP 和连接 Bosea 子公司路由器 R4 的相应网络；
- 确保总公司各路由器均能学习到相应子网的明细路由，总公司各子网间可以互相通信；
- 配置总公司接入 ISP 的路由器 R1 使其可以通过 ISP 接入 Internet，同时配置 ISP 使其能为 Changda 总公司及 BoSea 子公司转发来自 Internet 访问该公司的数据流；
- 在总公司的边界路由器 R1 上传播默认路由使总公司各子网均可以访问 Internet；
- 禁止 RIP 的更新信息发送到公司各业务网段，以免影响网络性能；
- 修改 RIP 的计时器以加快网络收敛，基本更新周期不变，刷新时间改为 4 倍更新周期，其余计时器改为 3 倍更新周期；
- 配置总公司的路由器 R3 使其与 BoSea 子公司的路由器 R4 通过静态路由实现互连，并使 BoSea 子公司能通过总部访问 Internet，同时为使公司总部其他子网可以访问子公司，需要通过 R3 将去往子公司的静态路由注入到公司总部各路由器；
- 对网络进行连通性测试，实现公司总部各子网间、总部和分部间能相互通信，且保证总部和分部均可以访问 Internet。

4. 项目实施

- 任务一：配置各路由器接口地址，确保直连路由安装到路由表；
- 任务二：配置 RIPv2，在总公司 3 台路由器上配置 RIPv2 协议；
- 任务三：关闭自动汇总，确保公司总部各路由器均可学习到子网明细路由；
- 任务四：配置静态路由，使公司总部和分部均能接入 Inetenet，且总部边界路由器与分部边界路由器实现互通；
- 任务五：传播默认路由，确保公司总部各子网均可以接入 Internet；
- 任务六：配置路由注入，使公司总部各网段均可以访问公司分部；
- 任务七：配置被动接口，避免 RIP 更新报文被发送到各业务网段；
- 任务八：配置计时参数，加快网络收敛；
- 任务九：测试网络连通性，实现全网互通。

7.2.1 任务一：配置接口地址

确保路由器接口配置正确的 IP 地址且都处于 up 状态。

```
ISP#show ip interface brief | include manual
GigabitEthernet0/0      200.200.200.254      YES manual up                    up
Serial0/0/0             216.11.5.6           YES manual up                    up

R1#show ip interface brief | include manual
GigabitEthernet0/0      200.5.19.62          YES manual up                    up
GigabitEthernet0/1      10.0.12.1            YES manual up                    up
Serial0/0/0             216.11.5.5           YES manual up                    up

R2#show ip interface brief | include manual
GigabitEthernet0/0      10.0.12.2            YES manual up                    up
GigabitEthernet0/1      200.5.19.126         YES manual up                    up
GigabitEthernet0/2      10.0.23.2            YES manual up                    up

R3#show ip interface brief | include manual
GigabitEthernet0/0      10.0.23.3            YES manual up                    up
GigabitEthernet0/1      200.5.19.190         YES manual up                    up
GigabitEthernet0/2      200.5.19.254         YES manual up                    up
Serial0/0/0             10.0.34.3            YES manual up                    up

R4#show ip interface brief | include manual
GigabitEthernet0/1      213.9.16.254         YES manual up                    up
Serial0/0/0             10.0.34.4            YES manual up                    up
```

7.2.2 任务二：配置路由协议

在路由器 R1 上配置 RIPv2。

R1>enable
R1#**configure terminal**
R1(config)#**router rip**
R1(config-router)#**version 2**
R1(config-router)#**network 10.0.0.0**
R1(config-router)#**network 200.5.19.0**

在路由器 R2 上配置 RIPv2。

R2>enable

```
R2#configure terminal
R2(config)#router rip
R2(config-router)#version 2
R2(config-router)#network 10.0.0.0
R2(config-router)#network 200.5.19.0
```

在路由器 R3 上配置 RIPv2

```
R3>enable
R3#configure terminal
R3(config)#router rip
R3(config-router)#version 2
R3(config-router)#network 10.0.0.0
R3(config-router)#network 200.5.19.0
```

检验 R1 的路由表，查看是否学习到远程 RIP 域内的路由。

```
R1#show ip route | begin Gateway
Gateway of last resort is not set

      10.0.0.0/8 is variably subnetted, 4 subnets, 2 masks
C        10.0.12.0/24 is directly connected, GigabitEthernet0/1
L        10.0.12.1/32 is directly connected, GigabitEthernet0/1
R        10.0.23.0/24 [120/1] via 10.0.12.2, 00:00:20, GigabitEthernet0/1
R        10.0.34.0/24 [120/2] via 10.0.12.2, 00:00:20, GigabitEthernet0/1
      200.5.19.0/24 is variably subnetted, 3 subnets, 3 masks
R        200.5.19.0/24 [120/1] via 10.0.12.2, 00:00:20, GigabitEthernet0/1
C        200.5.19.0/26 is directly connected, GigabitEthernet0/0
L        200.5.19.62/32 is directly connected, GigabitEthernet0/0
      216.11.5.0/24 is variably subnetted, 2 subnets, 2 masks
C        216.11.5.4/30 is directly connected, Serial0/0/0
L        216.11.5.5/32 is directly connected, Serial0/0/0
```

通过查看以上 R1 路由表的输出，我们可以看到除了路由器 R1 的 3 个直连路由，路由器通过 RIP 协议仅学习到 3 条路由 10.0.23.0/24、10.0.34.0/24 和 200.5.19.0/24，实际上 R1 所在 RIP 域中除了 200.5.19.0/26 子网外，并没有学到其他 3 个子网 200.5.19.64/26、200.5.19.128/26 和 200.5.19.192/26，学到的却是主网路由，我们再来看一下 R2 的路由表，如下所示：

```
R2#show ip route | begin Gateway
Gateway of last resort is not set

      10.0.0.0/8 is variably subnetted, 5 subnets, 2 masks
C        10.0.12.0/24 is directly connected, GigabitEthernet0/0
```

```
L       10.0.12.2/32 is directly connected, GigabitEthernet0/0
C       10.0.23.0/24 is directly connected, GigabitEthernet0/2
L       10.0.23.2/32 is directly connected, GigabitEthernet0/2
R       10.0.34.0/24 [120/1] via 10.0.23.3, 00:00:17, GigabitEthernet0/2
        200.5.19.0/24 is variably subnetted, 3 subnets, 3 masks
R       200.5.19.0/24 [120/1] via 10.0.23.3, 00:00:17, GigabitEthernet0/2
                      [120/1] via 10.0.12.1, 00:00:01, GigabitEthernet0/0
C       200.5.19.64/26 is directly connected, GigabitEthernet0/1
L       200.5.19.126/32 is directly connected, GigabitEthernet0/1
```

从路由器 R2 的路由表输出可以发现，R2 共学习到 3 条路由，即 10.0.34.0/24 和 200.5.19.0/24。其中通过 R2 左接口 GigabitEthernet0/0，经邻居 R1 10.0.12.1 学习到网络地址 200.5.19.0/24；同时又通过 R2 右接口 GigabitEthernet0/2，经邻居 R3 10.0.23.3 也学习到网络地址 200.5.19.0/24，因度量都是 1，路由等价，所以都被写进路由表。

```
R3>show ip route | begin Gateway
Gateway of last resort is not set

        10.0.0.0/8 is variably subnetted, 5 subnets, 2 masks
R       10.0.12.0/24 [120/1] via 10.0.23.2, 00:00:29, GigabitEthernet0/0
C       10.0.23.0/24 is directly connected, GigabitEthernet0/0
L       10.0.23.3/32 is directly connected, GigabitEthernet0/0
C       10.0.34.0/24 is directly connected, Serial0/0/0
L       10.0.34.3/32 is directly connected, Serial0/0/0
        200.5.19.0/24 is variably subnetted, 5 subnets, 3 masks
R       200.5.19.0/24 [120/1] via 10.0.23.2, 00:00:29, GigabitEthernet0/0
C       200.5.19.128/26 is directly connected, GigabitEthernet0/1
L       200.5.19.190/32 is directly connected, GigabitEthernet0/1
C       200.5.19.192/26 is directly connected, GigabitEthernet0/2
L       200.5.19.254/32 is directly connected, GigabitEthernet0/2
```

从 R3 的路由表我们发现 R3 学习到了两条路由，其中 200.5.19.0/24 是主网路由，而非子网路由。为什么 3 台路由器都学习不到远程的有关主网 200.5.19.0/24 的子网路由呢？因为 3 台路由器都是网络边界路由器（接口），属于不同网络，当发送更新信息时，子网路由会发生自动汇总，因此邻居路由器学到的是汇总的主网路由。

7.2.3　任务三：关闭自动汇总

在 3 台路由器上关闭自动汇总功能。

```
R1(config)#router rip
```

R1(config-router)#**no auto-summary**

R2(config)#**router rip**
R2(config-router)#**no auto-summary**

R3(config)#**router rip**
R3(config-router)#**no auto-summary**

查看路由器 R1 的路由表。

R1>**show ip route | begin Gateway**
Gateway of last resort is not set

      10.0.0.0/8 is variably subnetted, 4 subnets, 2 masks
C   10.0.12.0/24 is directly connected, GigabitEthernet0/1
L   10.0.12.1/32 is directly connected, GigabitEthernet0/1
R   10.0.23.0/24 [120/1] via 10.0.12.2, 00:00:03, GigabitEthernet0/1
R   10.0.34.0/24 [120/2] via 10.0.12.2, 00:00:03, GigabitEthernet0/1
      200.5.19.0/24 is variably subnetted, 5 subnets, 2 masks
C   200.5.19.0/26 is directly connected, GigabitEthernet0/0
L   200.5.19.62/32 is directly connected, GigabitEthernet0/0
R   **200.5.19.64/26 [120/1] via 10.0.12.2, 00:00:03, GigabitEthernet0/1**
R   **200.5.19.128/26 [120/2] via 10.0.12.2, 00:00:03, GigabitEthernet0/1**
R   **200.5.19.192/26 [120/2] via 10.0.12.2, 00:00:03, GigabitEthernet0/1**
      216.11.5.0/24 is variably subnetted, 2 subnets, 2 masks
C   216.11.5.4/30 is directly connected, Serial0/0/0
L   216.11.5.5/32 is directly connected, Serial0/0/0

我们发现 R1 路由器学习到了 200.5.19.0 主网的三条子路由。

R2>**show ip route | begin Gateway**
Gateway of last resort is not set

      10.0.0.0/8 is variably subnetted, 5 subnets, 2 masks
C   10.0.12.0/24 is directly connected, GigabitEthernet0/0
L   10.0.12.2/32 is directly connected, GigabitEthernet0/0
C   10.0.23.0/24 is directly connected, GigabitEthernet0/2
L   10.0.23.2/32 is directly connected, GigabitEthernet0/2
R   10.0.34.0/24 [120/1] via 10.0.23.3, 00:00:09, GigabitEthernet0/2
      200.5.19.0/24 is variably subnetted, 5 subnets, 2 masks
R   **200.5.19.0/26 [120/1] via 10.0.12.1, 00:00:16, GigabitEthernet0/0**
C   200.5.19.64/26 is directly connected, GigabitEthernet0/1

L	200.5.19.126/32 is directly connected, GigabitEthernet0/1	
R	200.5.19.128/26 [120/1] via 10.0.23.3, 00:00:09, GigabitEthernet0/2	
R	200.5.19.192/26 [120/1] via 10.0.23.3, 00:00:09, GigabitEthernet0/2	

我们发现 R2 路由器也学习到了 200.5.19.0 主网的 3 条子路由。

R3>**show ip route | begin Gateway**
Gateway of last resort is not set

```
     10.0.0.0/8 is variably subnetted, 5 subnets, 2 masks
R       10.0.12.0/24 [120/1] via 10.0.23.2, 00:00:03, GigabitEthernet0/0
C       10.0.23.0/24 is directly connected, GigabitEthernet0/0
L       10.0.23.3/32 is directly connected, GigabitEthernet0/0
C       10.0.34.0/24 is directly connected, Serial0/0/0
L       10.0.34.3/32 is directly connected, Serial0/0/0
     200.5.19.0/24 is variably subnetted, 6 subnets, 2 masks
R       200.5.19.0/26 [120/2] via 10.0.23.2, 00:00:03, GigabitEthernet0/0
R       200.5.19.64/26 [120/1] via 10.0.23.2, 00:00:03, GigabitEthernet0/0
C       200.5.19.128/26 is directly connected, GigabitEthernet0/1
L       200.5.19.190/32 is directly connected, GigabitEthernet0/1
C       200.5.19.192/26 is directly connected, GigabitEthernet0/2
L       200.5.19.254/32 is directly connected, GigabitEthernet0/2
```

由以上输出可知，在路由器 R3 的路由表里，也显示了学到的主网 200.5.19.0 的两条子网路由。

Fire	Last Status	Source	Destination	Type	Color	Time(sec)	Periodic	Num	Edit	Delete
●	Successful	PC1_1	PC2_1	ICMP		0.000	N	0	(edit)	
●	Successful	PC1_2	PC2_2	ICMP		0.000	N	1	(edit)	
●	Successful	PC2_1	PC3_1	ICMP		0.000	N	2	(edit)	
●	Successful	PC2_2	PC3_2	ICMP		0.000	N	3	(edit)	
●	Successful	PC3_1	PC1_1	ICMP		0.000	N	4	(edit)	
●	Successful	PC3_2	PC1_2	ICMP		0.000	N	5	(edit)	

图 7-3　总公司各子网间 ping 测试结果

至此，3 台路由器直连的局域网网段间可以实现互访，即主网为 200.5.19.0 的 4 个子网彼此可以通信。

7.2.4　任务四：配置静态路由

在路由器 ISP 上配置静态路由。

ISP(config)#**ip route 200.5.19.0 255.255.255.0 s0/0/0**
ISP(config)#**ip route 213.9.16.0 255.255.255.0 s0/0/0**

查看路由器 ISP 的路由表，我们发现已经添加了两条静态路由。

```
ISP#show ip route | begin Gateway
Gateway of last resort is not set

S       200.5.19.0/24 is directly connected, Serial0/0/0
        200.200.200.0/24 is variably subnetted, 2 subnets, 2 masks
C       200.200.200.0/24 is directly connected, GigabitEthernet0/0
L       200.200.200.254/32 is directly connected, GigabitEthernet0/0
S       213.9.16.0/24 is directly connected, Serial0/0/0
        216.11.5.0/24 is variably subnetted, 2 subnets, 2 masks
C       216.11.5.4/30 is directly connected, Serial0/0/0
L       216.11.5.6/32 is directly connected, Serial0/0/0
```

在企业边界路由器 R1 上配置静态默认路由。

```
R1(config)#ip route 0.0.0.0 0.0.0.0 s0/0/0
```

查看路由器 R1 的路由表，发现静态默认路由已经配置成功，如下所示：

```
R1>show ip route static
S*      0.0.0.0/0 is directly connected, Serial0/0/0
```

在边界路由器 R4 上配置静态默认路由。

```
R4(config)#ip route 0.0.0.0 0.0.0.0 s0/0/0
```

查看路由器 R4 的路由表，发现静态默认路由已经配置成功，如下所示：

```
R4>show ip route | begin Gateway
Gateway of last resort is 0.0.0.0 to network 0.0.0.0

        10.0.0.0/8 is variably subnetted, 2 subnets, 2 masks
C       10.0.34.0/24 is directly connected, Serial0/0/0
L       10.0.34.4/32 is directly connected, Serial0/0/0
        213.9.16.0/24 is variably subnetted, 2 subnets, 2 masks
C       213.9.16.0/24 is directly connected, GigabitEthernet0/1
L       213.9.16.254/32 is directly connected, GigabitEthernet0/1
S*      0.0.0.0/0 is directly connected, Serial0/0/0
```

在路由器 R3 上配置静态路由。

```
R3(config)#ip route 213.9.16.0 255.255.255.0 s0/0/0
```

查看路由器 R3 的路由表，发现静态路由配置成功。

```
R3>show ip route static
S       213.9.16.0/24 is directly connected, Serial0/0/0
```

我们对全网主机做一下连通性测试，测试结果如图 7-4 所示。

Fire	Last Status	Source	Destination	Type	Color	Time(sec)	Periodic	Num	Edit	Delete
●	Successful	PC1_1	200.200.20···	ICMP		0.000	N	0	(edit)	
●	Successful	PC1_2	200.200.20···	ICMP		0.000	N	1	(edit)	
●	Failed	PC2_1	200.200.20···	ICMP		0.000	N	2	(edit)	
●	Failed	PC3_1	200.200.20···	ICMP		0.000	N	3	(edit)	
●	Failed	PC4_1	200.200.20···	ICMP		0.000	N	4	(edit)	
●	Successful	PC3_1	PC4_1	ICMP		0.000	N	5	(edit)	
●	Successful	PC3_2	PC4_1	ICMP		0.000	N	6	(edit)	
●	Successful	PC3_3	PC4_1	ICMP		0.000	N	7	(edit)	
●	Failed	PC2_1	PC4_1	ICMP		0.000	N	8	(edit)	
●	Failed	PC1_1	PC4_1	ICMP		0.000	N	9	(edit)	

图 7-4 全网 ping 测试结果

路由器 R1 的局域网段可以访问 Internet，R2 和 R3 的局域网段尚不能访问 Internet，因为它们没有访问 Internet 的默认路由。路由器 R3 的局域网段与 R4 的局域网段间可实现互通，但 R1 和 R2 的局域网段与 R4 的局域网段不能互通，因为 R3 有去 R4 的静态路由，但 R1 和 R2 没有。

7.2.5 任务五：传播默认路由

在路由器 R1 上，将默认路由通过 RIP 协议传播给其他两台路由器。

R1(config)#**router rip**

R1(config-router)#**default-information originate**

查看邻居路由器 R2 的路由表

R2#**show ip route | begin Gateway**

Gateway of last resort is 10.0.12.1 to network 0.0.0.0

```
      10.0.0.0/8 is variably subnetted, 5 subnets, 2 masks
C        10.0.12.0/24 is directly connected, GigabitEthernet0/0
L        10.0.12.2/32 is directly connected, GigabitEthernet0/0
C        10.0.23.0/24 is directly connected, GigabitEthernet0/2
L        10.0.23.2/32 is directly connected, GigabitEthernet0/2
R        10.0.34.0/24 [120/1] via 10.0.23.3, 00:00:22, GigabitEthernet0/2
      200.5.19.0/24 is variably subnetted, 5 subnets, 2 masks
R        200.5.19.0/26 [120/1] via 10.0.12.1, 00:00:12, GigabitEthernet0/0
C        200.5.19.64/26 is directly connected, GigabitEthernet0/1
L        200.5.19.126/32 is directly connected, GigabitEthernet0/1
R        200.5.19.128/26 [120/1] via 10.0.23.3, 00:00:22, GigabitEthernet0/2
R        200.5.19.192/26 [120/1] via 10.0.23.3, 00:00:22, GigabitEthernet0/2
R*    0.0.0.0/0 [120/1] via 10.0.12.1, 00:00:12, GigabitEthernet0/0
```

从以上 R2 路由表的输出可见，路由器 R2 从邻居 R1 学习到了默认路由，度量为 1 跳，我们再查看 R3 的路由表。

```
R3#show ip route | begin Gateway
Gateway of last resort is 10.0.23.2 to network 0.0.0.0

     10.0.0.0/8 is variably subnetted, 5 subnets, 2 masks
R       10.0.12.0/24 [120/1] via 10.0.23.2, 00:00:18, GigabitEthernet0/0
C       10.0.23.0/24 is directly connected, GigabitEthernet0/0
L       10.0.23.3/32 is directly connected, GigabitEthernet0/0
C       10.0.34.0/24 is directly connected, Serial0/0/0
L       10.0.34.3/32 is directly connected, Serial0/0/0
     200.5.19.0/24 is variably subnetted, 6 subnets, 2 masks
R       200.5.19.0/26 [120/2] via 10.0.23.2, 00:00:18, GigabitEthernet0/0
R       200.5.19.64/26 [120/1] via 10.0.23.2, 00:00:18, GigabitEthernet0/0
C       200.5.19.128/26 is directly connected, GigabitEthernet0/1
L       200.5.19.190/32 is directly connected, GigabitEthernet0/1
C       200.5.19.192/26 is directly connected, GigabitEthernet0/2
L       200.5.19.254/32 is directly connected, GigabitEthernet0/2
S    213.9.16.0/24 is directly connected, Serial0/0/0
R*   0.0.0.0/0 [120/2] via 10.0.23.2, 00:00:18, GigabitEthernet0/0
```

由上可知，R3 从其邻居 R2 那里也学习到了默认路由，度量为 2 跳。

7.2.6　任务六：配置路由注入

将 R3 路由器的静态路由 213.9.16.0/24 注入到 RIP 域。

```
R3>show ip route static
S    213.9.16.0/24 is directly connected, Serial0/0/0
```

```
R3(config)#router rip
R3(config-router)#redistribute static metric 10
```

查看路由器 R3 直连的邻居 R2 的路由表。

```
R2>show ip route | begin Gate
Gateway of last resort is 10.0.12.1 to network 0.0.0.0

     10.0.0.0/8 is variably subnetted, 5 subnets, 2 masks
C       10.0.12.0/24 is directly connected, GigabitEthernet0/0
L       10.0.12.2/32 is directly connected, GigabitEthernet0/0
C       10.0.23.0/24 is directly connected, GigabitEthernet0/2
L       10.0.23.2/32 is directly connected, GigabitEthernet0/2
```

```
R          10.0.34.0/24 [120/1] via 10.0.23.3, 00:00:18, GigabitEthernet0/2
           200.5.19.0/24 is variably subnetted, 5 subnets, 2 masks
R          200.5.19.0/26 [120/1] via 10.0.12.1, 00:00:27, GigabitEthernet0/0
C          200.5.19.64/26 is directly connected, GigabitEthernet0/1
L          200.5.19.126/32 is directly connected, GigabitEthernet0/1
R          200.5.19.128/26 [120/1] via 10.0.23.3, 00:00:18, GigabitEthernet0/2
R          200.5.19.192/26 [120/1] via 10.0.23.3, 00:00:18, GigabitEthernet0/2
R          213.9.16.0/24 [120/10] via 10.0.23.3, 00:00:18, GigabitEthernet0/2
R*         0.0.0.0/0 [120/1] via 10.0.12.1, 00:00:27, GigabitEthernet0/0
```

由上可知，我们发现 R3 已经成功将其静态路由 213.9.16.0/24 注入到了 R2 的路由表中，度量为 10。我们再查看 R2 直连邻居 R1 的路由表。

```
R1>show ip route | begin Gateway
Gateway of last resort is 0.0.0.0 to network 0.0.0.0

           10.0.0.0/8 is variably subnetted, 4 subnets, 2 masks
C          10.0.12.0/24 is directly connected, GigabitEthernet0/1
L          10.0.12.1/32 is directly connected, GigabitEthernet0/1
R          10.0.23.0/24 [120/1] via 10.0.12.2, 00:00:17, GigabitEthernet0/1
R          10.0.34.0/24 [120/2] via 10.0.12.2, 00:00:17, GigabitEthernet0/1
           200.5.19.0/24 is variably subnetted, 5 subnets, 2 masks
C          200.5.19.0/26 is directly connected, GigabitEthernet0/0
L          200.5.19.62/32 is directly connected, GigabitEthernet0/0
R          200.5.19.64/26 [120/1] via 10.0.12.2, 00:00:17, GigabitEthernet0/1
R          200.5.19.128/26 [120/2] via 10.0.12.2, 00:00:17, GigabitEthernet0/1
R          200.5.19.192/26 [120/2] via 10.0.12.2, 00:00:17, GigabitEthernet0/1
R          213.9.16.0/24 [120/11] via 10.0.12.2, 00:00:17, GigabitEthernet0/1
           216.11.5.0/24 is variably subnetted, 2 subnets, 2 masks
C          216.11.5.4/30 is directly connected, Serial0/0/0
L          216.11.5.5/32 is directly connected, Serial0/0/0
S*         0.0.0.0/0 is directly connected, Serial0/0/0
```

通过以上 R1 的路由表输出我们发现，R3 通过 R2 将其静态路由成功注入到了 R1 的路由表中，度量为 11。

7.2.7 任务七：配置被动接口

在 3 台路由器上配置被动接口。

```
R1(config)#router rip
```

R1(config-router)#**passive-interface g0/0**

R2(config)#**router rip**
R2(config-router)#**passive-interface g0/1**

R3(config)#**router rip**
R3(config-router)#**passive-interface g0/1**
R3(config-router)#**passive-interface g0/2**
R3(config-router)#**passive-interface s0/0/0**

查看路由器 R3 的协议信息。

R3>**show ip protocols**
Routing Protocol is "rip"
Sending updates every 30 seconds, next due in 8 seconds
Invalid after 180 seconds, hold down 180, flushed after 240
Outgoing update filter list for all interfaces is not set
Incoming update filter list for all interfaces is not set
Redistributing: rip
Default version control: send version 2, receive 2

Interface	Send	Recv	Triggered RIP	Key-chain
GigabitEthernet0/0	2	2		

Automatic network summarization is not in effect
Maximum path: 4
Routing for Networks:
   10.0.0.0
   200.5.19.0
Passive Interface(s):
   GigabitEthernet0/1
   GigabitEthernet0/2
   Serial0/0/0
Routing Information Sources:

Gateway	Distance	Last Update
10.0.23.2	120	00:00:21

Distance: (default is 120)

我们看到路由器 R3 的 3 个接口 G0/1、G0/2 和 S0/0/0 已经成功被设置为被动接口。

7.2.8 任务八：设置计时参数

设置 3 台路由器的计时器。

R1(config)#**router rip**
R1(config-router)#**Timers Basic 30 90 90 120**

R2(config)#**router rip**
R2(config-router)#**Timers Basic 30 90 90 120**

R3(config)#**router rip**
R3(config-router)#**Timers Basic 30 90 90 120**

查看路由器 R3 的协议信息，我们发现 4 个计时器第 3 个的值都被重新设置了。

R3>**show ip protocols**
Routing Protocol is "rip"
Sending updates every 30 seconds, next due in 17 seconds
Invalid after 90 seconds, hold down 90, flushed after 120
Outgoing update filter list for all interfaces is not set
Incoming update filter list for all interfaces is not set
Redistributing: rip
Default version control: send version 2, receive 2

Interface	Send	Recv	Triggered RIP	Key-chain
GigabitEthernet0/0	2	2		

Automatic network summarization is not in effect
Maximum path: 4
Routing for Networks:
    10.0.0.0
    200.5.19.0
Passive Interface(s):
    GigabitEthernet0/1
    GigabitEthernet0/2
    Serial0/0/0
Routing Information Sources:

Gateway	Distance	Last Update
10.0.23.2	120	00:00:23

Distance: (default is 120)

7.2.9 任务九：测试网络连通

查看主机 PC1_1 的 IP 地址。

PC>**ipconfig**
FastEthernet0 Connection:(default port)

```
Link-local IPv6 Address.........: FE80::230:F2FF:FEC1:DD3
IP Address......................: 200.5.19.1
Subnet Mask.....................: 255.255.255.192
Default Gateway.................: 200.5.19.62
```

测试主机 PC1_1 与 PC2_1 主机的连通性。

```
PC>ping 200.5.19.65

Pinging 200.5.19.65 with 32 bytes of data:

Reply from 200.5.19.65: bytes=32 time=1ms TTL=126
Reply from 200.5.19.65: bytes=32 time=11ms TTL=126
Reply from 200.5.19.65: bytes=32 time=13ms TTL=126
Reply from 200.5.19.65: bytes=32 time=11ms TTL=126

Ping statistics for 200.5.19.65:
    Packets: Sent = 4, Received = 4, Lost = 0 (0% loss),
Approximate round trip times in milli-seconds:
    Minimum = 1ms, Maximum = 13ms, Average = 9ms
```

子网 1 和子网 2 之间实现了互通，再测试 PC1_1 和 PC3_1 主机的连通性。

```
PC>ping 200.5.19.129

Pinging 200.5.19.129 with 32 bytes of data:

Reply from 200.5.19.129: bytes=32 time=0ms TTL=125
Reply from 200.5.19.129: bytes=32 time=12ms TTL=125
Reply from 200.5.19.129: bytes=32 time=15ms TTL=125
Reply from 200.5.19.129: bytes=32 time=15ms TTL=125

Ping statistics for 200.5.19.129:
    Packets: Sent = 4, Received = 4, Lost = 0 (0% loss),
Approximate round trip times in milli-seconds:
    Minimum = 0ms, Maximum = 15ms, Average = 10ms
```

子网 1 和子网 3 也实现了互通，继续测试 PC1_1 和 PC3_3 的连通性。

```
PC>ping 200.5.19.193

Pinging 200.5.19.193 with 32 bytes of data:
```

Reply from 200.5.19.193: bytes=32 time=2ms TTL=125
Reply from 200.5.19.193: bytes=32 time=15ms TTL=125
Reply from 200.5.19.193: bytes=32 time=11ms TTL=125
Reply from 200.5.19.193: bytes=32 time=13ms TTL=125

Ping statistics for 200.5.19.193:
    Packets: Sent = 4, Received = 4, Lost = 0 (0% loss),
Approximate round trip times in milli-seconds:
  Minimum = 2ms, Maximum = 15ms, Average = 10ms

子网 1 和子网 4 也同样实现了互通，再测试 PC1_1 和分公司服务器的连通性。

PC>**ping 213.9.16.1**

Pinging 213.9.16.1 with 32 bytes of data:

Reply from 213.9.16.1: bytes=32 time=2ms TTL=124
Reply from 213.9.16.1: bytes=32 time=16ms TTL=124
Reply from 213.9.16.1: bytes=32 time=11ms TTL=124
Reply from 213.9.16.1: bytes=32 time=11ms TTL=124

Ping statistics for 213.9.16.1:
    Packets: Sent = 4, Received = 4, Lost = 0 (0% loss),
Approximate round trip times in milli-seconds:
    Minimum = 2ms, Maximum = 16ms, Average = 10ms

由测试结果可知，总部子网 1 可以访问分公司服务器，继续测试 PC1_1 对公网的访问。

PC>**ping 200.200.200.200**

Pinging 200.200.200.200 with 32 bytes of data:

Reply from 200.200.200.200: bytes=32 time=1ms TTL=126
Reply from 200.200.200.200: bytes=32 time=1ms TTL=126
Reply from 200.200.200.200: bytes=32 time=1ms TTL=126
Reply from 200.200.200.200: bytes=32 time=10ms TTL=126

Ping statistics for 200.200.200.200:
    Packets: Sent = 4, Received = 4, Lost = 0 (0% loss),
Approximate round trip times in milli-seconds:
    Minimum = 1ms, Maximum = 10ms, Average = 3ms

由测试结果可知,子网 1 已经可以成功访问 Internet,再进一步从公网到公司内部进行 ping 测试,可知全网均实现了互连互通,如图 7-5 所示。

Fire	Last Status	Source	Destination	Type	Color	Time(sec)	Periodic	Num	Edit	Delete
●	Successful	200.200.200.200	PC1_1	ICMP		0.000	N	0	(edit)	
●	Successful	200.200.200.200	PC2_1	ICMP		0.000	N	1	(edit)	
●	Successful	200.200.200.200	PC3_1	ICMP		0.000	N	2	(edit)	
●	Successful	200.200.200.200	213.9.16.1	ICMP		0.000	N	3	(edit)	
●	Successful	PC1_1	PC2_1	ICMP		0.000	N	4	(edit)	
●	Successful	PC2_1	PC3_1	ICMP		0.000	N	5	(edit)	
●	Successful	PC2_2	PC3_2	ICMP		0.000	N	6	(edit)	
●	Successful	PC2_2	PC3_3	ICMP		0.000	N	7	(edit)	
●	Successful	PC1_1	213.9.16.1	ICMP		0.000	N	8	(edit)	
●	Successful	PC2_1	213.9.16.1	ICMP		0.000	N	9	(edit)	
●	Successful	PC3_1	213.9.16.1	ICMP		0.000	N	10	(edit)	
●	Successful	Server	213.9.16.1	ICMP		0.000	N	11	(edit)	

图 7-5 全网互通测试结果

7.3 园区网络规划案例

背景描述:MT 服装有限公司是一家致力于休闲服饰设计和生产的知名公司,目前公司已具规模,拥有 A、B、C、D 4 个园区。为提高生产效率,公司要求对企业园区网实现互连,请用所学知识配置网络。以下有 4 种网络规划方案:方案 A 采用 A 类网络地址进行规划,方案 B 采用 B 类网络地址进行规划,方案 C 和 D 均采用 C 类网络地址进行规划,全网采用 RIPv2 协议。

- 请你为 MT 公司选择较好的两个设计方案;
- 请分析这 4 种方案各自的特点说明能否给出优化方案;
- 请你与项目组成员一起探讨方案,以确定最优设计方案并进行实施。

7.3.1 园区网络规划方案 A

方案 A:本方案采用 A 类网络地址来规划园区网,A、B、C 和 D 4 个园区采用的地址段分别为 10.1.0.0/16、10.2.0.0/16、10.3.0.0/16 和 10.4.0.0/16,具体网络拓扑和地址规划如图 7-6 和表 7-2 所示。

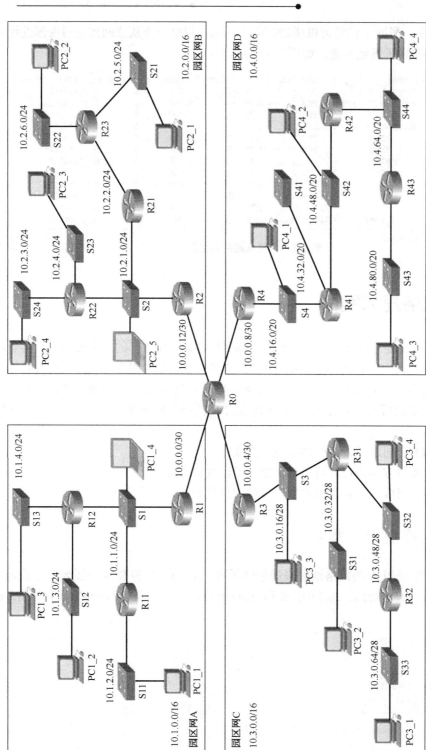

图 7-6 园区网络规划方案 A 拓扑

表7-2 方案A地址规划表

区　域	地址规划	备　注
园区网 A	10.1.0.0/16	10.1.x.0/24（x=1,2,3,4）
园区网 B	10.2.0.0/16	10.2.x.0/24（x=1,2,3,4,5,6）
园区网 C	10.3.0.0/16	10.3.0.x/28（x=16,32,48,64）
园区网 D	10.4.0.0/16	10.4.x.0/20 （x=,16,32,48,64,80）
核心区域	10.0.0.x/30	x=0,4,8,12

全网已经收敛，终端实现互通，我们来查看核心路由器R0的路由表，如下所示：

```
R0#show ip route
Codes: C - connected, S - static, I - IGRP, R - RIP, M - mobile, B - BGP
Gateway of last resort is not set

     10.0.0.0/8 is variably subnetted, 23 subnets, 4 masks
C       10.0.0.0/30 is directly connected, GigabitEthernet0/0        //核心的直连路由（4条）
C       10.0.0.4/30 is directly connected, GigabitEthernet2/0
C       10.0.0.8/30 is directly connected, GigabitEthernet3/0
C       10.0.0.12/30 is directly connected, GigabitEthernet1/0
R       10.1.1.0/24 [120/1] via 10.0.0.2, 00:00:12, GigabitEthernet0/0    //园区A的路由（4条）
R       10.1.2.0/24 [120/2] via 10.0.0.2, 00:00:12, GigabitEthernet0/0
R       10.1.3.0/24 [120/2] via 10.0.0.2, 00:00:12, GigabitEthernet0/0
R       10.1.4.0/24 [120/2] via 10.0.0.2, 00:00:12, GigabitEthernet0/0
R       10.2.1.0/24 [120/1] via 10.0.0.14, 00:00:11, GigabitEthernet1/0   //园区B的路由（6条）
R       10.2.2.0/24 [120/2] via 10.0.0.14, 00:00:11, GigabitEthernet1/0
R       10.2.3.0/24 [120/2] via 10.0.0.14, 00:00:11, GigabitEthernet1/0
R       10.2.4.0/24 [120/2] via 10.0.0.14, 00:00:11, GigabitEthernet1/0
R       10.2.5.0/24 [120/3] via 10.0.0.14, 00:00:11, GigabitEthernet1/0
R       10.2.6.0/24 [120/3] via 10.0.0.14, 00:00:11, GigabitEthernet1/0
R       10.3.0.16/28 [120/1] via 10.0.0.6, 00:00:07, GigabitEthernet2/0   //园区C的路由（4条）
R       10.3.0.32/28 [120/2] via 10.0.0.6, 00:00:07, GigabitEthernet2/0
R       10.3.0.48/28 [120/2] via 10.0.0.6, 00:00:07, GigabitEthernet2/0
R       10.3.0.64/28 [120/3] via 10.0.0.6, 00:00:07, GigabitEthernet2/0
R       10.4.16.0/20 [120/1] via 10.0.0.10, 00:00:28, GigabitEthernet3/0  //园区D的路由（5条）
R       10.4.32.0/20 [120/2] via 10.0.0.10, 00:00:28, GigabitEthernet3/0
R       10.4.48.0/20 [120/2] via 10.0.0.10, 00:00:28, GigabitEthernet3/0
R       10.4.64.0/20 [120/3] via 10.0.0.10, 00:00:28, GigabitEthernet3/0
R       10.4.80.0/20 [120/4] via 10.0.0.10, 00:00:28, GigabitEthernet3/0
```

通过查看R0路由表的输出，我们发现核心路由器学习到了全网的明细路由（子网路由）。接下来，我们分别查看4个园区网内部路由器的路由表，结果所有路由器都与核心路由器的路由数量一致，例如，园区A的路由器R11的路由表如下所示：

```
R11>show ip route
Codes: C – connected, S – static, I – IGRP, R – RIP, M – mobile, B – BGP
Gateway of last resort is not set

     10.0.0.0/8 is variably subnetted, 23 subnets, 4 masks
R       10.0.0.0/30 [120/1] via 10.1.1.254, 00:00:25, FastEthernet0/0
R       10.0.0.4/30 [120/2] via 10.1.1.254, 00:00:25, FastEthernet0/0
R       10.0.0.8/30 [120/2] via 10.1.1.254, 00:00:25, FastEthernet0/0
R       10.0.0.12/30 [120/2] via 10.1.1.254, 00:00:25, FastEthernet0/0
C       10.1.1.0/24 is directly connected, FastEthernet0/0
C       10.1.2.0/24 is directly connected, FastEthernet0/1
R       10.1.3.0/24 [120/1] via 10.1.1.2, 00:00:03, FastEthernet0/0
R       10.1.4.0/24 [120/1] via 10.1.1.2, 00:00:03, FastEthernet0/0
R       10.2.1.0/24 [120/3] via 10.1.1.254, 00:00:25, FastEthernet0/0
R       10.2.2.0/24 [120/4] via 10.1.1.254, 00:00:25, FastEthernet0/0
R       10.2.3.0/24 [120/4] via 10.1.1.254, 00:00:25, FastEthernet0/0
R       10.2.4.0/24 [120/4] via 10.1.1.254, 00:00:25, FastEthernet0/0
R       10.2.5.0/24 [120/5] via 10.1.1.254, 00:00:25, FastEthernet0/0
R       10.2.6.0/24 [120/5] via 10.1.1.254, 00:00:25, FastEthernet0/0
R       10.3.0.16/28 [120/3] via 10.1.1.254, 00:00:25, FastEthernet0/0
R       10.3.0.32/28 [120/4] via 10.1.1.254, 00:00:25, FastEthernet0/0
R       10.3.0.48/28 [120/4] via 10.1.1.254, 00:00:25, FastEthernet0/0
R       10.3.0.64/28 [120/5] via 10.1.1.254, 00:00:25, FastEthernet0/0
R       10.4.16.0/20 [120/3] via 10.1.1.254, 00:00:25, FastEthernet0/0
R       10.4.32.0/20 [120/4] via 10.1.1.254, 00:00:25, FastEthernet0/0
R       10.4.48.0/20 [120/4] via 10.1.1.254, 00:00:25, FastEthernet0/0
R       10.4.64.0/20 [120/5] via 10.1.1.254, 00:00:25, FastEthernet0/0
R       10.4.80.0/20 [120/6] via 10.1.1.254, 00:00:25, FastEthernet0/0
```

该规划方案的特点是全网使用一个 A 类网络地址 10.0.0.0/8，各园区采用/16 的一个子网，园区内的网段根据其网络规模继续划分子网。这种规划配置非常简单，全网路由器宣告的网络地址一样，有足够的地址空间，便于今后网络的扩展，其缺点是全网路由器的路由表数量一样且都有明细路由，降低了路由查询效率。方案 A 的优化方案是在园区边界路由器的出口采用命令 **ip summary-address rip** *network-address mask* 将各个园区网络的路由表进行手动汇总，从而实现路由表的优化。

7.3.2　园区网络规划方案 B

方案 B：本方案采用 B 类地址来规划园区网，A、B、C 和 D 4 个园区采用的地址段分别为 172.16.0.0/16、172.17.0.0/16、172.18.0.0/16 和 172.19.0.0/16，具体网络拓扑和地址规划如图 7-7 和表 7-3 所示。

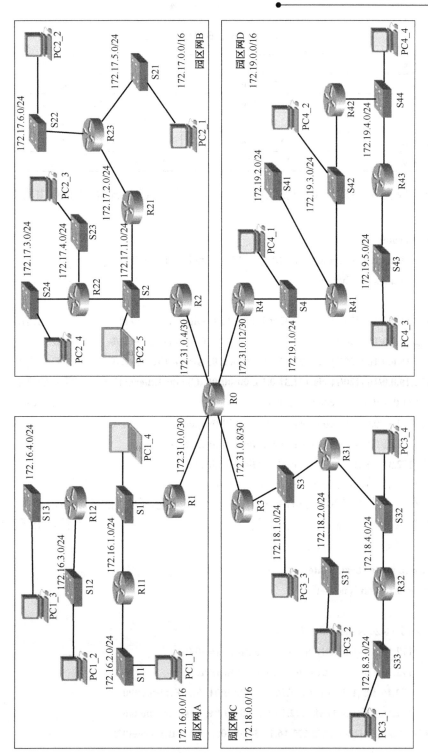

图 7-7 园区网络规划方案 B 拓扑

表 7-3 方案 B 地址规划表

区　域	地址规划	备　注
园区网 A	172.16.0.0/16	172.16.x.0/24 （x=1,2,3,4）
园区网 B	172.17.0.0/16	172.17.x.0/24 （x=1,2,3,4,5,6）
园区网 C	172.18.0.0/16	172.18.x.0/24 （x=1,2,3,4）
园区网 D	172.19.0.0/16	172.19.x.0/24 （x=1,2,3,4,5）
核心区域	172.31.0.x/30	x=0,4,8,12

全网已完成收敛，终端 PC 间彼此可以互访。下面，我们来查看核心路由器 R0 的路由表，如下所示：

```
R0#show ip route
Codes: C - connected, S - static, I - IGRP, R - RIP, M - mobile, B - BGP
Gateway of last resort is not set

R    172.16.0.0/16 [120/1] via 172.31.0.1, 00:00:21, GigabitEthernet0/0     //园区 A 的路由（1 条）
R    172.17.0.0/16 [120/1] via 172.31.0.5, 00:00:21, GigabitEthernet1/0     //园区 B 的路由（1 条）
R    172.18.0.0/16 [120/1] via 172.31.0.9, 00:00:10, GigabitEthernet2/0     //园区 C 的路由（1 条）
R    172.19.0.0/16 [120/1] via 172.31.0.13, 00:00:05, GigabitEthernet3/0    //园区 D 的路由（1 条）
     172.31.0.0/30 is subnetted, 4 subnets                                  //核心的直连路由（4 条）
C       172.31.0.0 is directly connected, GigabitEthernet0/0
C       172.31.0.4 is directly connected, GigabitEthernet1/0
C       172.31.0.8 is directly connected, GigabitEthernet2/0
C       172.31.0.12 is directly connected, GigabitEthernet3/0
```

通过查看核心路由器 R0 的路由表输出，我们发现它学习到的 4 条路由是各园区的主网路由，相比方案 A 少了 15 条路由，大大简化了路由表。接下来，我们查看 4 个园区网内部路由器的路由表，例如，园区 A 的路由器 R11 的路由表如下所示：

```
R11>show ip route | begin Gateway
Gateway of last resort is not set

     172.16.0.0/24 is subnetted, 4 subnets
C       172.16.1.0 is directly connected, FastEthernet0/0
C       172.16.2.0 is directly connected, FastEthernet0/1
R       172.16.3.0 [120/1] via 172.16.1.1, 00:00:01, FastEthernet0/0
R       172.16.4.0 [120/1] via 172.16.1.1, 00:00:01, FastEthernet0/0
R    172.17.0.0/16 [120/3] via 172.16.1.254, 00:00:03, FastEthernet0/0
```

```
R       172.18.0.0/16 [120/3] via 172.16.1.254, 00:00:03, FastEthernet0/0
R       172.19.0.0/16 [120/3] via 172.16.1.254, 00:00:03, FastEthernet0/0
R       172.31.0.0/16 [120/1] via 172.16.1.254, 00:00:03, FastEthernet0/0
```

由上我们发现 R11 路由表里记录了本园区网的 4 条明细路由和来自其他 3 个园区 B、C、D 及核心的主网路由，路由条目也简化了很多。下面是园区网 B 的路由器 R22 的路由表，如下所示：

```
R22>show ip route | begin Gateway
Gateway of last resort is not set

R       172.16.0.0/16 [120/3] via 172.17.1.254, 00:00:26, FastEthernet0/0
        172.17.0.0/24 is subnetted, 6 subnets
C       172.17.1.0 is directly connected, FastEthernet0/0
R       172.17.2.0 [120/1] via 172.17.1.1, 00:00:24, FastEthernet0/0
C       172.17.3.0 is directly connected, FastEthernet1/0
C       172.17.4.0 is directly connected, FastEthernet0/1
R       172.17.5.0 [120/2] via 172.17.1.1, 00:00:24, FastEthernet0/0
R       172.17.6.0 [120/2] via 172.17.1.1, 00:00:24, FastEthernet0/0
R       172.18.0.0/16 [120/3] via 172.17.1.254, 00:00:26, FastEthernet0/0
R       172.19.0.0/16 [120/3] via 172.17.1.254, 00:00:26, FastEthernet0/0
R       172.31.0.0/16 [120/1] via 172.17.1.254, 00:00:26, FastEthernet0/0
```

R22 的路由表里记录了园区 B 的 6 条明细路由 1/24 以及来自其他 3 个园区 A、C、D 和核心的主网路由 1/16。下面是园区网 C 的路由器 R32 的路由表，如下所示：

```
R32>show ip route | begin Gateway
Gateway of last resort is not set

R       172.16.0.0/16 [120/4] via 172.18.4.1, 00:00:19, FastEthernet0/0
R       172.17.0.0/16 [120/4] via 172.18.4.1, 00:00:19, FastEthernet0/0
        172.18.0.0/24 is subnetted, 4 subnets
R       172.18.1.0 [120/1] via 172.18.4.1, 00:00:19, FastEthernet0/0
R       172.18.2.0 [120/1] via 172.18.4.1, 00:00:19, FastEthernet0/0
C       172.18.3.0 is directly connected, FastEthernet0/1
C       172.18.4.0 is directly connected, FastEthernet0/0
R       172.19.0.0/16 [120/4] via 172.18.4.1, 00:00:19, FastEthernet0/0
R       172.31.0.0/16 [120/2] via 172.18.4.1, 00:00:19, FastEthernet0/0
```

R32 的路由表里记录了园区 C 的 4 条明细路由（/24）及 4 条来自园区 A、B、D 和核心的主网路由（/16）。

下面是园区网 D 的路由器 R42 的路由表，如下所示：

```
R42>show ip route | begin Gateway
Gateway of last resort is not set

R      172.16.0.0/16 [120/4] via 172.19.3.1, 00:00:05, FastEthernet0/0
R      172.17.0.0/16 [120/4] via 172.19.3.1, 00:00:05, FastEthernet0/0
R      172.18.0.0/16 [120/4] via 172.19.3.1, 00:00:05, FastEthernet0/0
       172.19.0.0/24 is subnetted, 5 subnets
R         172.19.1.0 [120/1] via 172.19.3.1, 00:00:05, FastEthernet0/0
R         172.19.2.0 [120/1] via 172.19.3.1, 00:00:05, FastEthernet0/0
C         172.19.3.0 is directly connected, FastEthernet0/0
C         172.19.4.0 is directly connected, FastEthernet0/1
R         172.19.5.0 [120/1] via 172.19.4.2, 00:00:00, FastEthernet0/1
R      172.31.0.0/16 [120/2] via 172.19.3.1, 00:00:05, FastEthernet0/0
```

R42 的路由表里记录了园区 D 的 5 条明细路由（/24）及 4 条来自园区 A、B、C 和核心的主网路由（/16）。

规划方案 B 的特点是各园区分别采用了一个 B 类的网络地址，园区内的网段根据其网络规模继续划分子网。这种方案配置相对简单，各园区内的路由器宣告的网络地址一样，园区边界路由器宣告两个主网地址，该方案也有足够的地址空间，便于今后网络的扩展，路由表条目简化，每台路由器的路由表中存储了本区域的明细路由及其他园区的汇总路由，极大地提高了网络的查询效率，是一种理想的网络规划方案。

7.3.3　园区网络规划方案 C

方案 C：本方案采用 C 类地址来规划园区网，A、B、C 和 D 4 个园区采用的地址段分别为 192.168.1x.0/24、192.168.2x.0/24、192.168.3x.0/24 和 192.168.4x.0/24，具体网络拓扑和地址规划如图 7-8 和表 7-4 所示。

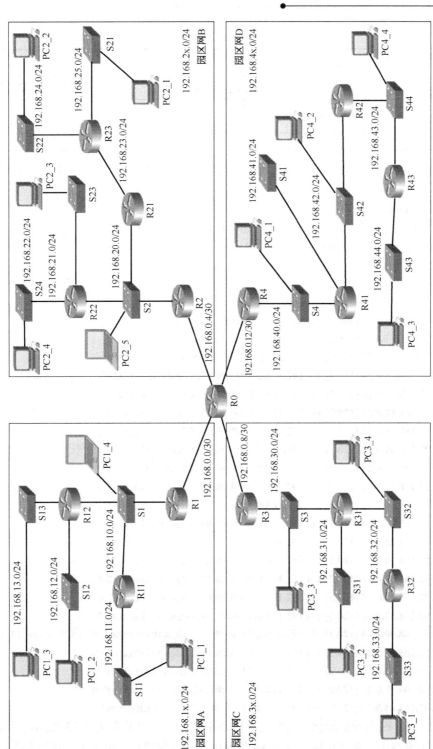

图 7-8 园区网络规划方案 C 拓扑

表 7-4　方案 C 地址规划表

区　域	地址规划	备　注
园区网 A	192.168.1x.0/24	x =0,1,2,3
园区网 B	192.168.2x.0/24	x =0,1,2,3,4,5
园区网 C	192.168.3x.0/24	x =0,1,2,3
园区网 D	192.168.4x.0/24	x =0,1,2,3,4
核心区域	192.168.0.x/30	x =0,4,8,12

全网已完成收敛，终端 PC 间彼此可以互访。下面，我们来查看核心路由器 R0 的路由表，如下所示：

```
R0#show ip route
Codes: C - connected, S - static, I - IGRP, R - RIP, M - mobile, B - BGP
Gateway of last resort is not set

        192.168.0.0/30 is subnetted, 4 subnets                           //核心的直连路由（4 条）
C        192.168.0.0 is directly connected, GigabitEthernet0/0
C        192.168.0.4 is directly connected, GigabitEthernet1/0
C        192.168.0.8 is directly connected, GigabitEthernet2/0
C        192.168.0.12 is directly connected, GigabitEthernet3/0
R     192.168.10.0/24 [120/1] via 192.168.0.1, 00:00:25, GigabitEthernet0/0   //园区 A 的路由（4 条）
R     192.168.11.0/24 [120/2] via 192.168.0.1, 00:00:25, GigabitEthernet0/0
R     192.168.12.0/24 [120/2] via 192.168.0.1, 00:00:25, GigabitEthernet0/0
R     192.168.13.0/24 [120/2] via 192.168.0.1, 00:00:25, GigabitEthernet0/0
R     192.168.20.0/24 [120/1] via 192.168.0.5, 00:00:04, GigabitEthernet1/0   //园区 B 的路由（6 条）
R     192.168.21.0/24 [120/2] via 192.168.0.5, 00:00:04, GigabitEthernet1/0
R     192.168.22.0/24 [120/2] via 192.168.0.5, 00:00:04, GigabitEthernet1/0
R     192.168.23.0/24 [120/2] via 192.168.0.5, 00:00:04, GigabitEthernet1/0
R     192.168.24.0/24 [120/3] via 192.168.0.5, 00:00:04, GigabitEthernet1/0
R     192.168.25.0/24 [120/3] via 192.168.0.5, 00:00:04, GigabitEthernet1/0
R     192.168.30.0/24 [120/1] via 192.168.0.9, 00:00:01, GigabitEthernet2/0   //园区 C 的路由（4 条）
R     192.168.31.0/24 [120/2] via 192.168.0.9, 00:00:01, GigabitEthernet2/0
R     192.168.32.0/24 [120/2] via 192.168.0.9, 00:00:01, GigabitEthernet2/0
R     192.168.33.0/24 [120/3] via 192.168.0.9, 00:00:01, GigabitEthernet2/0
R     192.168.40.0/24 [120/1] via 192.168.0.13, 00:00:07, GigabitEthernet3/0  //园区 D 的路由（5 条）
R     192.168.41.0/24 [120/2] via 192.168.0.13, 00:00:07, GigabitEthernet3/0
R     192.168.42.0/24 [120/2] via 192.168.0.13, 00:00:07, GigabitEthernet3/0
R     192.168.43.0/24 [120/3] via 192.168.0.13, 00:00:07, GigabitEthernet3/0
R     192.168.44.0/24 [120/4] via 192.168.0.13, 00:00:07, GigabitEthernet3/0
```

通过查看 R0 路由表的输出，我们发现核心路由器学习到了全网的 23 条路由。只有核心的 4 条路由是子网路由，来自其他园区的 19 条路由全是主网路由。接下来，我们查看 4 个园区网内部路由器的路由表，发现除了园区边界路由器的路由数量和核心路由器的数量一致外，其他路由器的路由表里的记录数都是 20 条，因为核心的 4 条子路由经园区边界路由器自动汇总成一条主网路由传进了园区内。例如，园区 C 路由器 R31 的路由表如下所示：

```
R31>show ip route
Codes: C - connected, S - static, I - IGRP, R - RIP, M - mobile, B - BGP
Gateway of last resort is not set

     R    192.168.0.0/24 [120/1] via 192.168.30.1, 00:00:08, FastEthernet0/0
     R    192.168.10.0/24 [120/3] via 192.168.30.1, 00:00:08, FastEthernet0/0
     R    192.168.11.0/24 [120/4] via 192.168.30.1, 00:00:08, FastEthernet0/0
     R    192.168.12.0/24 [120/4] via 192.168.30.1, 00:00:08, FastEthernet0/0
     R    192.168.13.0/24 [120/4] via 192.168.30.1, 00:00:08, FastEthernet0/0
     R    192.168.20.0/24 [120/3] via 192.168.30.1, 00:00:08, FastEthernet0/0
     R    192.168.21.0/24 [120/4] via 192.168.30.1, 00:00:08, FastEthernet0/0
     R    192.168.22.0/24 [120/4] via 192.168.30.1, 00:00:08, FastEthernet0/0
     R    192.168.23.0/24 [120/4] via 192.168.30.1, 00:00:08, FastEthernet0/0
     R    192.168.24.0/24 [120/5] via 192.168.30.1, 00:00:08, FastEthernet0/0
     R    192.168.25.0/24 [120/5] via 192.168.30.1, 00:00:08, FastEthernet0/0
     C    192.168.30.0/24 is directly connected, FastEthernet0/0
     C    192.168.31.0/24 is directly connected, FastEthernet0/1
     C    192.168.32.0/24 is directly connected, FastEthernet1/0
     R    192.168.33.0/24 [120/1] via 192.168.32.2, 00:00:11, FastEthernet1/0
     R    192.168.40.0/24 [120/3] via 192.168.30.1, 00:00:08, FastEthernet0/0
     R    192.168.41.0/24 [120/4] via 192.168.30.1, 00:00:08, FastEthernet0/0
     R    192.168.42.0/24 [120/4] via 192.168.30.1, 00:00:08, FastEthernet0/0
     R    192.168.43.0/24 [120/5] via 192.168.30.1, 00:00:08, FastEthernet0/0
     R    192.168.44.0/24 [120/6] via 192.168.30.1, 00:00:08, FastEthernet0/0
```

规划方案 C 的特点是各园区分别采用了多个 C 类/24 的网络地址，每个园区分配的网络地址为 192.168.xy.0/24（x=1,2,3,4; y=0,1,2,3,…,9，即 4 个园区，每个园区最多分配 10 个网络地址），园区内网段都采用 C 类主网地址。这种规划方案配置相对复杂，每台路由器宣告的网络地址不一样，相比方案 A 和方案 B，可扩展的地址空间小。其缺点是路由表条目多，不能通过精确汇总来优化路由表，因此该方案不是理想的网络规划方案。

7.3.4　园区网络规划方案 D

方案 D：本方案采用 C 类网络地址来规划园区网，A、B、C 和 D 4 个园区采用的地址段分别为 192.168.0.0/22、192.168.8.0/21、192.168.4.0/22 和 192.168.16.0/21，具体网络拓扑和地址规划如图 7-9 和表 7-5 所示。

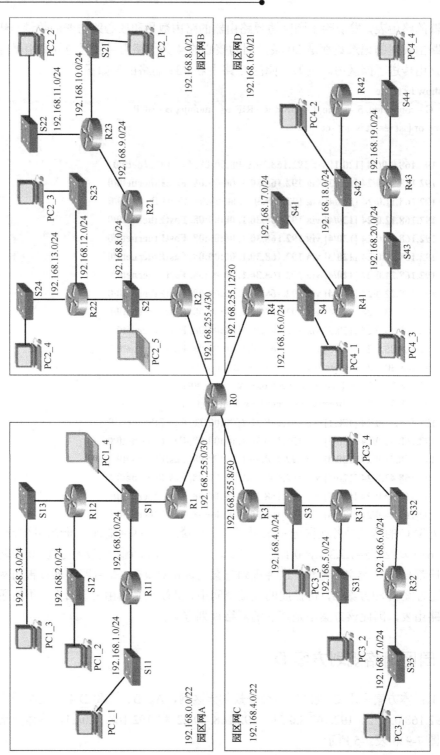

图 7-9　园区网络规划方案 D 拓扑

表 7-5　方案 D 地址规划表

区　　域	地址规划	备　　注
园区网 A	192.168.0.0/22	192.168.x.0/24（x=0,1,2,3）
园区网 B	192.168.8.0/21	192.168.x.0/24（x=8,9,10,11,12,13）
园区网 C	192.168.4.0/22	192.168.x.0/24（x=4,5,6,7）
园区网 D	192.168.16.0/21	192.168.x.0/24（x=16,17,18,19,20）
核心区域	192.168.255.x/30	x=0,4,8,12

全网已完成收敛，终端 PC 间彼此可以互访。下面，我们来查看核心路由器 R0 的路由表，如下所示：

```
R0#show ip route
Codes: C - connected, S - static, I - IGRP, R - RIP, M - mobile, B - BGP
Gateway of last resort is not set

R    192.168.0.0/24 [120/1] via 192.168.255.2, 00:00:03, GigabitEthernet0/0    //园区 A 的路由（4 条）
R    192.168.1.0/24 [120/2] via 192.168.255.2, 00:00:03, GigabitEthernet0/0
R    192.168.2.0/24 [120/2] via 192.168.255.2, 00:00:03, GigabitEthernet0/0
R    192.168.3.0/24 [120/2] via 192.168.255.2, 00:00:03, GigabitEthernet0/0
R    192.168.4.0/24 [120/1] via 192.168.255.10, 00:00:00, GigabitEthernet2/0   //园区 C 的路由（4 条）
R    192.168.5.0/24 [120/2] via 192.168.255.10, 00:00:00, GigabitEthernet2/0
R    192.168.6.0/24 [120/2] via 192.168.255.10, 00:00:00, GigabitEthernet2/0
R    192.168.7.0/24 [120/3] via 192.168.255.10, 00:00:00, GigabitEthernet2/0
R    192.168.8.0/24 [120/1] via 192.168.255.6, 00:00:00, GigabitEthernet1/0    //园区 B 的路由（6 条）
R    192.168.9.0/24 [120/2] via 192.168.255.6, 00:00:00, GigabitEthernet1/0
R    192.168.10.0/24 [120/3] via 192.168.255.6, 00:00:00, GigabitEthernet1/0
R    192.168.11.0/24 [120/3] via 192.168.255.6, 00:00:00, GigabitEthernet1/0
R    192.168.12.0/24 [120/2] via 192.168.255.6, 00:00:00, GigabitEthernet1/0
R    192.168.13.0/24 [120/2] via 192.168.255.6, 00:00:00, GigabitEthernet1/0
R    192.168.16.0/24 [120/1] via 192.168.255.14, 00:00:04, GigabitEthernet3/0  //园区 D 的路由（5 条）
R    192.168.17.0/24 [120/2] via 192.168.255.14, 00:00:04, GigabitEthernet3/0
R    192.168.18.0/24 [120/2] via 192.168.255.14, 00:00:04, GigabitEthernet3/0
R    192.168.19.0/24 [120/3] via 192.168.255.14, 00:00:04, GigabitEthernet3/0
R    192.168.20.0/24 [120/4] via 192.168.255.14, 00:00:04, GigabitEthernet3/0
     192.168.255.0/30 is subnetted, 4 subnets                                  //核心的直连路由（4 条）
C    192.168.255.0 is directly connected, GigabitEthernet0/0
C    192.168.255.4 is directly connected, GigabitEthernet1/0
C    192.168.255.8 is directly connected, GigabitEthernet2/0
```

C   192.168.255.12 is directly connected, GigabitEthernet3/0

核心路由器 R0 的路由表共有 23 条路由，同方案 C。4 台区域边界路由器与核心路由器的路由数量一致，其他园区内路由器的路由条数为 20 条，路由器 R41 的路由表如下所示：

R41>**show ip route**
Codes: C - connected, S - static, I - IGRP, R - RIP, M - mobile, B - BGP
Gateway of last resort is not set

R   192.168.0.0/24 [120/3] via 192.168.16.1, 00:00:21, FastEthernet0/0
R   192.168.1.0/24 [120/4] via 192.168.16.1, 00:00:21, FastEthernet0/0
R   192.168.2.0/24 [120/4] via 192.168.16.1, 00:00:21, FastEthernet0/0
R   192.168.3.0/24 [120/4] via 192.168.16.1, 00:00:21, FastEthernet0/0
R   192.168.4.0/24 [120/3] via 192.168.16.1, 00:00:21, FastEthernet0/0
R   192.168.5.0/24 [120/4] via 192.168.16.1, 00:00:21, FastEthernet0/0
R   192.168.6.0/24 [120/4] via 192.168.16.1, 00:00:21, FastEthernet0/0
R   192.168.7.0/24 [120/5] via 192.168.16.1, 00:00:21, FastEthernet0/0
R   192.168.8.0/24 [120/3] via 192.168.16.1, 00:00:21, FastEthernet0/0
R   192.168.9.0/24 [120/4] via 192.168.16.1, 00:00:21, FastEthernet0/0
R   192.168.10.0/24 [120/5] via 192.168.16.1, 00:00:21, FastEthernet0/0
R   192.168.11.0/24 [120/5] via 192.168.16.1, 00:00:21, FastEthernet0/0
R   192.168.12.0/24 [120/4] via 192.168.16.1, 00:00:21, FastEthernet0/0
R   192.168.13.0/24 [120/4] via 192.168.16.1, 00:00:21, FastEthernet0/0
C   192.168.16.0/24 is directly connected, FastEthernet0/0
C   192.168.17.0/24 is directly connected, FastEthernet0/1
C   192.168.18.0/24 is directly connected, FastEthernet1/0
R   192.168.19.0/24 [120/1] via 192.168.18.2, 00:00:14, FastEthernet1/0
R   192.168.20.0/24 [120/2] via 192.168.18.2, 00:00:14, FastEthernet1/0
R   192.168.255.0/24 [120/1] via 192.168.16.1, 00:00:21, FastEthernet0/0

从园区内路由器的路由表输出来看，核心路由器的 4 个子网路由被汇总成了 1 条主网路由 192.168.255.0/24。

规划方案 D 的特点是依据网络规模大小为各园区分配了 C 类网络的聚合地址，因为分配的地址是连续可汇总的。园区 A 和 C 分配的是 /22 的网络地址，即 192.168.0.0/22 和 192.168.4.0/22；园区 B 和 D 分配的是 /21 的网络地址，即 192.168.8.0/21 和 192.168.16.0/21。这种规划方案和方案 C 一样，配置都相对复杂，每台路由器宣告的网络地址都不一样，园区 A 和 C 已经用尽了分配的网络地址，无法适应园区网络规模的扩展；园区 C 和 D 可扩展的地址空间也很有限。尽管方案 D 也可以在边界路由器出口通过手动汇总来优化路由表，但地址空间缺乏扩展性，因此也不是理想的网络规划方案。

7.3.5 园区网规划方案总结

MT 服装有限公司共提出了 4 种园区网规划方案，其中方案 A 采用 A 类私有网络地址 10.0.0.0/8 进行规划，4 个园区地址分别为 10.1.0.0/16～10.4.0.0/16；方案 B 采用 B 类私有网络地址进行规划，园区地址分别为 172.16.0.0/16～172.19.0.0/16；方案 C 和方案 D 都采用 C 类私有网络地址进行规划，其中方案 C 园区地址分别为 192.168.1x.0/24～192.168.4x.0/24，方案 D 的园区地址分别为 192.168.0.0/22、192.168.4.0/22、192.168.8.0/21 和 192.168.16.0/21。

方案 A：全网使用一个私有网络地址 10.0.0.0/8，进行子网划分后再分配给 4 个园区，每个园区网地址进一步划分子网后分配给每个网段。方案 A 在配置方面非常简单，且扩展性好，当某个园区新增某个网段时，除非新增路由器，否则无须宣告就可自动参与 RIP 更新。缺点是核心路由器和各园区内部路由器的路由条目数量大，定期更新会占 CPU、内存并且链路开销大。我们可以采用相应措施来优化路由表，比如可以在园区边界路由器上开启手动汇总等办法进行优化。

方案 B：4 个园区采用 4 个 B 类的私有网络地址，每个园区内部对同一主网地址进行子网划分后再分配给相应网段，这种地址规划使得园区连接核心的路由器就成为网络边界路由器，即边界路由器会将子路由自动汇总成主网路由宣告给核心路由器。因此方案 B 中核心路由器学习到的园区网都是有类的网络地址，各园区内路由器的路由表里都安装的是其他园区的主网路由，因而极大简化了路由表，降低了资源的占用率，从而提高了查询效率。从配置方面来讲，方案 B 比较简单，是比较理想的设计方案。

方案 C：采用了 C 类私有网地址进行规划，全网除了核心网段用了子网，其余全是主网，尽管地址规划比较直观，但不利于路由表的优化，该方案利用了我们比较熟悉的十进制特点，从视觉上来看是一种比较好的设计方案，但从路由表的优化、地址空间的扩展以及相应配置方面来看是相对较差的规划方案。

方案 D：和方案 C 一样都采用了 C 类私有网络地址进行规划，4 个园区得到的是聚合地址（22<24，21<24），显然分配的是连续地址空间。根据网络规模的大小进行地址规划，园区 A 和 C 规模等同，分配的是/22 的网络地址空间；园区 B 和园区 D 相对较大，分配的是/21 的网络地址空间。尽管路由器的路由表比较庞大，但我们可以和方案 A 一样在园区边界路由器上采取手动汇总等方案对路由表进行优化。本方案的缺点是可扩展性差，尤其园区 A 和园区 C 目前已经将分配的地址空间全部用完，若要新增加网段，需要额外分配地址，因此不是理想的规划方案。

7.3.6 园区网络优化方案 A

实际上，MT 公司全网采用 RIPv2 协议实现互通并非是最优设计方案。从网络扩展角度来看，方案 A 和方案 B 都是可行设计方案；从配置简单方面来讲，方案 A 占很大优势；从路由表的优化来看，方案 B 占优势。要想取得最优网络规划方案，采用动态路由协议和静态路由相结合的方式，采用方案 A 的地址规划，配置最简单，且扩展性好，便于网络管理和维护。最终采用的方案是在方案 A 的基础上进行进一步优化，具体操作如下。

在核心路由器 R0 上，禁用 RIP 协议，配置到 4 个园区的静态路由。

R0(config)#**no router rip**
R0(config)#**ip route 10.1.0.0 255.255.0.0 10.0.0.2**
R0(config)#**ip route 10.2.0.0 255.255.0.0 10.0.0.14**
R0(config)#**ip route 10.3.0.0 255.255.0.0 10.0.0.6**
R0(config)#**ip route 10.4.0.0 255.255.0.0 10.0.0.10**

在园区 A 边界路由器 R1 上配置静态默认路由，传播默认路由并设置被动接口，避免不必要更新。

R1(config)#**ip route 0.0.0.0 0.0.0.0 10.0.0.1**
R1(config)#**router rip**
R1(config-router)#**default-information originate**
R1(config-router)#**passive-interface g0/0**

在园区 B 边界路由器 R2 上配置静态默认路由，传播默认路由并设置被动接口。

R2(config)#**ip route 0.0.0.0 0.0.0.0 10.0.0.13**
R2(config)#**router rip**
R2(config-router)#**default-information originate**
R2(config-router)#**passive-interface g0/0**

在园区 C 边界路由器 R3 上配置静态默认路由，传播默认路由并设置被动接口。

R3(config)#**ip route 0.0.0.0 0.0.0.0 10.0.0.5**
R3(config)#**router rip**
R3(config-router)#**default-information originate**
R3(config-router)#**passive-interface g0/0**

在园区 D 边界路由器 R4 上配置静态默认路由，传播默认路由并设置被动接口。

R4(config)#**ip route 0.0.0.0 0.0.0.0 10.0.0.9**
R4(config)#**router rip**
R4(config-router)#**default-information originate**
R4(config-router)#**passive-interface g0/0**

查看核心路由器 R0 的路由表。

R0>**show ip route | begin Gateway**
Gateway of last resort is not set

  10.0.0.0/8 is variably subnetted, 8 subnets, 2 masks
C 10.0.0.0/30 is directly connected, GigabitEthernet0/0

C	10.0.0.4/30 is directly connected, GigabitEthernet2/0
C	10.0.0.8/30 is directly connected, GigabitEthernet3/0
C	10.0.0.12/30 is directly connected, GigabitEthernet1/0
S	10.1.0.0/16 [1/0] via 10.0.0.2
S	10.2.0.0/16 [1/0] via 10.0.0.14
S	10.3.0.0/16 [1/0] via 10.0.0.6
S	10.4.0.0/16 [1/0] via 10.0.0.10

查看园区 A 内部路由器的路由表，路由器 R11 的路由表如下所示：

R11>**show ip route | begin Gateway**
Gateway of last resort is 10.1.1.254 to network 0.0.0.0

      10.0.0.0/8 is variably subnetted, 5 subnets, 2 masks
R   10.0.0.0/30 [120/1] via 10.1.1.254, 00:00:09, FastEthernet0/0
C   10.1.1.0/24 is directly connected, FastEthernet0/0
C   10.1.2.0/24 is directly connected, FastEthernet0/1
R   10.1.3.0/24 [120/1] via 10.1.1.2, 00:00:18, FastEthernet0/0
R   10.1.4.0/24 [120/1] via 10.1.1.2, 00:00:18, FastEthernet0/0
R*   **0.0.0.0/0 [120/1] via 10.1.1.254, 00:00:09, FastEthernet0/0**

由上我们看到路由器 R11 已经学习到了路由器 R1 传播来的默认路由。查看园区 B 内部路由器的路由表，路由器 R21 的显示如下：

R21>**show ip route | begin Gateway**
Gateway of last resort is 10.2.1.254 to network 0.0.0.0

      10.0.0.0/8 is variably subnetted, 7 subnets, 2 masks
R   10.0.0.12/30 [120/1] via 10.2.1.254, 00:00:07, FastEthernet0/0
C   10.2.1.0/24 is directly connected, FastEthernet0/0
C   10.2.2.0/24 is directly connected, FastEthernet0/1
R   10.2.3.0/24 [120/1] via 10.2.1.2, 00:00:08, FastEthernet0/0
R   10.2.4.0/24 [120/1] via 10.2.1.2, 00:00:08, FastEthernet0/0
R   10.2.5.0/24 [120/1] via 10.2.2.2, 00:00:08, FastEthernet0/1
R   10.2.6.0/24 [120/1] via 10.2.2.2, 00:00:08, FastEthernet0/1
R*   **0.0.0.0/0 [120/1] via 10.2.1.254, 00:00:07, FastEthernet0/0**

路由器 R21 也学习到了 R2 路由器传播来的默认路由。查看园区 C 内部路由器的路由表，路由器 R31 的显示如下：

R31>**show ip route | begin Gateway**
Gateway of last resort is 10.3.0.30 to network 0.0.0.0

```
           10.0.0.0/8 is variably subnetted, 5 subnets, 2 masks
R          10.0.0.4/30 [120/1] via 10.3.0.30, 00:00:17, FastEthernet0/0
C          10.3.0.16/28 is directly connected, FastEthernet0/0
C          10.3.0.32/28 is directly connected, FastEthernet0/1
C          10.3.0.48/28 is directly connected, FastEthernet1/0
R          10.3.0.64/28 [120/1] via 10.3.0.62, 00:00:14, FastEthernet1/0
R*         0.0.0.0/0 [120/1] via 10.3.0.30, 00:00:17, FastEthernet0/0
```

从以上输出我们看到，路由器 R31 也学习到了默认路由。查看园区 D 内部路由器的路由表，路由器 R41 的路由表显示如下：

```
R41>show ip route | begin Gateway
Gateway of last resort is 10.4.31.254 to network 0.0.0.0

           10.0.0.0/8 is variably subnetted, 6 subnets, 2 masks
R          10.0.0.8/30 [120/1] via 10.4.31.254, 00:00:06, FastEthernet0/0
C          10.4.16.0/20 is directly connected, FastEthernet0/0
C          10.4.32.0/20 is directly connected, FastEthernet0/1
C          10.4.48.0/20 is directly connected, FastEthernet1/0
R          10.4.64.0/20 [120/1] via 10.4.48.2, 00:00:13, FastEthernet1/0
R          10.4.80.0/20 [120/2] via 10.4.48.2, 00:00:13, FastEthernet1/0
R*         0.0.0.0/0 [120/1] via 10.4.31.254, 00:00:06, FastEthernet0/0
```

从以上路由表的输出来看，园区网内部采用的是动态路由协议，园区网与核心之间采用静态路由，避免了园区网与核心间的路由定期更新，从而也优化了路由表，提高了查询效率。

7.4 课外拓展训练

7.4.1 训练一：分析能力技能挑战

任务描述：如图 7-10 所示网络拓扑，所有路由器与 PC 接口的 IP 地址均已配置，路由器接口的状态均为 up，4 台路由器均已配置路由协议 RIPv1。但是 10.1.1.0/24 网络的用户不能访问 10.2.1.0/24 的用户。

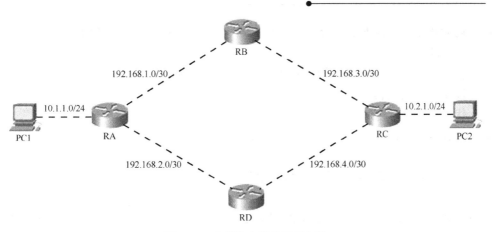

图 7-10 分析能力技能挑战拓扑

- 请分析：什么问题导致两个网段的用户间不能互访？
- 试分析：4 台路由器的路由表中各有多少条记录？
- 请回答：如何解决两个网段的用户间互访问题？

7.4.2 训练二：综合应用技能挑战

任务描述：如图 7-11 所示，ABC 公司企业网的 3 个分支分别通过路由器 BR1、BR2 和 BR3 的 WAN 接口连接到了公司总部路由器 HQ，并通过 HQ 的 WAN 接口接到 ISP。请按照如下要求完成对网络的配置并实现全网互通。

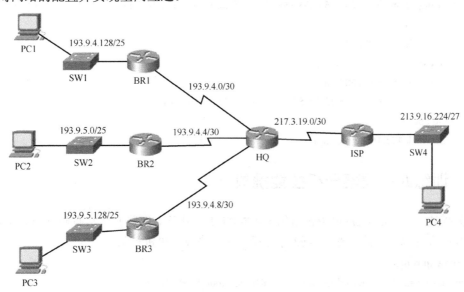

图 7-11 综合应用技能挑战拓扑

- 请为核心路由器 HQ 和分支路由器 BR1、BR2 和 BR3 配置 RIPv2 协议，实现企业网间互通；
- 请采用最优设计方案配置路由器 HQ 和 ISP 实现企业网接入 ISP，使企业网主机可以访问公网资源并可以通过公网主机访问企业网；
- 优化企业网，限制不必要的路由更新；
- 配置 SSH，使网络管理员可以远程管理网络设备，用户名为 ADMIN，口令为 cisco123。

7.4.3 训练三：奇思妙想技能挑战

任务描述：XYZ 公司是一家小型的网络公司，目前公司规模在迅速扩大，急需招聘多名有工作经验的网络工程师。为确保招聘进来的工程师对业务尽快上手，公司任命技术部经理对面试考试题目严格把关，题库需要不断完善。你作为本次参加面试的人员，下面是你抽取到的面试考题：

- 请你依据如下 RIP 协议的 debug 输出信息，推断出网络拓扑；
- 请打开 Packet Tracer 模拟器，搭建网络拓扑并完成相应配置；
- 请在已完成配置的路由器 RtrA 上打开 debug 命令输出与下面一致的结果。

```
RtrA#debug ip rip
RIP: received v1 update from 10.0.15.2 on Serial0/0/0
      192.168.1.0 in 1 hops
      192.168.168.0 in 1 hops
RIP: sending v1 update to 255.255.255.255 via FastEthernet0/0 (172.16.1.1)
RIP: build update entries
      network 10.0.0.0 metric 1
      network 192.168.1.0 metric 2
      network 192.168.168.0 metric 2
RIP: sending v1 update to 255.255.255.255 via Serial0/0/0 (10.0.8.1)
RIP: build update entries
      network 172.16.0.0 metric 1
```

7.4.4 训练四：洞察分析技能挑战

任务描述：以下是某位同学提供的路由器 R1 的路由表，你能发现它有破绽吗？请找出来，用两种方法进行修改，确保路由表没有问题，并对修改结果进行验证。

```
R1#show ip route
Codes: L - local, C - connected, S - static, R - RIP, M - mobile, B - BGP
Gateway of last resort is 0.0.0.0 to network 0.0.0.0
```

```
         10.0.0.0/8 is variably subnetted, 4 subnets, 2 masks
C        10.0.12.0/24 is directly connected, GigabitEthernet0/1
L        10.0.12.1/32 is directly connected, GigabitEthernet0/1
R        10.0.23.0/24 [120/1] via 10.0.12.2, 00:00:12, GigabitEthernet0/1
         200.5.19.0/24 is variably subnetted, 3 subnets, 3 masks
R        200.5.19.0/24 [120/1] via 10.0.12.2, 00:00:12, GigabitEthernet0/1
C        200.5.19.0/26 is directly connected, GigabitEthernet0/0
L        200.5.19.62/32 is directly connected, GigabitEthernet0/0
         216.11.5.0/24 is variably subnetted, 2 subnets, 2 masks
C        216.11.5.4/30 is directly connected, Serial0/0/0
L        216.11.5.5/32 is directly connected, Serial0/0/0
```

"RIP 网络实战"一章，到此即将结束。本章设计了 1 个故障排错场景，1 个企业网综合配置案例，4 个园区网规划方案以及 1 个网络优化方案，外加 4 个课外拓展训练。基于实验室分组实验的场景，展现了同学们主动学习的热情，通过两个小组的故障案例，带你完成排错学习，引导你如何发现并解决问题。1 个企业网综合配置案例，教你如何对所学知识进行综合应用。任务实施过程体现任务之间逻辑分层的实施理念，向你展示了整体的实施思路，承载的知识包括接口配置、协议配置、关闭自动汇总、配置静态路由、默认路由传播以及静态路由注入等。园区网规划案例，通过 4 个规划方案和 1 个优化方案，展示了合理地址规划在网络规划设计中的重要性，优化方案让我们能深入理解了在某些应用场合，采用动态路由协议和静态路由相结合的方式才更加完美。课外拓展训练可以进一步巩固所学知识，开拓思维，培养创新能力。

第8章

Multiuser 分布式多用户案例

本章要点

- PT Multiuser 实验案例
- PT Multiuser 综合案例
- PT Multiuser 拓展案例

8.1　PT Multiuser 实验案例

本章"Multiuser 分布式多用户案例",是基于 PT 的 Multiuser 扩展功能开发的 LAN 多用户联机实验教学案例,在课堂实施过程中产生了意想不到的、非常好的教学效果。有的同学反馈这是在课堂上玩联机游戏,有的同学说 Multiuser 实验是对传统实验课的颠覆,也有的同学说从此让他爱上了网络技术,爱上了 Packet Tracer。本章介绍了 PT Multiuser 的功能特点,总结了常见的 Multiuser 连接故障,通过设计的两用户、三用户以及多用户的教学案例,将全书所学知识进行了融合,是对全书内容的一个概括总结。通过 Teacher 和 Student 的双机 PT 实例连接实验,展现了 Multiuser 的神奇功能;通过设计的综合教学案例,让三个用户协作完成一个综合实验,增强了团队间的合作与交流,增进了学生间的友谊,是对现实生活真实项目的模拟体验;最后通过师生联合开发的拓展案例,巩固所学知识,让更多用户参与进来,进而扩大游戏规模,挖掘 Multiuser 在团队竞技方面的潜能,展现网络技术无穷大的魔力。

8.1.1　PT Multiuser功能简介

Multiuser(多用户)是 Cisco Packet Tracer(简称 PT)的扩展功能,它允许在多个运行的 PT 实例间建立点对点连接。Multiuser 功能可以为建立连接的两个用户提供虚拟的通信信道。这种虚拟通道可以连接多个 PT 实例,构成分布式实验环境,从而扩展网络规模。Multiuser 对课堂展开网络教学,为实现师生互动提供了一个非常好的实验平台。它可以极大地活跃课堂氛围,增强团队协作能力,提高课堂的趣味性和挑战性,达到完美的教学效果。

Multiuser 打破了传统的集中式实验理念,采用多用户分布式模式,将一个复杂的任务进行分解,由多个用户协同完成,这不仅能增强学生的团队意识和竞争意识,同时可以培养学生的学习兴趣,积累排错经验,提高协作能力,进一步提升专业技术水平。

Packet Tracer Multiuser注意事项如下:
- Multiuser 多 PT 实例间的连接基于 TCP 的 PTMP(Packet Tracer Multiuser Protocol)协议。
- PT Multiuser 功能可以转发远程设备的 PTMP 广播或组播流量。
- PT Multiuser 默认端口为 38000,若主机同时开启多个 PT 实例,那么实例各自占用的端口会依次递增。
- PT Multiuser 功能默认开启,可通过点击菜单栏 Extensions→Multiuser → Listen →Stop listening 来禁用此功能。
- PT Multiuser 尝试使用 UPnP(Universal Plug and Play,通用即插即用)协议来建立端口转发,以使 PT 多实例设备间彼此能自动连接并协同工作。

- PT Multiuser 网络通信可以在 PT 实例间进行，Console 线缆允许在 Multiuser 中使用。
- 要使 PT Multiuser 间能正常通信，必须确保采用相同的 PT 版本创建不同的实例，例如，PT 6.3 下的实例无法与 PT7.0 下的实例进行通信。
- PT Multiuser 多实例间的连接需要设置监听参数，包括监听地址和端口、连接口令以及对远程连接采取的接受方式（Always Accept、Always Deny、Prompt）。连接口令默认是 cisco，建议配置时自行修改。
- PT Multiuser 的连接口令可以采用"全局口令"或"专用口令"与远程 PT 建立连接，默认采用全局口令。例如，三个 PT 实例 B、C 和 D 与实例 PT A 建立连接，可以统一采用全局口令；也可以采用专用口令，如实例 B 和 A 的连接采用口令 1，实例 C 和 A 的连接采用口令 2，实例 D 和 A 的连接采用口令 3。
- PT Multiuser 实例间连接的监听方可以随意指定，例如，实例 A 和实例 B 连接，我们可以指定实例 A 为监听方，也可指定实例 B 为监听方，则另一方需要主动发起连接请求。
- PT Multiuser 监听方需要指定设备的可见端口，以便于远程请求方可以正确连接到监听方设备的端口，确保物理层的正确连接。例如，监听方的交换机使用 Fa0/24 端口连接远程主机，则监听方需要点击菜单栏的 Extensions → Multiuser →Port Visibility，来勾选交换机的指定端口 Fa0/24。
- PT Multiuser 实例间的连接，需要通过 PT 的设备选区将 Multiuser 的云图标拖拽至需要建立连接的模拟环境，该图标就代表远程网络，我们可以设置名字（默认为 Peer0），以便接入方以该名称进行连接请求，此时图标呈现灰色，表示还没有与其他模拟网络建立连接。
- 在 PT Multiuser 实例连接的过程中，Multiuser 的云连接图标会发生一定的颜色变化，例如，初始未连接的状态是"灰色"，连接成功是"蓝色"等。

8.1.2 PT Multiuser 案例背景

案例背景：今天的网络课，CFL 老师将带领大家一起做 Multiuser 实验。上节课给大家布置的作业是 Multiuser 实验的准备工作，具体要求如下：
- 要求全班同学安装相同版本的 Packet Tracer 模拟器，建议升级到最新版本；
- 要求全班同学的笔记本电脑能通过局域网互连在一起，确保彼此间可以 ping 通；
- 为老师的笔记本电脑准备一根网线，分配一个局域网的 IP 地址（也可无线连接）。

有关今天的 Multiuser 实验，老师课前已经将实验拓扑投放在大屏幕上，如图 8-1 所示，地址规划见表 8-1。本案例的 Multiuser 只有两个用户，一个用户创建 Teacher 端 PT 实例，另一个创建 Student 端 PT 实例，两个 PT 实例通过 PT Multiuser 功能互连起来，使 Student 端 PC 能通过远端 Teacher 的 DHCP Server 成功获取地址，并实现主机间的通信。简单地说，就是将图 8-2 所示的单机单用户实验分解成图 8-1 所示的由两个用户协同完成的实验。

第 8 章 Multiuser 分布式多用户案例

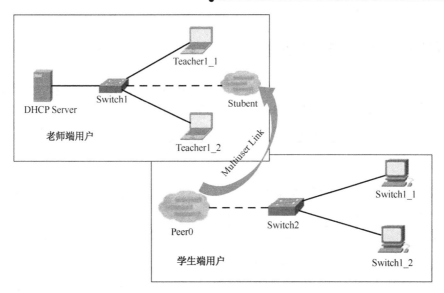

图 8-1 多用户实验拓扑

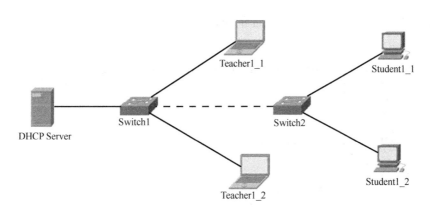

图 8-2 单机单用户实验拓扑

表 8-1 地址规划表

设 备 名 称	IP 地址	子网掩码	设 备 描 述
DHCP Server	192.168.1.250	255.255.255.0	Link to Switch1
Teacher1_1	DHCP	DHCP	Link to Switch1
Teacher1_2	DHCP	DHCP	Link to Switch1
Student1_1	DHCP	DHCP	Link to Switch2
Student1_2	DHCP	DHCP	Link to Switch2

8.1.3 PT Multiuser 实例创建

接下来，CFL 老师用自己的笔记本电脑创建了如图 8-3 所示的 PT 实例，她要求同学们用各自的笔记本电脑创建如图 8-4 所示的 PT 实例，看哪位同学最先能和老师的 PT 实例成功连接。

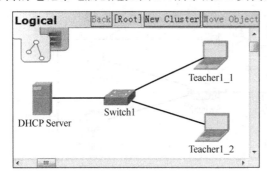 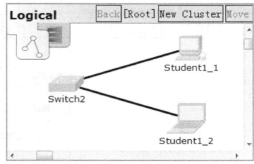

图 8-3 Teacher.pkt　　　　　　　　　　图 8-4 Student.pk

CFL 老师给大家演示了 DHCP Server 的配置，如图 8-5 所示。

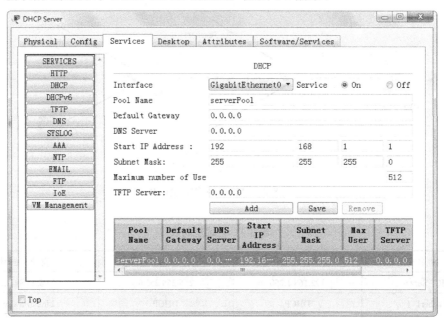

图 8-5 设置教师端 DHCP 服务

CFL 老师让大家观察教师端的主机能否通过 DHCP 从 DHCP Server 上动态获取地址。
我们发现 Teacher1_1 主机从 DHCP Server 上成功获取到地址 192.168.1.1/24，如图 8-6 所示。

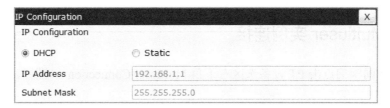

图 8-6　教师端主机 Teacher1_1 动态获取到 IP 地址

Teacher1_2 主机也从 DHCP Server 上成功获取到地址 192.168.1.2/24，如图 8-7 所示。

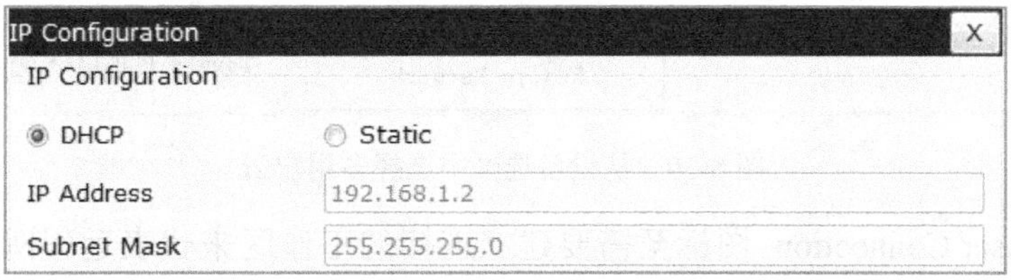

图 8-7　教师端主机 Teacher1_2 动态获取到 IP 地址

CFL 老师让全班同学将各自 PT 中的两台主机的 IP 配置设置为 DHCP 获取，并观察和记录实验结果。同学们惊讶地发现 PT 中的两台 PC 都获取到 169.254.x.x/16 的 IP 地址，SJX 同学的 PC 获取到的地址如图 8-8 和 8-9 所示。老师告诉大家这个地址被称为自动专用地址，该地址专门服务于 DHCP 协议。当 DHCP 服务器没有开机或出现故障使得 DHCP 客户端无法与其联系时，客户端会自动得到 169.254.x.x/16 的随机地址，供局域网临时通信使用。

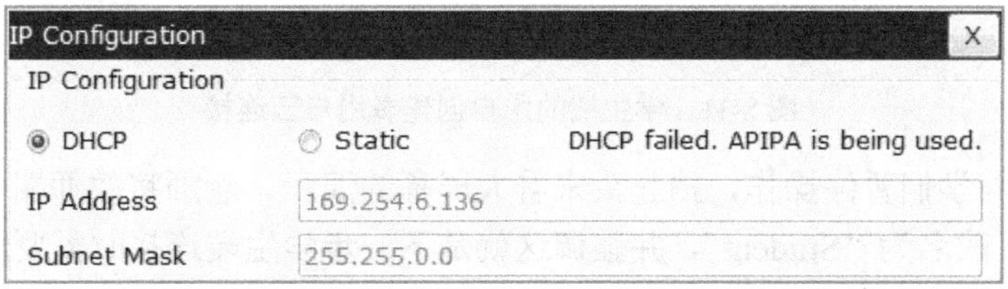

图 8-8　设置学生端主机 Student1_1 动态获取 IP 地址

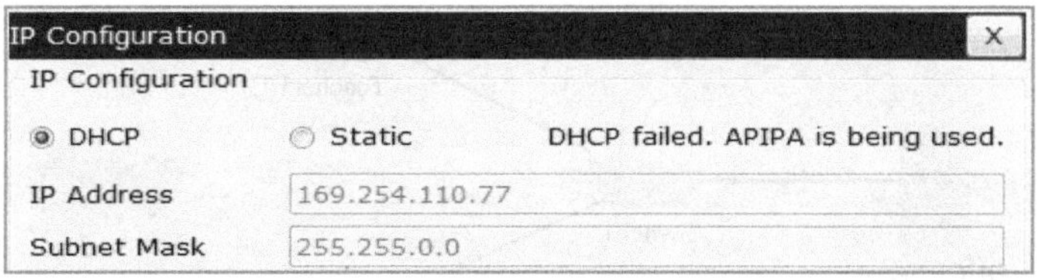

图 8-9　设置学生端主机 Student1_2 动态获取 IP 地址

CFL 老师向大家解释目前学生的 PT 实例和教师的 PT 实例之间没有进行连接，也就是学生端主机与 DHCP Server 无法通信，因此仅能获取到上述的临时自动专用 IP 地址。如何实现两个 PT 实例间的互连呢，我们将在 8.1.4 节学习。

8.1.4　PT Multiuser 实例连接

CFL 老师让同学们点击 PT 设备选区右下角 Multiuser Connection 图标，如图 8-10 所示。

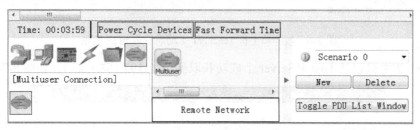

图 8-10　从设备选区中选择多用户云

拖拽 Multiuser Connection 图标至需要建立连接的工作区来代表远程网络，默认名称为 Peer0，如图 8-11 所示。

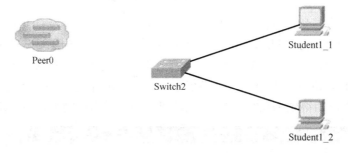

图 8-11　学生端的用户创建多用户云连接

CFL 老师让同学们暂停操作，抬起头来看大屏幕的演示。老师将教师端 PT 实例的远程网络名称"Peer0"，改名为"Student"，并强调这就是下一步学生端连接时需要输入的参数"Peer Network Name"的值，如图 8-12 所示。

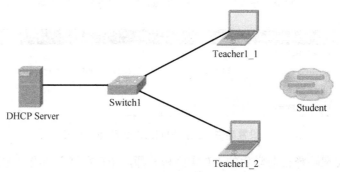

图 8-12　教师端的用户创建多用户云连接

CFL 老师告诉大家下一步要设置 PT 实例连接的参数，包括监听地址、端口号、对端的网络名称及口令等。两个 PT 实例要选出一方为监听端，另一方为接入端。CFL 老师把教师端设置成监听端，她想让学生端向她的 PT 实例发出连接请求。CFL 老师让同学们观看大屏幕上监听端的参数设置方法。PT Multiuser 的监听功能默认开启，通过点击菜单栏 Extensions→Multiuser→Listen 打开如图 8-13 所示窗口。教师端监听地址是 192.168.1.104，端口为 38000，口令设为 ytvc。老师对远程网络接入请求采取的接受方式是 Prompt，若选择 Always Accept，则对远程接入总是允许；若选择 Always Deny，则总是拒绝，永远无法建立连接。

接下来，CFL 老师要求全班同学依据图 8-13 所设置的监听参数完成相关设置，学生端接入参数设置如图 8-14 所示，看哪位同学能成功与教师端 PT 实例连接，CFL 老师话音未落，就见大屏幕上弹出了一个窗口，如图 8-15 所示。

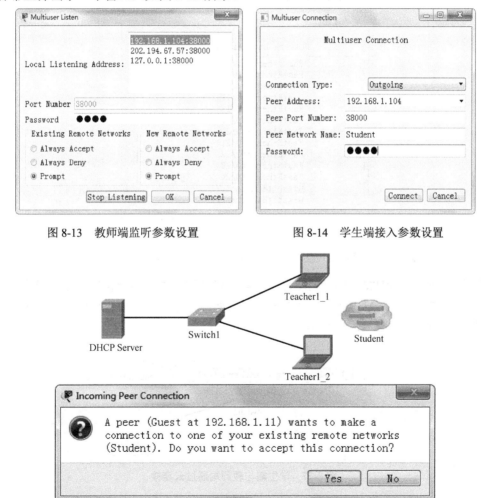

图 8-13　教师端监听参数设置　　　　图 8-14　学生端接入参数设置

图 8-15　远程的学生端成功接入

CFL 老师高兴地向大家宣布："恭喜 IP 地址为 192.168.1.11 的同学成功接入！"同学们都抬起头来看着大屏幕，有的同学在问谁是.11，原来是 ZHWQ 同学，还有同学问老师是否还可以再接入，老师回答说当然可以，全班同学都可以接进来。CFL 老师让大家先不要着急，ZHWQ 同学还没有真正接入进来，要全班同学看大屏幕，观察接下来会发生什么。只见老师点击了"Yes"按钮，然后奇迹发生了，老师的云由"灰色"变成了"蓝色"，ZHWQ 同学说："老师，我的云也变色了！变成了'蓝色'。"同学们目瞪口呆，老师笑着说就应该变成蓝色，蓝色表明我们两个 PT 实例已经连通。但是，我们的主机怎么连通呢？同学们回答要用交叉线连接。CFL 老师说回答正确，还需最关键的一步，那就是"端口可见"设置。只见老师点击 **Extensions→Multiuser →Port Visibility**，打开了如图 8-16 所示的窗口，并勾选了 Gig0/1 端口。老师通知远端的 ZHWQ 同学，选择交叉线连接到老师交换机 Switch1 的 Gig0/1 端口。

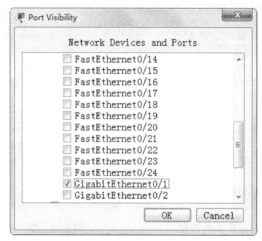

图 8-16 设置教师端交换机 G0/1 口为端口可见

好多同学跑到了 ZHWQ 同学的座位上观察，有的同学说确实能看到老师那端的交换机名称和端口，如图 8-17 所示；瞪着屏幕的同学也开始惊叫起来，老师这端自动地出现一根交叉线，连接在 Switch1 和云上，如图 8-18 所示。

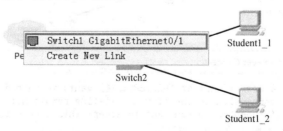

图 8-17 学生端与教师端通过云连接

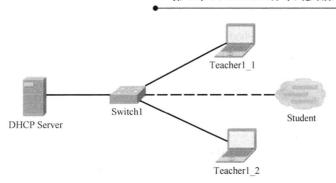

图 8-18 教师端自动连接至云端

看着连接云端的链路指示灯变成绿色,全班同学都鼓起掌来。CFL 老师高兴地对全班同学说实验还没有结束,还需要做进一步的测试。

8.1.5 PT Multiuser 连通测试

CFL 老师让 ZHWQ 同学查看一下他 PT 中的两台主机是否通过 DHCP 成功获取到地址。

ZHWQ 反馈他的两台主机分别获取到地址 192.168.1.3/24 和 192.168.1.4/24,如图 8-19 和图 8-20 所示。

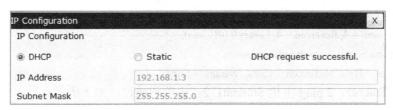

图 8-19 学生端主机 Student1_1 成功获取地址

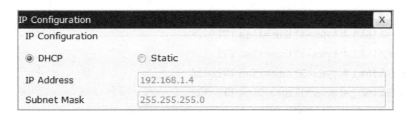

图 8-20 学生端主机 Student1_2 成功获取地址

ZHWQ 同学查看了一下主机 Student1_2 的地址,该地址确实是来自老师端的 DHCP Server,如下所示:

```
C:\>ipconfig /all
FastEthernet0 Connection:(default port)
    Connection-specific DNS Suffix..:
```

```
Physical Address.................: 0001.970A.45A3
Link-local IPv6 Address.........: FE80::201:97FF:FE0A:45A3
IP Address........................: 192.168.1.4
Subnet Mask......................: 255.255.255.0
Default Gateway.................: 0.0.0.0
DNS Servers......................: 0.0.0.0
DHCP Servers....................: 192.168.1.250
DHCPv6 Client DUID...........: 00-01-00-01-13-61-BE-3A-00-01-97-0A-45-A3
```

CFL 老师通过大屏幕向全班同学展示了最终的实验结果，她的主机 Teacher1_1 ping 通了 ZHWQ 同学的主机 Student1_1，如下所示：

```
C:\>ping 192.168.1.3

Pinging 192.168.1.3 with 32 bytes of data:
Reply from 192.168.1.3: bytes=32 time=15ms TTL=128
Reply from 192.168.1.3: bytes=32 time=12ms TTL=128
Reply from 192.168.1.3: bytes=32 time=11ms TTL=128
Reply from 192.168.1.3: bytes=32 time=15ms TTL=128

Ping statistics for 192.168.1.3:
    Packets: Sent = 4, Received = 4, Lost = 0 (0% loss),
Approximate round trip times in milli-seconds:
    Minimum = 11ms, Maximum = 15ms, Average = 13ms
```

换成主机 Teacher1_2 ping 主机 Student1_2，也顺利通过，如下所示：

```
C:\>ping 192.168.1.4

Pinging 192.168.1.4 with 32 bytes of data:
Reply from 192.168.1.4: bytes=32 time=33ms TTL=128
Reply from 192.168.1.4: bytes=32 time=13ms TTL=128
Reply from 192.168.1.4: bytes=32 time=15ms TTL=128
Reply from 192.168.1.4: bytes=32 time=25ms TTL=128

Ping statistics for 192.168.1.4:
    Packets: Sent = 4, Received = 4, Lost = 0 (0% loss),
Approximate round trip times in milli-seconds:
    Minimum = 13ms, Maximum = 33ms, Average = 21ms
```

至此，本节课的实验已经顺利完成，CFL 老师让同学们第二节课分组比赛完成该实验。同桌俩为一个组，一个为教师端，另一个为学生端，看看哪个小组能拿到冠军。

8.1.6　PT Multiuser 故障排错

基于 Packet Tracer 的多用户案例故障现象时有发生，多用户 PT 互连不通的现象非常多，首先要确保多用户 PT 的版本必须一致，其次再观察多用户连接时的云颜色是否发生改变，并按以下罗列的几点依次排除。

- Connection Type（连接类型）：确保监听端的连接类型为"incoming"，接入端为"outgoing"。
- Peer Address（对端地址）：若不同物理主机运行多个 PT 实例连接，首先要确保物理主机在同一网段，输入的对端地址必须为可达的物理网卡 IP 地址。
- Peer Port Number（对端端口号）：监听端的端口号，也是接入端主机需要输入的端口号。若监听端主机同时打开多个 PT 实例，端口会发生相应变化，需要记录下正确端口号，本端口号可能会因多个 PT 实例运行顺序而发生改变。
- Peer Network Name（对端网络名称）：大小写不敏感，但必须与"监听端"网络名称保护一致。
- Password（口令）：必须保证接入端口令与监听端一致；建议输入口令至记事本，再粘贴，避免输入错误。
- Connection Status（连接状态）：查看网络云连接时颜色的变化，确保本地与远程的云连接状态为 connected。

为顺利排除 Multiuser 故障，了解 Multiuser 的 5 种连接状态是非常有必要的，如表 8-2 所示。

表 8-2　Multiuser 的五种连接状态

连接类型编号	云状态图	连接状态	详细说明
1	Peer0 灰色	Disconnected（未连接）	现有网络与远程网络没有进行 Multiuser 连接时的连接状态
2	Peer0 浅黄	Prompt（连接变化）	提醒你远端连接内容有变化了，可以检查连接的变化
3	Peer0 蓝灰	Connecting（连接中）	云端用户发起连接请求，尝试和远程云端用户建立虚拟连接，检验 IP，端口号，连接口令等参数

续表

连接类型编号	云状态图	连接状态	详细说明
4	Peer0 蓝色	Connected（已连接）	云端用户之间已经成功完成了连接，最终的理想连接状态
5	Peer0 灰色	Error（连接出错）	表示连接中出现了故障，在连接成功之后出现的故障，往往是另一端 PT 实例关闭导致，也可能是再次修改连接参数产生的故障

本 Multiuser 实验案例，两个 PT 实例成功连接后，若先关闭教师端的 PT 实例，学生端 PT 的云连接图标中就会显示"!"，表示连接中断，如图 8-21 所示。

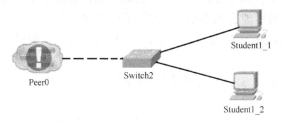

图 8-21　学生端 PT 显示云连接中断

若先关闭学生端 PT 实例，教师端 PT 实例的云连接图标中也会显示"!"，表示连接中断，如图 8-22 所示。

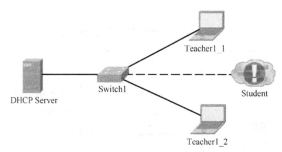

图 8-22　教师端 PT 显示云连接中断

此时，若重新打开已经关闭的 PT 实例，再次连接，就会显示正常的连接状态。

8.2 PT Multiuser 综合案例

8.2.1 PT Multiuser 任务背景

任务背景： 上节课 CFL 老师带同学们一起做的 Multiuser 实验，大家做得很开心也很投入，下课铃声响了，同学们似乎都没有听见。那专注劲，让一旁的老师感到非常欣慰。Multiuser 让同学们收获很多，也体会到了实验的乐趣和成就感。好多同学反映对计算机网络课程的学习兴趣越来越浓了，没想到实验还可以这么做，真的特别好玩。为了巩固所学知识，CFL 老师打算下次课让同学们三人一组完成如图 8-23 所示的综合案例。为确保下次课能顺利进行，CFL 老师邀请到 XL 和 LX 两位同学和她一起按照具体要求事先演练一遍，以便检验综合案例的难易程度，做到心中有数。老师给两位同学大致介绍了网络的拓扑结构，主要分三层：接入层（包括 S1、S2、S3、S4）、汇聚层（包括 MS1 和 MS2）以及核心层（R1），整个拓扑的特点是采用冗余链路提高网络的可靠性。实验最终目的是实现全网互通，并且要求网络设备间的任何一条链路出现故障，都不会中断网络的正常通信。

1. 网络拓扑

图 8-23 为综合案例网络拓扑。

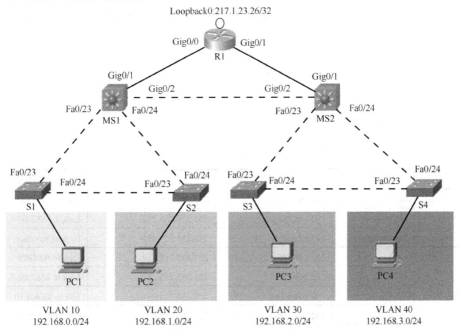

图 8-23 综合案例网络拓扑

2. 具体要求

- 设备主机名的配置。
- 设备 IP 地址的配置：参照 8-3 所示的地址规划表。
- 对所有明文口令进行加密。
- 特权密文口令的配置：密文口令为 **ytvc**。
- 远程 Telnet 的配置：最多允许 4 个用户同时登录，登录口令为 **e509**。
- VLAN 和 Trunk 的配置：VLAN 配置参照表 8-4 的 VLAN 规划表，并将相应端口配置成 Trunk。
- 配置 DHCP 服务器：在 MS1 上创建地址池 14net1 和 14net2 分别为 VLAN 10 和 VLAN 20 的主机分配地址，在 MS2 上创建地址池 15net1 和 15net2 分别为 VLAN 30 和 VLAN 40 的主机分配地址。
- 配置 SVI 接口：在 MS1 和 MS2 上配置 SVI 接口作为各 VLAN 的网关。
- 配置静态路由：不允许采用 ARP 代理。
- 在 MS1 和 MS2 上各配置一条静态默认路由，正常情况下通过上行链路转发数据；当上行链路出现故障时则通过水平链路转发数据。
- 在 MS1 和 MS2 上分别配置到对方 VLAN 的静态路由，不允许经路由器转发。
- 在路由器 R1 上配置到相应 VLAN 的静态路由，其中 VLAN 10、VLAN 20 及 VLAN 99 的流量正常情况下从 G0/0 接口转发，若该接口出现故障，则从 G0/1 口转发；VLAN 30、VLAN 40 及 VLAN 199 的流量正常情况下从 G0/1 接口转发，若该接口出现故障，则从 G0/0 口转发。
- 保存设备的配置文件。

3. 地址规划

表 8-3 为 Multiuser 综合案例地址规划表。

表 8-3　Multiuser 综合案例地址规划表

设备名称	接口	IP 地址	子网掩码	接口描述/网关
R1	G0/0	192.168.255.2	255.255.255.252	Link to MS1 G0/1
	G0/1	192.168.255.10	255.255.255.252	Link to MS2 G0/1
	Loopback0	217.1.23.26	255.255.255.255	Logical interface
MS1	G0/1	192.168.255.1	255.255.255.252	Link to R1 G0/0
	G0/2	192.168.255.5	255.255.255.252	Link to MS2 G0/2
	VLAN 10	192.168.0.254	255.255.255.0	SVI to VLAN 10
	VLAN 20	192.168.1.254	255.255.255.0	SVI to VLAN 20
	VLAN 99	192.168.99.254	255.255.255.0	SVI to VLAN 99

续表

设备名称	接口	IP 地址	子网掩码	接口描述/网关
MS2	G0/1	192.168.255.9	255.255.255.252	Link to R1 G0/1
	G0/2	192.168.255.6	255.255.255.252	Link to MS1 G0/2
	VLAN 30	192.168.2.254	255.255.255.0	SVI to VLAN 30
	VLAN 40	192.168.3.254	255.255.255.0	SVI to VLAN 40
	VLAN 199	192.168.199.254	255.255.255.0	SVI to VLAN 199
S1	VLAN 99	192.168.99.10	255.255.255.0	192.168.99.254
S2	VLAN 99	192.168.99.20	255.255.255.0	192.168.99.254
S3	VLAN 199	192.168.199.30	255.255.255.0	192.168.199.254
S4	VLAN 199	192.168.199.40	255.255.255.0	192.168.199.254
PC1	NIC	DHCP	DHCP	DHCP
PC2	NIC	DHCP	DHCP	DHCP
PC3	NIC	DHCP	DHCP	DHCP
PC4	NIC	DHCP	DHCP	DHCP

表 8-4 为 Multiuser 综合案例 VLAN 规划表。

表 8-4　Multiuser 综合案例 VLAN 规划表

VLAN ID	VLAN 名称	交换机端口	地址规划
10	14NET1	S1：Fa0/1-22	192.168.0.0/24
20	14NET2	S2：Fa0/1-22	192.168.1.0/24
30	15NET1	S3：Fa0/1-22	192.168.2.0/24
40	15NET2	S4：Fa0/1-22	192.168.3.0/24
99	Admin	S1&S2	192.168.99.10/24 & 192.168.99.20/24
199	Admin	S3&S4	192.168.199.30/24 & 192.168.199.40/24

8.2.2　PT Multiuser 任务分解

该综合案例要求采用 Multiuser 来实现。本案例的实施由 CF、XL 和 LX 3 个用户在 3 台主机上协同完成，这就是所谓的分布式多用户实验。3 个用户分别负责核心层、汇聚层以及接入层的 PT 搭建与实施。最后通过 Multiuser 云将 3 个用户的 PT 实例进行连接，并实现全网互通。3 个用户的具体分工如表 8-5 所示。

表 8-5　Multiuser 任务分解表

用户名	分配任务	用户主机地址	分层
CF	R1	192.168.1.1/24	核心层
XL	G0/2 ———— G0/2 MS1　　　　　　MS2	192.168.1.2/24	汇聚层
LX	F0/24 F0/23　　　　　F0/24 F0/23 S1 —— S2　　　　　S3 —— S4 PC1　　PC2　　PC3　　PC4 VLAN 10　VLAN 20　VLAN 30　VLAN 40 192.168.0.0/24　192.168.1.0/24　192.168.2.0/24　192.168.3.0/24	192.168.1.3/24	接入层

下面让我们 3 个人一起来完成这个有趣的实验吧。

8.2.3　PT Multiuser 任务实施

任务一：接入层交换机任务实施

（1）用户 LX 配置接入层交换机 S1

```
Switch>enable
Switch#configure terminal
Switch(config)#hostname S1
S1(config)#service password-encryption
S1(config)#vlan 10
S1(config-vlan)#name 14NET1
S1(config-vlan)#vlan 20
S1(config-vlan)#name 14NET2
S1(config-vlan)#vlan 99
S1(config-vlan)#name Admin
S1(config-vlan)#interface vlan 99
S1(config-if)#ip address 192.168.99.10 255.255.255.0
S1(config-if)#no shutdown
S1(config-if)#ip default-gateway 192.168.99.254
S1(config)#enable secret ytvc
```

```
S1(config)#line vty 0 3
S1(config-line)#password e509
S1(config-line)#login
S1(config-line)#exit
S1(config)#interface range f0/1-22
S1(config-if-range)#switchport mode access
S1(config-if-range)#switchport access vlan 10
S1(config-if-range)#exit
S1(config)#interface range f0/23-24
S1(config-if-range)#switchport mode trunk
S1(config-if-range)#switchport trunk allowed vlan all
S1(config-if-range)#end
S1#write
```

（2）用户 LX 配置接入层交换机 S2

```
Switch>enable
Switch#configure terminal
Switch(config)#hostname S2
S2(config)#service password-encryption
S2(config)#vlan 10
S2(config-vlan)#name 14NET1
S2(config-vlan)#vlan 20
S2(config-vlan)#name 14NET2
S2(config-vlan)#vlan 99
S2(config-vlan)#name Admin
S2(config-vlan)#interface vlan 99
S2(config-if)#ip address 192.168.99.20 255.255.255.0
S2(config-if)#no shutdown
S2(config-if)#ip default-gateway 192.168.99.254
S2(config)#enable secret ytvc
S2(config)#line vty 0 3
S2(config-line)#password e509
S2(config-line)#login
S2(config-line)#interface range f0/1-22
S2(config-if-range)#switchport mode access
S2(config-if-range)#switchport access vlan 20
S2(config-if-range)#interface range f0/23-24
S2(config-if-range)#switchport mode trunk
S2(config-if-range)#switchport trunk allowed vlan all
```

```
S2(config-if-range)#end
S2#write
```

(3)用户 LX 配置接入层交换机 S3

```
Switch>enable
Switch#configure terminal
Switch(config)#hostname S3
S3(config)#service password-encryption
S3(config)#vlan 30
S3(config-vlan)#name 15NET1
S3(config-vlan)#vlan 40
S3(config-vlan)#name 15NET2
S3(config-vlan)#vlan 199
S3(config-vlan)#name Admin
S3(config-vlan)#interface vlan 199
S3(config-if)#ip address 192.168.199.30 255.255.255.0
S3(config-if)#no shutdown
S3(config-if)#ip default-gateway 192.168.199.254
S3(config)#enable secret ytvc
S3(config)#line vty 0 3
S3(config-line)#password e509
S3(config-line)#login
S3(config-line)#interface range f0/1-22
S3(config-if-range)#switchport mode access
S3(config-if-range)#switchport access vlan 30
S3(config-if-range)#interface range f0/23-24
S3(config-if-range)#switchport mode trunk
S3(config-if-range)#switchport trunk allowed vlan all
S3(config-if-range)#end
S3#write
```

(4)用户 LX 配置接入层交换机 S4

```
Switch>enable
Switch#configure terminal
Switch(config)#hostname S4
S4(config)#service password-encryption
S4(config)#vlan 30
S4(config-vlan)#name 15NET1
```

```
S4(config-vlan)#vlan 40
S4(config-vlan)#name 15NET2
S4(config-vlan)#vlan 199
S4(config-vlan)#name Admin
S4(config-vlan)#interface vlan 199
S4(config-if)#ip address 192.168.199.40 255.255.255.0
S4(config-if)#no shutdown
S4(config-if)#ip default-gateway 192.168.199.254
S4(config)#enable secret ytvc
S4(config)#line vty 0 3
S4(config-line)#password e509
S4(config-line)#login
S4(config-line)#interface range f0/1-22
S4(config-if-range)#switchport mode access
S4(config-if-range)#switchport access vlan 40
S4(config-if-range)#interface range f0/23-24
S4(config-if-range)#switchport mode trunk
S4(config-if-range)#switchport trunk allowed vlan all
S4(config-if-range)#end
S4#write
```

任务二：汇聚层交换机任务实施

（1）用户 XL 配置汇聚层交换机 MS1

```
Switch>enable
Switch#configure terminal
Enter configuration commands, one per line.    End with CNTL/Z.
Switch(config)#hostname MS1
MS1(config)#ip routing
MS1(config)#vlan 10
MS1(config-vlan)#name 14NET1
MS1(config-vlan)#vlan 20
MS1(config-vlan)#name 14NET2
MS1(config-vlan)#vlan 99
MS1(config-vlan)#name Admin
MS1(config-vlan)#interface vlan 10
MS1(config-if)#ip address 192.168.0.254 255.255.255.0
MS1(config-if)#no shutdown
```

```
MS1(config-if)#interface vlan 20
MS1(config-if)#ip address 192.168.1.254 255.255.255.0
MS1(config-if)#no shutdown
MS1(config-if)#interface vlan 99
MS1(config-if)#ip address 192.168.99.254 255.255.255.0
MS1(config-if)#no shutdown
MS1(config-if)#ip dhcp excluded-address 192.168.0.254
MS1(config)#ip dhcp excluded-address 192.168.1.254
MS1(config)#ip dhcp pool 14net1
MS1(dhcp-config)#network 192.168.0.0 255.255.255.0
MS1(dhcp-config)#default-router 192.168.0.254
MS1(dhcp-config)#ip dhcp pool 14net2
MS1(dhcp-config)#network 192.168.1.0 255.255.255.0
MS1(dhcp-config)#default-router 192.168.1.254
MS1(dhcp-config)#exit
MS1(config)#interface range f0/23-24
MS1(config-if-range)#switchport trunk encapsulation dot1q
MS1(config-if-range)#switchport mode trunk
MS1(config-if-range)#interface g0/1
MS1(config-if)#no switchport
MS1(config-if)#ip address 192.168.255.1 255.255.255.252
MS1(config-if)#no shutdown
MS1(config-if)#interface g0/2
MS1(config-if)#no switchport
MS1(config-if)#ip address 192.168.255.5 255.255.255.252
MS1(config-if)#no shutdown
MS1(config-if)#ip route 192.168.2.0 255.255.254.0 192.168.255.6
MS1(config)#ip route 192.168.199.0 255.255.255.0 192.168.255.6
MS1(config)#ip route 0.0.0.0 0.0.0.0 192.168.255.2
MS1(config)#ip route 0.0.0.0 0.0.0.0 192.168.255.6 250
MS1(config)#enable secret ytvc
MS1(config)#line vty 0 3
MS1(config-line)#password e509
MS1(config-line)#login
MS1(config-line)#end
MS1#write
```

（2）用户 XL 配置汇聚层交换机 MS2

```
Switch>enable
Switch#configure terminal
Switch(config)#hostname MS2
MS2(config)#ip routing
MS2(config)#vlan 30
MS2(config-vlan)#name 15NET1
MS2(config-vlan)#vlan 40
MS2(config-vlan)#name 15NET2
MS2(config-vlan)#vlan 199
MS2(config-vlan)#name Admin
MS2(config-vlan)#interface vlan 30
MS2(config-if)#ip address 192.168.2.254 255.255.255.0
MS2(config-if)#no shutdown
MS2(config-if)#interface vlan 40
MS2(config-if)#ip address 192.168.3.254 255.255.255.0
MS2(config-if)#no shutdown
MS2(config-if)#interface vlan 199
MS2(config-if)#ip address 192.168.199.254 255.255.255.0
MS2(config-if)#no shutdown
MS2(config-if)#ip dhcp excluded-address 192.168.2.254
MS2(config)#ip dhcp excluded-address 192.168.3.254
MS2(config)#ip dhcp pool 15net1
MS2(dhcp-config)#network 192.168.2.0 255.255.255.0
MS2(dhcp-config)#default-router 192.168.2.254
MS2(dhcp-config)#ip dhcp pool 15net2
MS2(dhcp-config)#network 192.168.3.0 255.255.255.0
MS2(dhcp-config)#default-router 192.168.3.254
MS2(dhcp-config)#exit
MS2(config)#interface range f0/23-24
MS2(config-if-range)#switchport trunk encapsulation dot1q
MS2(config-if-range)#switchport mode trunk
MS2(config-if-range)#interface g0/1
MS2(config-if)#no switchport
MS2(config-if)#ip address 192.168.255.9 255.255.255.252
MS2(config-if)#no shutdown
MS2(config-if)#interface g0/2
MS2(config-if)#no switchport
```

```
MS2(config-if)#ip add 192.168.255.6 255.255.255.252
MS2(config-if)#no shutdown
MS2(config-if)#ip route 192.168.0.0 255.255.254.0 192.168.255.5
MS2(config)#ip route 192.168.99.0 255.255.255.0 192.168.255.5
MS2(config)#ip route 0.0.0.0 0.0.0.0 192.168.255.10
MS2(config)#ip route 0.0.0.0 0.0.0.0 192.168.255.5 250
MS2(config)#enable secret ytvc
MS2(config)#line vty 0 3
MS2(config-line)#password e509
MS2(config-line)#login
MS2(config-line)#end
MS2#write
```

任务三：核心层路由器任务实施

用户 CF 配置核心层路由器 R1

```
Router>enable
Router#configure terminal
Router(config)#hostname R1
R1(config)#interface g0/0
R1(config-if)#ip address 192.168.255.2 255.255.255.252
R1(config-if)#no shutdown
R1(config-if)#interface g0/1
R1(config-if)#ip address 192.168.255.10 255.255.255.252
R1(config-if)#no shutdown
R1(config-if)#exit
R1(config)#ip route 192.168.0.0 255.255.254.0 192.168.255.1
R1(config)#ip route 192.168.0.0 255.255.254.0 192.168.255.9 250
R1(config)#ip route 192.168.99.0 255.255.255.0 192.168.255.1
R1(config)#ip route 192.168.99.0 255.255.255.0 192.168.255.9 250
R1(config)#ip route 192.168.2.0 255.255.254.0 192.168.255.9
R1(config)#ip route 192.168.2.0 255.255.254.0 192.168.255.1 250
R1(config)#ip route 192.168.199.0 255.255.255.0 192.168.255.9
R1(config)#ip route 192.168.199.0 255.255.255.0 192.168.255.1 250
R1(config)#enable secret ytvc
R1(config)#line vty 0 3
R1(config-line)#password e509
R1(config-line)#login
R1(config-line)#end
R1#write
```

8.2.4　PT Multiuser 实例连接

3 个用户已经完成了各自的分工，进入 PT 连接阶段，CFL 老师让大家按顺序先将接入层连接到汇聚层，待连通性测试通过后，再将汇聚层与核心层连接，并完成连通性测试，下面我们将按要求完成各部分对接。

任务一：接入层与汇聚层 PT 实例连接

首先，我们准备将 LX 负责的接入层 PT 实例（见图 8-24）与 XL 负责的汇聚层 PT 实例（见图 8-25）进行连接。

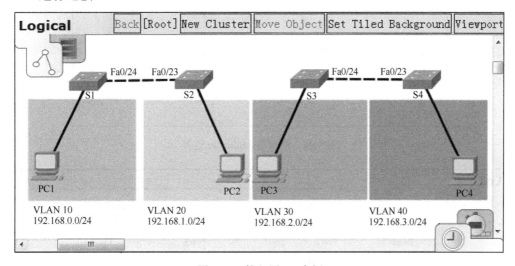

图 8-24　接入层 PT 实例

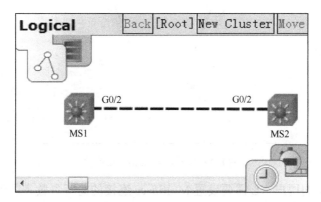

图 8-25　汇聚层 PT 实例

连接之前，需要确定监听方和接入方，在这里我们选汇聚层为监听方。两方都需要从设备选区中选择 Multiuser 连接，并拖至逻辑工作区。监听方修改远程网络的名称分别为 S1、S2、S3 和 S4，接入层则无须修改。图 8-26 所示为汇聚层 Multiuser 连接，图 8-27 所示为接入层 Multiuser 连接。

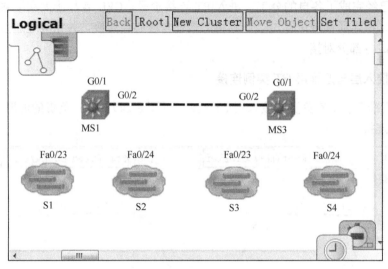

图 8-26　汇聚层 Multiuser 连接

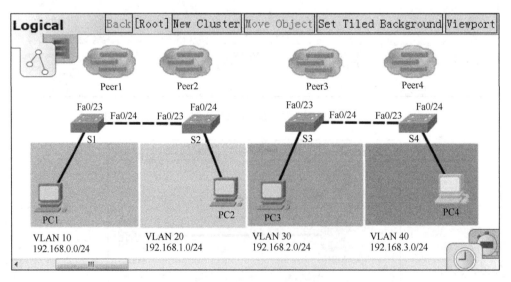

图 8-27　接入层 Multiuser 连接

接下来，汇聚层的用户 XL 需要设置 Multiuser 连接的监听参数，之后才可以使接入层的用户输入与监听参数一致的信息，请求远程接入。监听端的用户 XL 需要点击菜单栏 Extensions→Multiuser→Listen 打开如图 8-28 所示的监听窗口。

设置监听参数：

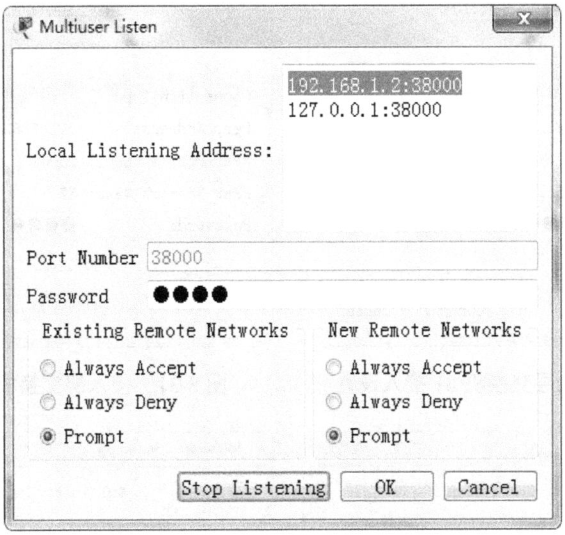

图 8-28 汇聚层 PT 监听参数设置

监听端的 IP 地址、端口号和口令分别为 192.168.1.2、38000 和 ytvc。

设置端口可见：选择菜单 **Extensions** → **Multiuser** →**Port Visibility**，打开如图 8-29 所示窗口，点击汇聚层交换机 MS1 和 MS2，通过拖动右面的滚动条，找到端口 Fa0/23 和 Fa0/24，并进行勾选，即将这 4 个端口设置为"端口可见"。

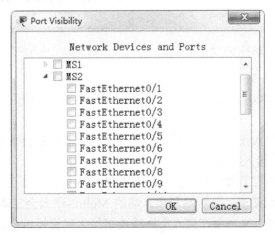

图 8-29 汇聚层 PT "端口可见"设置

监听端设置完毕，接下来需要接入层的用户 LX 根据 XL 提供的监听参数，向远程汇聚层 PT 发出连接请求。LX 同学依次点击接入层的 4 个 Multiuser 云，进行参数输入，如图 8-30～图 8-33 所示。

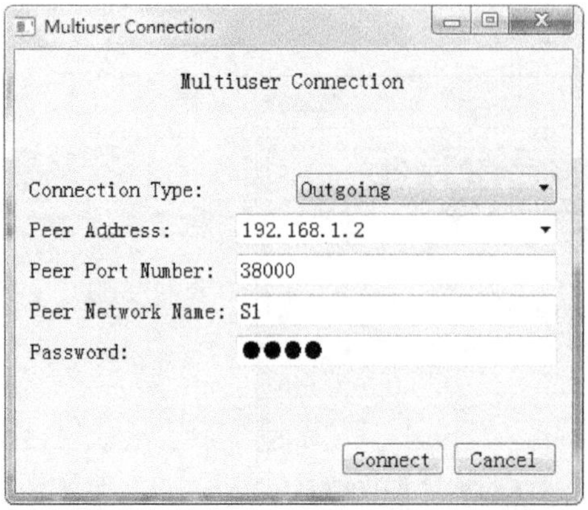

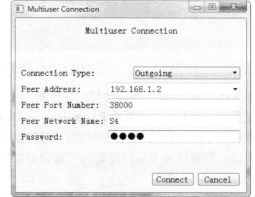

图 8-30　接入层交换机 S1 接入设置 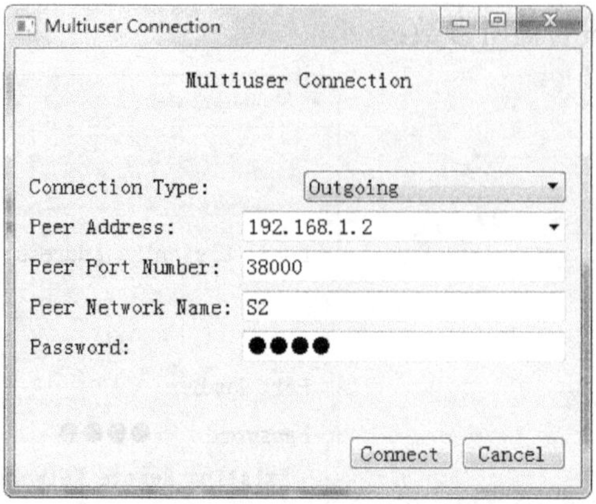　　　图 8-31　接入层交换机 S2 接入设置

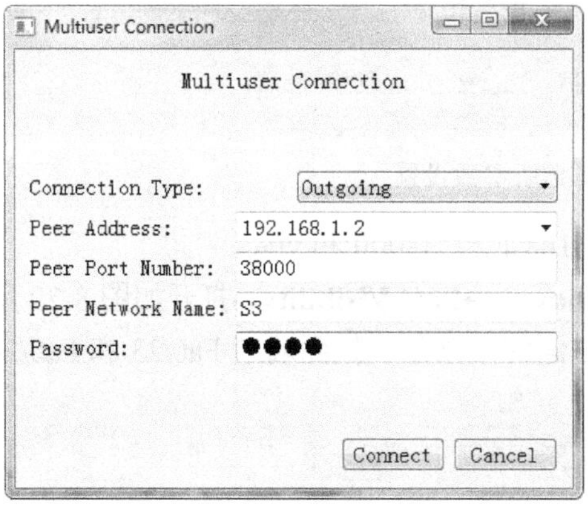

图 8-32　接入层交换机 S3 接入设置　　　图 8-33　接入层交换机 S4 接入设置

接入层的用户 LX 将 Multiuser 的连接类型设置为 **Outgoing**，设置输入对端的 IP 地址为 192.168.1.2、端口号为 38000、网络名称（各自的名称）以及连接口令为 ytvc，然后点击 **Connect**，接入方就发出了连接请求。

接入层的用户 LX 输入正确的参数后，汇聚层的用户 XL 就会看到其 PT 界面自动弹出如图 8-34 和图 8-35 所示的接入提示。

监听端的 XL 同学收到了来自同一台主机 192.168.1.3 发出的 4 个不同的连接请求，并依次点击 Yes 按钮，接受远程请求。监听端和接入端的 Multiuser 云会立即变成蓝色，表示 4 个虚拟通道已经成功建立。接下来需要在接入层选线连接接入层交换机至云端。注意选取正确的接口进行连接。

我们可以看到在接入层连接 S1 到云端，如图 8-36 所示；监听端的云端 S1 至汇聚层交换机 MS1 之间会自动出现连线，如图 8-37 所示。

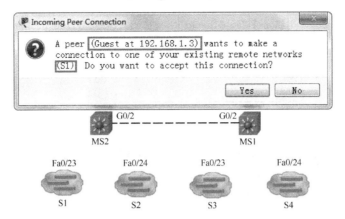

图 8-34　汇聚层 PT 收到接入层 S1 的连接请求

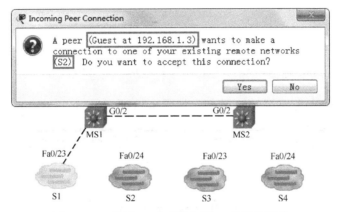

图 8-35　汇聚层 PT 收到接入层 S2 的连接请求

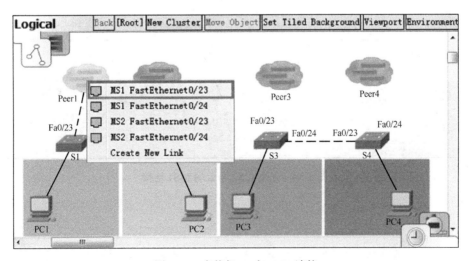

图 8-36　交换机 S1 与 MS1 连接

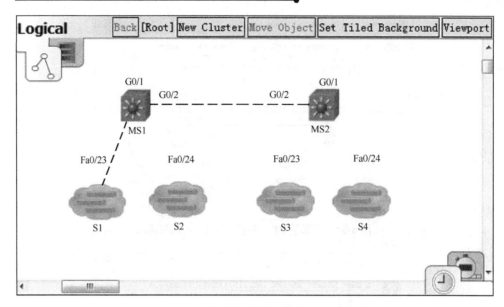

图 8-37 MS1 成功连接到远程云端 S1

接入层交换机 S2 与监听端 MS1 连接,如图 8-38 所示。

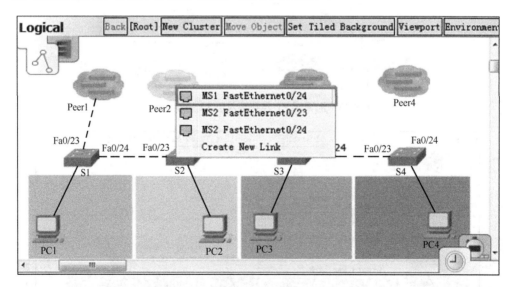

图 8-38 交换机 S2 与云端 MS1 连接

汇聚层会自动出现连线，如图 8-39 所示。

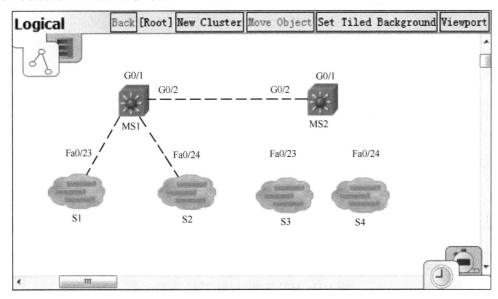

图 8-39　MS1 成功连接到远程云端 S2

接入层交换机 S3 至汇聚层交换机 MS2 物理连接成功，如图 8-40 和图 8-41 所示。

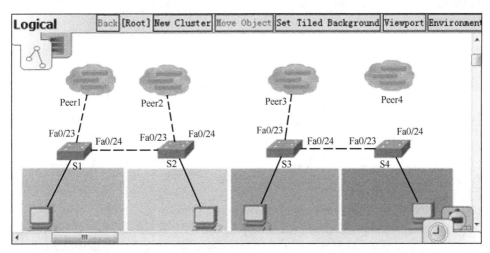

图 8-40　交换机 S3 与云端 MS2 连接

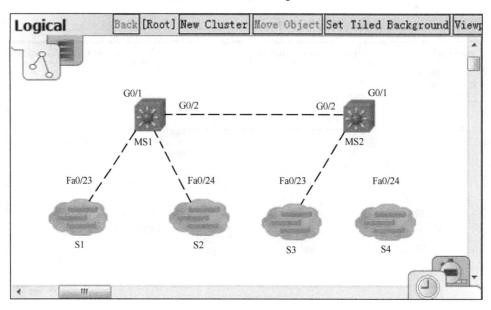

图 8-41　MS2 成功连接到远程云端 S3

接入层交换机 S4 至汇聚层交换机 MS2 物理连接成功，如图 8-42 和图 8-43 所示。

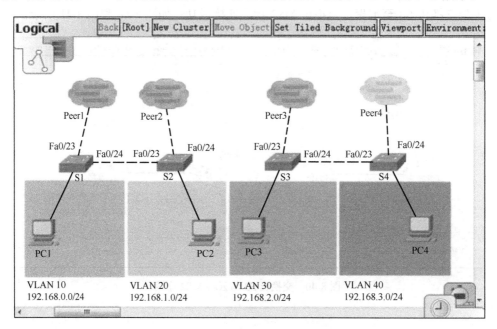

图 8-42　交换机 S4 与云端 MS2 连接

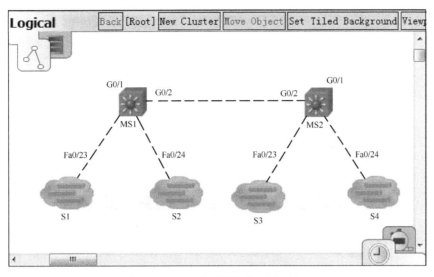

图 8-43　MS2 成功连接到远程云端 S4

至此，接入层 PT 和汇聚层 PT 之间已经成功实现对接。下面我们需要对接入层和汇聚层做连通性测试。

任务二：接入层与汇聚层 PT 连通性测试

首先，我们需要查看接入层 PC 是否能自动获取来自汇聚层交换机分配的 IP 地址，通过图 8-44～图 8-47 可知，4 台 PC 都成功获取到相应网段的 IP 地址。

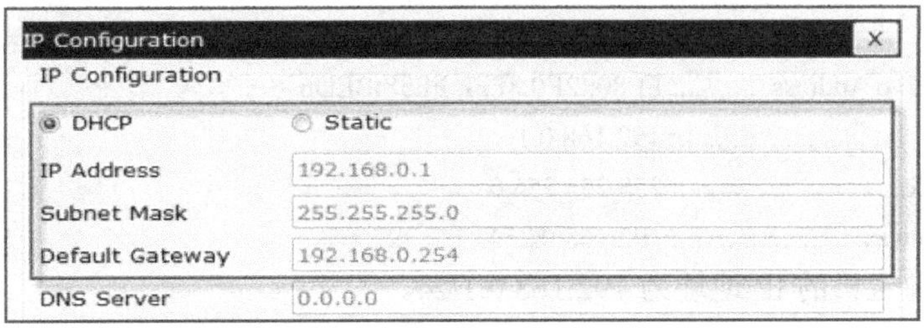

图 8-44　接入层主机 PC1 自动获取 MS1 分配的地址

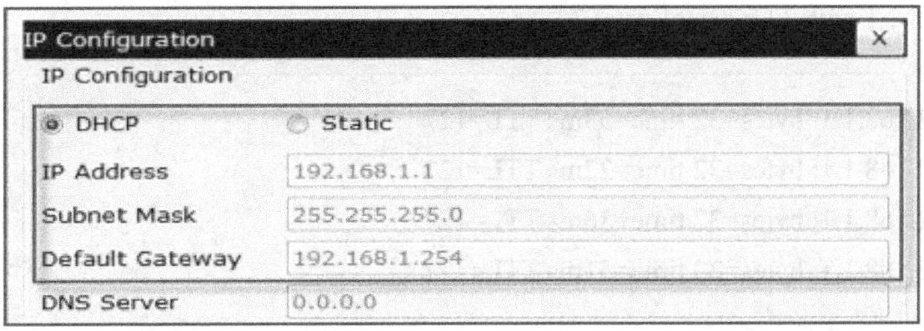

图 8-45　接入层主机 PC2 自动获取 MS1 分配的地址

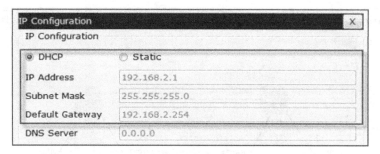

图 8-46　接入层主机 PC3 自动获取 MS2 分配的地址

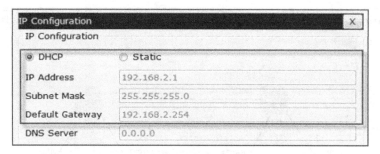

图 8-47　接入层主机 PC4 自动获取 MS2 分配的地址

在 PC1 上测试与其他网段主机的连通性。

C:\>ipconfig

FastEthernet0 Connection:(default port)

　　Link-local IPv6 Address............: FE80::2E0:8FFF:FE38:BED6
　　IP Address....................: 192.168.0.1
　　Subnet Mask..................: 255.255.255.0
　　Default Gateway...............: 192.168.0.254

检查发现当前主机的 IP 地址为 192.168.0.1/24。

C:\>ping 192.168.1.1

Pinging 192.168.1.1 with 32 bytes of data:

Reply from 192.168.1.1: bytes=32 time=53ms TTL=127
Reply from 192.168.1.1: bytes=32 time=22ms TTL=127
Reply from 192.168.1.1: bytes=32 time=36ms TTL=127
Reply from 192.168.1.1: bytes=32 time=51ms TTL=127

Ping statistics for 192.168.1.1:

Packets: Sent = 4, Received = 4, Lost = 0 (0% loss),
Approximate round trip times in milli-seconds:
Minimum = 22ms, Maximum = 53ms, Average = 40ms

PC1 与 PC2 主机的通信成功，TTL=127，说明两个网段之间跨越了 1 跳。

我们再来测试 PC1 和 PC3 主机的连通性，如下所示：

C:\>**ping 192.168.2.1**

Pinging 192.168.2.1 with 32 bytes of data:

Reply from 192.168.2.1: bytes=32 time=52ms TTL=126
Reply from 192.168.2.1: bytes=32 time=33ms TTL=126
Reply from 192.168.2.1: bytes=32 time=44ms TTL=126
Reply from 192.168.2.1: bytes=32 time=71ms TTL=126

Ping statistics for 192.168.2.1:
Packets: Sent = 4, Received = 4, Lost = 0 (0% loss),
Approximate round trip times in milli-seconds:
Minimum = 33ms, Maximum = 71ms, Average = 50ms

PC1 与 PC3 主机的通信成功，TTL=126，说明两个网段之间跨越了 2 跳。

我们继续测试 PC1 和 PC4 主机的连通性，如下所示：

C:\>**ping 192.168.3.1**

Pinging 192.168.3.1 with 32 bytes of data:

Reply from 192.168.3.1: bytes=32 time=44ms TTL=126
Reply from 192.168.3.1: bytes=32 time=38ms TTL=126
Reply from 192.168.3.1: bytes=32 time=30ms TTL=126
Reply from 192.168.3.1: bytes=32 time=49ms TTL=126

Ping statistics for 192.168.3.1:
Packets: Sent = 4, Received = 4, Lost = 0 (0% loss),
Approximate round trip times in milli-seconds:
Minimum = 30ms, Maximum = 49ms, Average = 40ms

PC1 与 PC4 主机的通信成功，TTL=126，说明两个网段之间也跨越了 2 跳，即经 MS1 和 MS2 到达目标主机，显然走的路径与我们设计规划的一致。

我们进一步查看汇聚层设备 MS1 的路由表，如下所示：

```
MS1>show ip route
Codes: C - connected, S - static, I - IGRP, R - RIP, M - mobile, B - BGP
Gateway of last resort is 192.168.255.2 to network 0.0.0.0

C    192.168.0.0/24 is directly connected, Vlan10
C    192.168.1.0/24 is directly connected, Vlan20
S    192.168.2.0/23 [1/0] via 192.168.255.6
C    192.168.99.0/24 is directly connected, Vlan99
S    192.168.199.0/24 [1/0] via 192.168.255.6
     192.168.255.0/30 is subnetted, 2 subnets
C       192.168.255.0 is directly connected, GigabitEthernet0/1
C       192.168.255.4 is directly connected, GigabitEthernet0/2
S*   0.0.0.0/0 [1/0] via 192.168.255.2
```

我们看到了汇聚层交换机 MS1 的路由表除了 5 条直连路由（3 个 SVI 接口，2 个物理接口）外，还有配置的 3 条静态路由，其中默认路由走上行链路，对其他业务数据段的访问均走水平链路。我们继续查看交换机 MS2 的路由表，如下所示：

```
MS2>show ip route
Codes: C - connected, S - static, I - IGRP, R - RIP, M - mobile, B - BGP
Gateway of last resort is 192.168.255.10 to network 0.0.0.0

S    192.168.0.0/23 [1/0] via 192.168.255.5
C    192.168.2.0/24 is directly connected, Vlan30
C    192.168.3.0/24 is directly connected, Vlan40
S    192.168.99.0/24 [1/0] via 192.168.255.5
C    192.168.199.0/24 is directly connected, Vlan199
     192.168.255.0/30 is subnetted, 2 subnets
C       192.168.255.4 is directly connected, GigabitEthernet0/2
C       192.168.255.8 is directly connected, GigabitEthernet0/1
S*   0.0.0.0/0 [1/0] via 192.168.255.10
```

MS2 路由表显示的结果类似 MS1，效果和预期规划配置的完全一致。至此，接入层 PT 到汇聚层 PT 已经成功连接。下面我们的任务是完成汇聚层与核心层 PT 实例的对接。

任务三：汇聚层与核心层 PT 实例连接

接下来，我们准备将 CF 负责的核心层 PT 实例（见图 8-48）与 XL 负责的汇聚层 PT 实例（见图 8-49）进行连接。

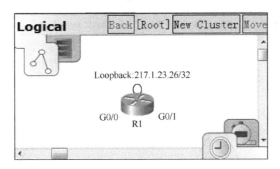

图 8-48　核心层 PT 实例

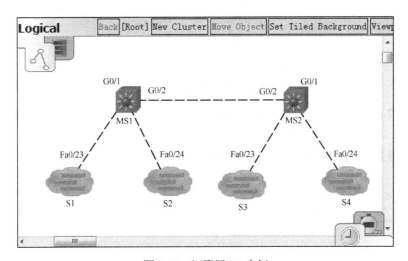

图 8-49　汇聚层 PT 实例

我们选定核心层 PT 为监听方。两端都要从设备选区中选择 Multiuser 云，并拖至逻辑工作区。监听方修改远程网络的名称分别为 MS1 和 MS2，汇聚层无须修改。如图 8-50 和图 8-51 所示。

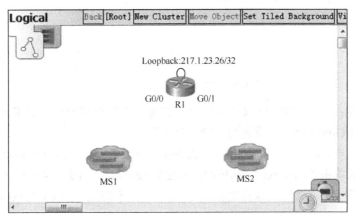

图 8-50　核心层 Multiuser 连接

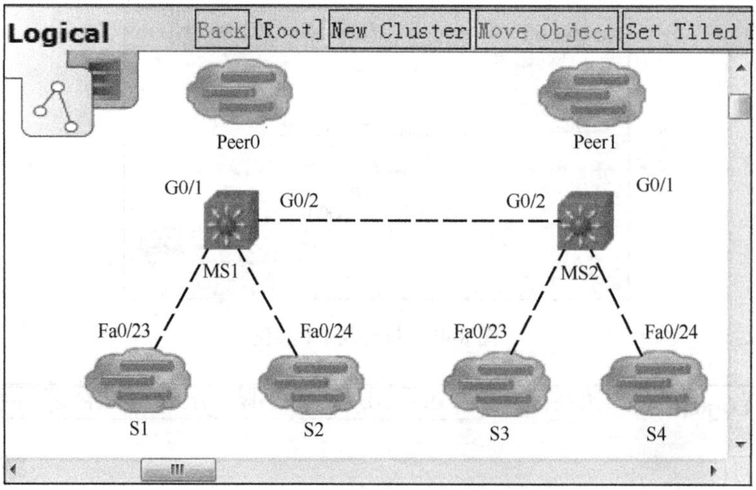

图 8-51 汇聚层 Multiuser 连接

接下来，核心层的用户 CF 需要设置 Multiuser 连接的监听参数，如图 8-52 所示，以便汇聚层的用户 XL 可以输入信息并发送连接请求。

设置监听参数：核心层的监听 IP 地址、端口号和口令分别为 192.168.1.1、38000 和 ytvc。

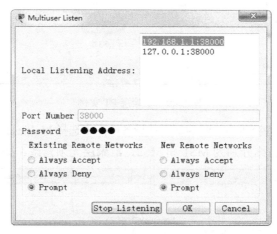

图 8-52 核心层 PT 监听参数设置

设置端口可见：监听端的用户 CF 打开"端口可见"的设置窗口，如图 8-53 所示，将路由器 R1 的两个端口 G0/0 和 G0/1 设置为"端口可见"。

监听端设置完毕，接下来汇聚层的用户 XL 根据用户 CF 提供的监听参数，点击两个 Multiuser 云输入信息，然后单击"Connect"，向核心层发出连接请求，如图 8-54 和图 8-55 所示。

核心层的用户 CF 需要处理来自主机 192.168.1.2 发来的汇聚层 PT 的连接请求，如图 8-56 所示。

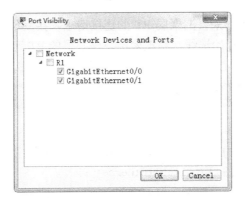

图 8-53　核心层 PT "端口可见"设置

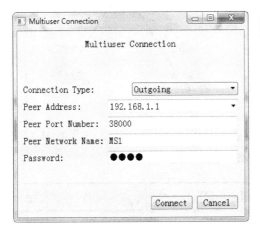

图 8-54　汇聚层交换机 MS1 接入设置

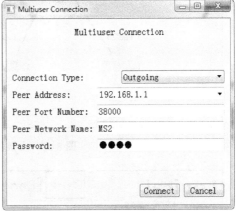

图 8-55　汇聚层交换机 MS2 接入设置

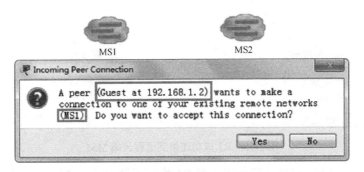

图 8-56　核心层 PT 收到汇聚层 MS1 的连接请求

核心层 PT 已经收到两个来自汇聚层 PT 的连接请求，接受请求后，两个虚拟通道就建立起来，接下来汇聚层的 XL 开始选线，将汇聚层与核心层的 PT 连接起来。

汇聚层交换机 MS1 与核心层的路由器 R1 物理连接成功，如图 8-57 和图 8-58 所示。

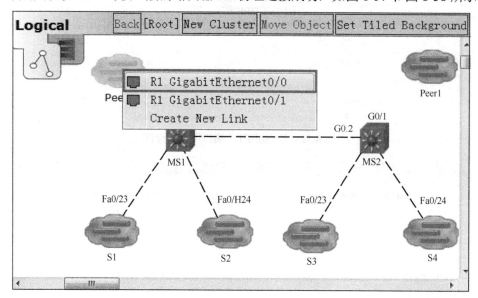

图 8-57　交换机 MS1 与 R1 连接

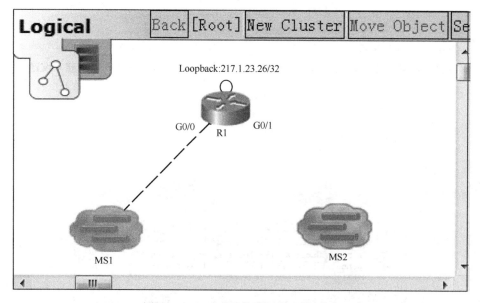

图 8-58　R1 成功连接到远程云端 MS1

汇聚层的交换机 MS2 与核心层的路由器 R1 物理连接成功，如图 8-59 和图 8-60 所示。

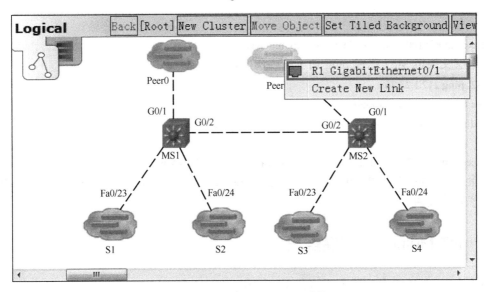

图 8-59　交换机 MS2 与 R1 连接

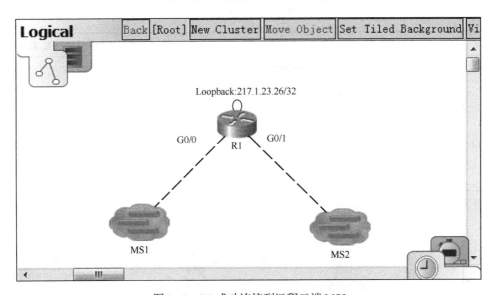

图 8-60　R1 成功连接到远程云端 MS2

至此，已经成功完成了汇聚层 PT 和核心层 PT 间的物理对接。

任务四：汇聚层与核心层 PT 连通测试

MS1>**ping 217.1.23.26**
Type escape sequence to abort.
Sending 5, 100-byte ICMP Echos to 217.1.23.26, timeout is 2 seconds:

!!!!!
Success rate is 100 percent (5/5), round-trip min/avg/max = 12/15/21 ms

汇聚层交换机 MS1 已经 ping 通了核心层的环回接口。

MS2>**ping 217.1.23.26**
Type escape sequence to abort.
Sending 5, 100-byte ICMP Echos to 217.1.23.26, timeout is 2 seconds:
!!!!!
Success rate is 100 percent (5/5), round-trip min/avg/max = 15/26/44 ms

汇聚层交换机 MS2 也 ping 通了核心层的环回接口，至此汇聚层 PT 与核心层 PT 实例间已经实现了互连互通。

8.2.5　PT Multiuser 互联结果

多用户 PT 互连结果一览表，如表 8-6 所示。

表 8-6　多用户 PT 互连结果一览表

用 户 名	分 配 任 务	互 联 关 系
CF	Loopback:217.1.23.26/32 G0/0　G0/1 R1 MS1　　MS2	核心层与汇聚层互连
XL	Peer0　　　　　Peer1 G0/1　G0/2　　G0/2　G0/1 MS1　　　　　MS2 Fa0/23　Fa0/24　Fa0/23　Fa0/24 S1　　S2　　S3　　S4	汇聚层与核心层互连，汇聚层与接入层互连

用户名	分配任务	互联关系
LX	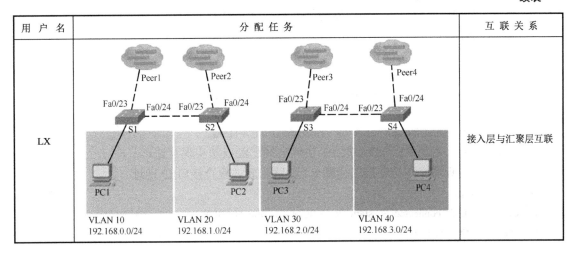	接入层与汇聚层互联

8.2.6 PT Multiuser 连通测试

为完成接入层 PT 与核心层 PT 实例的连通性测试，我们可以灵活地从核心路由器发出到接入层终端 PC 的 ping 包，如下所示：

R1>**ping 192.168.0.1**

Type escape sequence to abort.

Sending 5, 100-byte ICMP Echos to 192.168.0.1, timeout is 2 seconds:

!!!!!

Success rate is 100 percent (5/5), round-trip min/avg/max = 0/3/12 ms

以上结果表明，核心层路由器与接入层主机 PC1 实现了互通，我们继续 ping 主机 PC2。

R1>**ping 192.168.1.1**

Type escape sequence to abort.

Sending 5, 100-byte ICMP Echos to 192.168.1.1, timeout is 2 seconds:

!!!!!

Success rate is 100 percent (5/5), round-trip min/avg/max = 0/5/15 ms

核心层路由器与接入层主机 PC2 也实现了互通，接下来测试核心层路由器与主机 PC3 的连通性。

R1>**ping 192.168.2.1**

Type escape sequence to abort.

Sending 5, 100-byte ICMP Echos to 192.168.2.1, timeout is 2 seconds:

!!!!!

Success rate is 100 percent (5/5), round-trip min/avg/max = 0/2/12 ms

以上 ping 的结果表明核心层路由器与接入层主机 PC3 也实现了互通,继续 ping 主机 PC4。

R1>ping 192.168.3.1

Type escape sequence to abort.
Sending 5, 100-byte ICMP Echos to 192.168.3.1, timeout is 2 seconds:
!!!!!
Success rate is 100 percent (5/5), round-trip min/avg/max = 0/4/11 ms

以上结果表明接入层 PT 与核心层 PT 实例之间已经成功实现互通。
接下来,我们查看一下核心层路由器 R1 的路由表,检查接口 IP 地址。

R1#show ip route
Codes: L - local, C - connected, S - static, R - RIP, M - mobile, B - BGP

Gateway of last resort is not set

```
S    192.168.0.0/23 [1/0] via 192.168.255.1
S    192.168.2.0/23 [1/0] via 192.168.255.9
S    192.168.99.0/24 [1/0] via 192.168.255.1
S    192.168.199.0/24 [1/0] via 192.168.255.9
     192.168.255.0/24 is variably subnetted, 4 subnets, 2 masks
C       192.168.255.0/30 is directly connected, GigabitEthernet0/0
L       192.168.255.2/32 is directly connected, GigabitEthernet0/0
C       192.168.255.8/30 is directly connected, GigabitEthernet0/1
L       192.168.255.10/32 is directly connected, GigabitEthernet0/1
     217.1.23.0/32 is subnetted, 1 subnets
C       217.1.23.26/32 is directly connected, Loopback0
```

以上输出结果,我们发现路由器有 4 条静态路由,流量转发分摊在 G0/0 和 G0/1 两接口上。进一步查看核心路由器 R1 的 3 个接口的 IP 地址:

R1>show ip interface brief | include			up	up
GigabitEthernet0/0	192.168.255.2	YES manual	up	up
GigabitEthernet0/1	192.168.255.10	YES manual	up	up
Loopback0	217.1.23.26	YES manual	up	up

查看汇聚层交换机 MS1 接口的 IP 地址,了解到其上行链路接口地址为 192.168.255.1。

MS1>show ip interface brief | include			up	up
FastEthernet0/23	unassigned	YES unset	up	up
FastEthernet0/24	unassigned	YES unset	up	up
GigabitEthernet0/1	192.168.255.1	YES manual	up	up
GigabitEthernet0/2	192.168.255.5	YES manual	up	up

Vlan10	192.168.0.254	YES manual	up	up
Vlan20	192.168.1.254	YES manual	up	up
Vlan99	192.168.99.254	YES manual	up	up

查看汇聚层交换机MS2接口的IP地址，了解到其上行链路接口的IP地址为192.169.255.9。

```
MS2>show ip interface brief | include  up         up
FastEthernet0/23    unassigned      YES unset    up         up
FastEthernet0/24    unassigned      YES unset    up         up
GigabitEthernet0/1  192.168.255.9   YES manual   up         up
GigabitEthernet0/2  192.168.255.6   YES manual   up         up
Vlan30              192.168.2.254   YES manual   up         up
Vlan40              192.168.3.254   YES manual   up         up
Vlan199             192.168.199.254 YES manual   up         up
```

接下来我们从路由器R1上跟踪去往接入层的数据包，具体操作如下所示：

```
R1>traceroute 192.168.0.1
Type escape sequence to abort.
Tracing the route to 192.168.0.1

  1  192.168.255.1   1 msec    0 msec    0 msec
  2  192.168.0.1     13 msec   0 msec    0 msec

R1>traceroute 192.168.1.1
Type escape sequence to abort.
Tracing the route to 192.168.1.1

  1  192.168.255.1   2 msec    0 msec    0 msec
  2  192.168.1.1     13 msec   12 msec   10 msec

R1>traceroute 192.168.99.10
Type escape sequence to abort.
Tracing the route to 192.168.99.10

  1  192.168.255.1   1 msec    0 msec    0 msec
  2  192.168.99.10   0 msec    11 msec   11 msec

R1>traceroute 192.168.99.20
Type escape sequence to abort.
Tracing the route to 192.168.99.20

  1  192.168.255.1   2 msec    0 msec    0 msec
```

```
    2   192.168.99.20    13 msec    14 msec    0 msec
```

从以上输出可以验证路由器 R1 去 VLAN 10、VLAN 20、VLAN 99 的数据流量经其左分支，即经汇聚层交换机 MS1 转发至接入层交换机 S1 和 S2。我们继续跟踪发往接入层的数据包，如下所示：

```
R1>traceroute 192.168.2.1
Type escape sequence to abort.
Tracing the route to 192.168.2.1

    1   192.168.255.9    1 msec     0 msec     0 msec
    2   192.168.2.1      0 msec     0 msec     0 msec

R1>traceroute 192.168.3.1
Type escape sequence to abort.
Tracing the route to 192.168.3.1

    1   192.168.255.9    1 msec     0 msec     0 msec
    2   192.168.3.1      0 msec     5 msec     0 msec

R1>traceroute 192.168.199.30
Type escape sequence to abort.
Tracing the route to 192.168.199.30

    1   192.168.255.9    2 msec     0 msec     0 msec
    2   192.168.199.30   0 msec     0 msec     0 msec

R1>traceroute 192.168.199.40
Type escape sequence to abort.
Tracing the route to 192.168.199.40

    1   192.168.255.9    32 msec    0 msec     0 msec
    2   192.168.199.40   15 msec    12 msec    13 msec
```

以上输出表明路由器 R1 去 VLAN 30、VLAN 40、VLAN 199 的数据流量经其右分支，即经汇聚层交换机 MS2 转发至接入层交换机 S3 和 S4，数据包转发路径与规划设计的完全一致，路由配置正确。接下来，我们需要对冗余链路的可靠性进行测试。

8.2.7　PT Multiuser 冗余测试

1. 核心层冗余链路测试

我们先 down 掉路由器 R1 的左分支，即 down 掉 G0/0 接口，看路由器能否通过右分支转发数据包。

```
R1(config)#interface g0/0
R1(config-if)#shutdown
%LINK-5-CHANGED: Interface GigabitEthernet0/0, changed state to administratively down
%LINEPROTO-5-UPDOWN: Line protocol on Interface GigabitEthernet0/0, changed state to down
```

查看路由器 R1 的路由表：

```
R1#show ip route | begin Gateway
Gateway of last resort is not set

S        192.168.0.0/23 [250/0] via 192.168.255.9
S        192.168.2.0/23 [1/0] via 192.168.255.9
S        192.168.99.0/24 [250/0] via 192.168.255.9
S        192.168.199.0/24 [1/0] via 192.168.255.9
         192.168.255.0/24 is variably subnetted, 2 subnets, 2 masks
C           192.168.255.8/30 is directly connected, GigabitEthernet0/1
L           192.168.255.10/32 is directly connected, GigabitEthernet0/1
         217.1.23.0/32 is subnetted, 1 subnets
C           217.1.23.26/32 is directly connected, Loopback0
```

从以上路由器 R1 的路由表输出结果来看，我们发现两条管理距离为 250 的静态路由出现在路由表中，此时路由器去接入层的所有网段数据包，将通过路由器右分支转发，因此路由器没有中断数据包的转发。

接下来，我们进一步查看受到影响的汇聚层交换机 MS1 的路由表，如下所示：

```
MS1>show ip route | begin Gateway
Gateway of last resort is 192.168.255.6 to network 0.0.0.0

C        192.168.0.0/24 is directly connected, Vlan10
C        192.168.1.0/24 is directly connected, Vlan20
S        192.168.2.0/23 [1/0] via 192.168.255.6
C        192.168.99.0/24 is directly connected, Vlan99
S        192.168.199.0/24 [1/0] via 192.168.255.6
         192.168.255.0/30 is subnetted, 1 subnets
C           192.168.255.4 is directly connected, GigabitEthernet0/2
S*       0.0.0.0/0 [250/0] via 192.168.255.6
```

从以上 MS1 交换机路由表的输出结果来看，本来通过上行链路转发的默认路由切换到了水平链路，路由表中安装了管理距离是 250 的静态默认路由。

恢复路由器左分支，即开启 G0/0 口。我们继续 down 掉路由器的右分支，即 down 掉 G0/1 接口。

R1(config)#**interface g0/1**
R1(config-if)#**shutdown**
%LINK-5-CHANGED: Interface GigabitEthernet0/1, changed state to administratively down
%LINEPROTO-5-UPDOWN: Line protocol on Interface GigabitEthernet0/1, changed state to down

观察路由器 R1 的路由表，查看数据包的转发接口。

R1#**show ip route | begin Gateway**
Gateway of last resort is not set

```
S       192.168.0.0/23 [1/0] via 192.168.255.1
S       192.168.2.0/23 [250/0] via 192.168.255.1
S       192.168.99.0/24 [1/0] via 192.168.255.1
S       192.168.199.0/24 [250/0] via 192.168.255.1
        192.168.255.0/24 is variably subnetted, 2 subnets, 2 masks
C          192.168.255.0/30 is directly connected, GigabitEthernet0/0
L          192.168.255.2/32 is directly connected, GigabitEthernet0/0
        217.1.23.0/32 is subnetted, 1 subnets
C          217.1.23.26/32 is directly connected, Loopback0
```

以上输出结果表明，路由器去远程网络的流量将全部通过 G0/0 接口转发，即经右分支通过汇聚层交换机 MS1 转发，路由表中出现了两条管理距离为 250 的静态路由。接着我们进一步查看受到影响的汇聚层交换机 MS2 的路由表。

MS2>**show ip route | begin Gateway**
Gateway of last resort is 192.168.255.5 to network 0.0.0.0

```
S       192.168.0.0/23 [1/0] via 192.168.255.5
C       192.168.2.0/24 is directly connected, Vlan30
C       192.168.3.0/24 is directly connected, Vlan40
S       192.168.99.0/24 [1/0] via 192.168.255.5
C       192.168.199.0/24 is directly connected, Vlan199
        192.168.255.0/30 is subnetted, 1 subnets
C          192.168.255.4 is directly connected, GigabitEthernet0/2
S*      0.0.0.0/0 [250/0] via 192.168.255.5
```

从交换机 MS2 的路由表输出结果来看，本应通过上行链路转发的默认路由切换到水平链路经 MS1 交换机转发，该默认路由的管理距离为 250。

以上关于核心层冗余链路的可靠性测试，达到了预期目标。

2. 汇聚层和接入层冗余链路测试

我们通过设置不同网络环境，跟踪 PC1 至 PC4 的数据包，来观察其经过的路径，从而可以断定冗余链路是否能确保可靠通信。验证当前主机，如下所示：

```
C:\>ipconfig

FastEthernet0 Connection:(default port)
   Link-local IPv6 Address..........: FE80::2E0:8FFF:FE38:BED6
   IP Address...................: 192.168.0.1
   Subnet Mask..................: 255.255.255.0
   Default Gateway..............: 192.168.0.254
```

以上输出可以确认当前主机为 PC1。下面我们将设置多种网络环境来跟踪 PC1（192.168.0.1）至 PC4（192.168.3.1）的数据包。

（1）网络环境一：正常情况的网络环境（零处故障）

主机 PC1 跟踪 PC4 的结果如下：

```
C:\>tracert 192.168.3.1

Tracing route to 192.168.3.1 over a maximum of 30 hops:
  1   21 ms    6 ms     8 ms     192.168.0.254    //PC1 网关地址，好交换机 MS1 的 SVI 地址
  2   20 ms    20 ms    25 ms    192.168.255.6    //交换机 MS2 的 Gig0/2 接口地址
  3   44 ms    34 ms    47 ms    192.168.3.1
Trace complete.
```

从以上包跟踪的结果，我们推断出数据包经过的路径是 PC1→S1→MS1→MS2→S4→PC4（历经 4 台网络设备）。

（2）网络环境二：MS1 与 MS2 间的水平链路故障（一处故障）

```
MS1(config)#interface g0/2
MS1(config-if)#shutdown    //人为制造 down 掉接口故障
%LINK-5-CHANGED: Interface GigabitEthernet0/2, changed state to administratively down
%LINEPROTO-5-UPDOWN: Line protocol on Interface GigabitEthernet0/2, changed state to down
```

水平链路故障情况，包跟踪结果如下：

```
C:\>tracert 192.168.3.1
Tracing route to 192.168.3.1 over a maximum of 30 hops:
  1   10 ms    22 ms    10 ms    192.168.0.254
  2   27 ms    28 ms    20 ms    192.168.255.2    //路由器 R1 的 Gig0/2 接口地址
  3   22 ms    24 ms    29 ms    192.168.255.9    //交换机 MS2 的 Gig0/1 接口地址
  4   47 ms    41 ms    64 ms    192.168.3.1
Trace complete.
```

以上结果，我们可以推断数据包的转发路径为 PC1→S1→MS1→R1→MS2→S4→PC4（历经 5 台网络设备），此时数据包会通过上行链路经 R1 转发至 MS2。因此，转发路径改变，绕过了故障链路。

（3）网络环境三：MS1 与 MS2 间、MS1 与 S1 间链路故障（两处故障）

```
S1(config)#interface f0/23
S1(config-if)#shutdown
%LINK-5-CHANGED: Interface FastEthernet0/23, changed state to administratively down
%LINEPROTO-5-UPDOWN: Line protocol on Interface FastEthernet0/23, changed state to down
```

在两处故障情况下，主机 PC1 跟踪 PC4 的结果如下：

```
C:\>tracert 192.168.3.1

Tracing route to 192.168.3.1 over a maximum of 30 hops:

  1    13 ms    21 ms    11 ms    192.168.0.254
  2    17 ms    31 ms    34 ms    192.168.255.2
  3    50 ms    43 ms    34 ms    192.168.255.9
  4    59 ms    36 ms    53 ms    192.168.3.1

Trace complete.
```

我们进一步分析出包的转发路径是 PC1→S1→S2→MS1→R1→MS2→S4→PC4（历经 6 台网络设备），此时包的转发再一次发生改变，经 S2 到达 MS1。两条链路出现故障也没有中断网络的正常通信。

（4）网络环境四：MS1 与 MS2 间、MS1 与 S1 及 MS2 与 S4 间链路故障（三处故障）

```
S4(config)#interface f0/24
S4(config-if)#shutdown
%LINK-5-CHANGED: Interface FastEthernet0/24, changed state to administratively down
%LINEPROTO-5-UPDOWN: Line protocol on Interface FastEthernet0/24, changed state to down
```

包跟踪结果显示如下：

```
C:\>tracert 192.168.3.1

Tracing route to 192.168.3.1 over a maximum of 30 hops:

  1     8 ms    10 ms     8 ms    192.168.0.254
  2    15 ms    22 ms    11 ms    192.168.255.2
  3    26 ms    14 ms    41 ms    192.168.255.9
  4    45 ms    42 ms    29 ms    192.168.3.1

Trace complete.
```

结果表明 PC1 数据包的转发路径再一次发生了改变，本次转发路径是 PC1→S1→S2→MS1→R1→MS2→S3→S4→PC4（历经 7 台网络设备）。

由此可见，冗余链路真正做到了安全可靠，当主链路出现故障时，备用链路立即承担起转发数据的任务。

8.2.8　PT Multiuser 中断示意图

Multiuser 正常连接后，若云中出现了带叹号的圆圈，则表明 Multiuser 连接中断。中断的原因主要有两种情况：一种是远程 PT 实例关闭，另一种可能是修改了相关连接参数。

本案例中，若先关闭汇聚层的 PT 实例，则会同时影响到核心层 PT 和接入层 PT 的 Multiuser 连接，中断结果如图 8-61 和图 8-62 所示，因为汇聚层属于核心层和接入层通信的纽带。

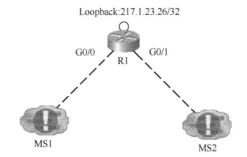

图 8-61　核心层出现 Multiuser 连接中断

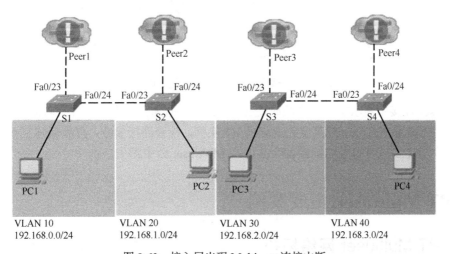

图 8-62　接入层出现 Multiuser 连接中断

若接入层 PT 实例先关闭，则会影响到汇聚层 PT 中的 Multiuser 连接，如图 8-63 所示。如果核心层 PT 和接入层 PT 都已关闭，则汇聚层的 Multiuser 连接中断，显示如图 8-64 所示。

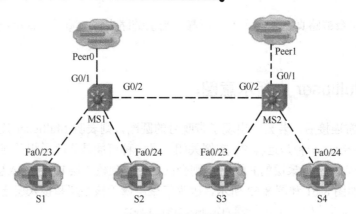

图 8-63 汇聚层出现 Multiuser 连接中断（A）

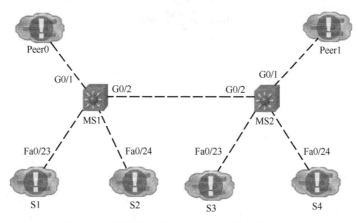

图 8-64 汇聚层出现 Multiuser 连接中断（B）

Multiuser 连接中断状态是一个临时变化的状态，不能被记录保存下来。若想改变当前的中断状态，则需要重新打开引起其状态变化的 PT 实例，检查连接参数，再重新建立 Multiuser 连接，一旦成功建立，带叹号的圆圈就会立即从 Multiuser 云中消失。

8.3　PT Multiuser 拓展案例

8.3.1　PT Multiuser 网络拓扑

> **案例背景**：PT Multiuser 确实激发了同学们的学习热情，活跃了课堂氛围，吸引全员参与。同学们反馈联机实验如同联机游戏，比单机版的好玩多了。HY 同学给 CFL 老师发送了一份电子邮件，附件是一个 PT 文件，如图 8-65 所示。他让老师帮忙看一下作为 Multiuser 拓展案例是否可行，并把要实现的功能告诉了老师。CFL 老师立即回复了 HY，

对他设计的案例给予了大大的赞赏。老师感叹"青出于蓝而胜于蓝"！这个案例简直就是 8.2 节 PT Multiuser 综合案例的姊妹篇，前者以交换为主，后者以路由为主；前者以静态路由为主，后者以动态路由为主；前者实现了 Telnet，后者采用 SSH 来补充；前者是老师设计，后者是学生设计。CFL 老师根据 HY 同学提供的拓扑稍做改动，如图 8-66 所示，老师制作了详细的地址规划表、VLAN 规划表以及设备一览表，CFL 老师就修改好的拓扑和 HY 同学商量项目需求，最终将这个拓展案例的项目需求罗列出来，供同学们课外拓展。请你根据实际情况，可以再约上几位小伙伴一起来完成这份师生共同开发的 PT Mutiuser 教学案例吧，人数不限哦！大家赶快行动起来吧！

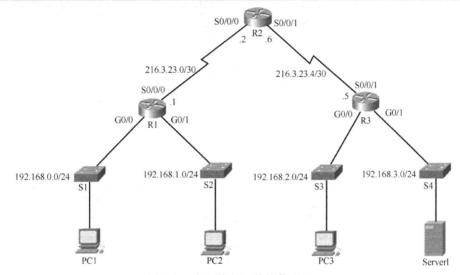

图 8-65　拓展案例网络拓扑（A）

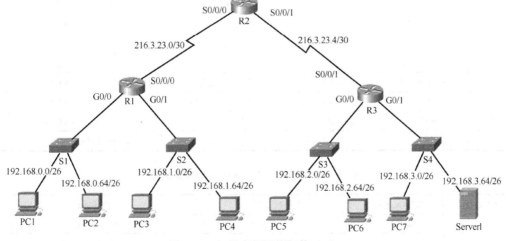

图 8-66　拓展案例网络拓扑（B）

8.3.2 PT Multiuser 地址规划

表 8-7 为 Multiuser 拓展案例地址规划表，表 8-8 为 Multiuser 拓展案例 VLAN 规划表。

表 8-7 Multiuser 拓展案例地址规划表

设备名称	接口	IP 地址	子网掩码	描述
R1	S0/0/0	216.3.23.1	255.255.255.252	Link to R2 S0/0/0
	G0/0	—	—	Link to S1 Gig0/1
	G0/0.10	192.168.0.62	255.255.255.192	Gateway to VLAN 10
	G0/0.20	192.168.0.126	255.255.255.192	Gateway to VLAN 20
	G0/0.99	192.168.0.190	255.255.255.192	Gateway to VLAN 99
	G0/1	—	—	Link to S2 Gig0/1
	G0/1.10	192.168.1.62	255.255.255.192	Gateway to VLAN 10
	G0/1.20	192.168.1.126	255.255.255.192	Gateway to VLAN 20
	G0/1.99	192.168.1.190	255.255.255.192	Gateway to VLAN 99
R2	S0/0/0	216.3.23.2	255.255.255.252	Link to R1 S0/0/0
	S0/0/1	216.3.23.6	255.255.255.252	Link to R3 S0/0/1
R3	S0/0/1	216.3.23.5	255.255.255.252	Link to R2 S0/0/1
	G0/0	—	—	Link to S3 Gig0/1
	G0/0.10	192.168.2.62	255.255.255.192	Gateway to VLAN 10
	G0/0.20	192.168.2.126	255.255.255.192	Gateway to VLAN 20
	G0/0.99	192.168.2.190	255.255.255.192	Gateway to VLAN 99
	G0/1	—	—	Link to S4 Gig0/1
	G0/1.10	192.168.3.62	255.255.255.192	Gateway to VLAN 10
	G0/1.20	192.168.3.126	255.255.255.192	Gateway to VLAN 20
	G0/1.99	192.168.3.190	255.255.255.192	Gateway to VLAN 99
S1	VLAN 99	192.168.0.129	255.255.255.192	Management VLAN
S2	VLAN 99	192.168.1.129	255.255.255.192	Management VLAN
S3	VLAN 99	192.168.2.129	255.255.255.192	Management VLAN
S4	VLAN 99	192.168.3.129	255.255.255.192	Management VLAN
PCx	NIC	DHCP	DHCP	Link to Each VLAN
Server1	NIC	192.168.3.100	255.255.255.192	Link to S4 Gig0/2

表 8-8 Multiuser 拓展案例 VLAN 规划表

VLAN ID	VLAN 名称	交换机端口	地址规划
10	14NET1A	S1：Fa0/1-12	192.168.0.0/26
20	14NET1B	S1：Fa0/13-24	192.168.0.64/26

续表

VLAN ID	VLAN 名称	交换机端口	地址规划
99	Admin	—	192.168.0.128/26
10	14NET2A	S2：Fa0/1-12	192.168.1.0/26
20	14NET2B	S2：Fa0/13-24	192.168.1.64/26
99	Admin	—	192.168.1.128/26
10	15NET1A	S3：Fa0/1-12	192.168.2.0/26
20	15NET1B	S3：Fa0/13-24	192.168.2.64/26
99	Admin	—	192.168.2.128/26
10	15NET2A	S4：Fa0/1-12	192.168.3.0/26
20	15NET2B	S4：Fa0/13-24，G0/2	192.168.3.64/26
99	Admin	—	192.168.3.128/26

8.3.3　PT Multiuser 项目需求

Multiuser 拓展案例设备一览表如表 8-9 所示。

表 8-9　Multiuser 拓展案例设备一览表

设备类型	设备数量	线缆类型	线缆数量	备　注
Cisco2911 Router	1 台	DCE 线缆	2 条	Console 线缆若干
Cisco1941 Router	2 台	DTE 线缆	2 条	
Cisco2960 Switch	4 台	UTP 线	8 条	

请参照图 8-66 所示拓扑，依据表 8-9 来进行设备选型，组建 Multiuser 技术团队，各自搭建自己的 PT 实例，按照如下要求实现全网互通。

基本要求如下所述。

- **网络基本配置**：包括主机名、特权口令、登录横幅（特权口令为 hycf）；
- **IP 地址的配置**：参照表 8-7 所示地址规划表，要求对网络设备间互连接口做详尽描述；
- **RIPV2 协议的配置**：避免子网明细路由出现在远程路由器的路由表中；
- **VLAN 的配置**：在交换机 S1 至 S4 配置相应的 VLAN，参照表 8-8 VLAN 规划表；
- **单臂路由的配置**：在路由器 R1 和 R3 上配置单臂路由，实现 VLAN 间互访；
- **SSH 远程管理配置**：用户名为 HY、XL、CF，口令为 ytvc，最多允许 3 个用户远程登录；
- **配置 DHCP 服务**：在路由器 R1 和 R3 上配置 DHCP 服务为各自的 VLAN 10 和 VLAN 20 分配地址，地址池采用各自 VLAN 的名称；

- **优化网络**：避免发送不必要的路由更新；
- **测试网络**：对全网进行互联互通测试

"Multiuser 分布式多用户案例"一章即将结束，本章详细介绍了 Multiuser 的功能特点，总结了常见的 Multiuser 故障，给出了 Multiuser 实验的具体步骤，可以带你轻松愉快地学习。设计了 1 个 Multiuser 实验案例（Teacher 与 Student 两用户），1 个 Multiuser 综合案例（3 用户）以及 1 个 Multiuser 综合拓展案例（多用户）。案例设计呈现螺旋上升的特点，两个综合案例几乎包含了全书所学知识，案例实施及测试环境灵活多变，可以拓展读者视野，让读者对所学知识达到活学活用。案例各有千秋，互为补充。知识涵盖了网络管理 Telnet、SSH、VLAN、Trunk、常见 VLAN 间路由（单臂路由，三层交换 SVI）、标准静态路由、静态默认路由、静态汇总路由、静态备份路由，以及动态路由协议 RIPv2 等技术。这些技术的融合通过本章设计的 Multiuser 多用户案例展现出来，让大家在玩中实践，在实践中探索，在探索中提高。Multiuser 是对传统单机实验课堂的颠覆，促进了师生互动，倡导团队协作，鼓励学生大胆创新，自主学习，迎接网络技能挑战。

参 考 文 献

[1] [美]格拉齐尼亚（Graziani, R.），[美]约翰逊（Johnson, A.）. CCNA Exploration：路由协议和概念. 思科系统公司，译. 北京：人民邮电出版社，2014.
[2] [美]刘易斯（Lewis, P.D）. CCNA Exploration：LAN 交换和无线. 思科系统公司，译. 北京：人民邮电出版社，2009.
[3] [美]瓦尚（Vachon, B.），[美]格拉齐尼亚（Graziani,R.）. CCNA Exploration：接入 WAN. 思科系统公司，译. 北京：人民邮电出版社，2009.
[4] [加]爱普森（Empson, S.），[美]施密特（Schmidt,C.）. 路由和交换基础. 思科系统公司，译. 北京：人民邮电出版社，2014.
[5] [美]戴伊（Dye, M.A.），[美]里德（Allan D.Reid）. 网络简介. 思科系统公司，译. 北京：人民邮电出版社，2014.
[6] [美]多伊尔（Doyle, J.），[美]卡罗尔（Carroll, J.）. TCP/IP 路由技术（第 1 卷）（第 2 版）. 葛建立，吴剑章，译. 北京：人民邮电出版社，2009.
[7] 李涤非. 网络故障分析——路由篇（上册）. 北京：电子工业出版社，2015.
[8] 李涤非. 网络故障分析——路由篇（下册）. 北京：电子工业出版社，2015.
[9] 雷震甲. 网络工程师教程（第 4 版）. 北京：清华大学出版社，2014.
[10] 梁广民，王隆杰. 思科网络实验室路由、交换实验指南（第 2 版）. 北京：电子工业出版社，2013.
[11] 张国清，车斌. 路由技术（IPV4 版）. 北京：电子工业出版社，2012.
[12] 何力. 网络实用技术项目教程. 武汉：华中科技大学出版社， 2010.